Gel Electrophoresis of Proteins (2nd edition)
Gene Targeting
Gene Transcription
Genome Analysis
Glycobiology
Growth Factors
Haemopoiesis
Histocompatibility Testing
HPLC of Macromolecules
HPLC of Small Molecules
Human Cytogenetics I and II (2nd edition)
Human Genetic Disease Analysis
Immobilised Cells and Enzymes
Immunocytochemistry
In Situ Hybridization
Iodinated Density Gradient Media
Light Microscopy in Biology
Lipid Analysis
Lipid Modification of Proteins
Lipoprotein Analysis
Liposomes
Lymphocytes
Mammalian Cell Biotechnology
Mammalian Development
Medical Bacteriology
Medical Mycology
Microcomputers in Biochemistry
Microcomputers in Biology
Microcomputers in Physiology
Mitochondria
Molecular Genetic Analysis of Populations
Molec[...] Neu[...]
Molec[...]
Molecular Plant Pathology I and II
Molecular Virology
Monitoring Neuronal Activity
Mutagenicity Testing
Neural Transplantation
Neurochemistry
Neuronal Cell Lines
NMR of Biological Macromolecules
Nucleic Acid and Protein Sequence Analysis
Nucleic Acid Hybridisation
Nucleic Acids Sequencing
Oligonucleotides and Analogues
Oligonucleotide Synthesis
PCR
Peptide Hormone Action
Peptide Hormone Secretion
Photosynthesis: Energy Transduction
Plant Cell Culture
Plant Molecular Biology
Plasmids
Pollination Ecology
Postimplantation Mammalian Embryos
Preparative Centrifugation
Prostaglandins and Related Substances
Protein Architecture
Protein Engineering
Protein Function

Protein Phosphorylation
Protein Purification Applications
Protein Purification Methods
Protein Sequencing
Protein Structure
Protein Targeting
Proteolytic Enzymes
Radioisotopes in Biology
Receptor Biochemistry
Receptor–Effector Coupling
Receptor–Ligand Interactions
Ribosomes and Protein Synthesis
Signal Transduction
Solid Phase Peptide Synthesis
Spectrophotometry and Spectrofluorimetry
Steroid Hormones
Teratocarcinomas and Embryonic Stem Cells
Transcription Factors
Transcription and Translation
Tumour Immunobiology
Virology
Yeast

Molecular Virology

A Practical Approach

Edited by

ANDREW J. DAVISON

MRC Virology Unit, Institute of Virology, Church Street, Glasgow G11 5JR

and

RICHARD M. ELLIOTT

Institute of Virology, University of Glasgow, Church Street, Glasgow G11 5JR

IRL PRESS
at
OXFORD UNIVERSITY PRESS
Oxford New York Tokyo

Oxford University Press, Walton Street, Oxford OX2 6DP
Oxford New York Toronto
Delhi Bombay Calcutta Madras Karachi
Kuala Lumpur Singapore Hong Kong Tokyo
Nairobi Dar es Salaam Cape Town
Melbourne Auckland Madrid
and associated companies in
Berlin Ibadan

Oxford is a trade mark of Oxford University Press

A Practical Approach is a registered trade mark of the Chancellor, Masters, and Scholars of the University of Oxford trading as Oxford University Press

Published in the United States
by Oxford University Press Inc., New York

A catalogue record for this book is available from the British Library

Library of Congress Cataloging in Publication Data
Molecular virology : a practical approach / edited by Andrew J. Davison and Richard M. Elliott.
(The Practical approach series)
Includes bibliographical references and index.
1. Molecular virology. I. Davison, Andrew J. II. Elliott, Richard M. III. Series.
QR389.M65 1993 576'.64—dc20 93–7052
ISBN 0–19–963358–4 (hbk.)
ISBN 0–19–963357–6 (pbk.)

Typeset by Footnote Graphics, Warminster, Wilts
Printed in Great Britain by Information Press Ltd, Eynsham, Oxon

Preface

Virology, like other biological disciplines, has been transformed in recent years by the availability of an ever-increasing range of techniques for molecular analysis. This has had a dual effect, expanding the number of virological specialities and widening the applicability of certain areas to studies of other viruses and to fields outside virology. The chapters in this volume centre on the latter aspect. The chapters on positive strand plant viruses, influenza viruses and paramyxoviruses, immunodeficiency viruses, and viral DNA replication cover areas that interest a significant number of virologists and include techniques that are applicable to viruses outside the immediate ambit of a particular chapter. The contributions describing vectors constructed from retroviruses, poliovirus, baculovirus, vaccinia virus, and herpes simplex virus provide detailed techniques that are of use to virologists who work with these agents and to those who wish to use vector technology in other fields. Also, the advent of PCR technology has revolutionized the study of viruses, even those which are inaccessible to classical methods, and so we have included a chapter on the use of PCR in analysing viral sequence variation.

This book contains a large number of techniques that were not available at the time of publication of a previous volume in this series, *Virology: a practical approach* edited by Brian Mahy. Nevertheless, many of the basic procedures in virology contained in the earlier publication are fundamental to our book, and we encourage readers to consider the two volumes as complementary. Needless to say, we have not covered some viruses which have been discussed adequately in other recent publications (for example, adenovirus vectors), nor have we dealt with fields presently at an early stage of development (for example, genetic engineering of negative strand RNA virus genomes).

Different areas are, of course, at different stages of development, and we have given the authors the freedom to pitch their chapters as they see fit. This has resulted in a varying level of detail, and some authors have presumed a greater degree of background expertise than others. All the chapters assume experience in the basic techniques of molecular biology and cell culture, and in some instances methods are presented in the form of guidance rather than as detailed protocols. Some methods are described in more than one chapter (for example, transfection and PCR protocols), and we have included them as supplied by the authors, since the minor variations reflect the practical flexibility possible in these systems. We have followed the practice of listing unusual or complex materials at the head of each protocol where they first occur, on the understanding that the contents of each chapter are considered as a whole. Most materials are available from major suppliers of chemicals and equipment, but we have indicated suppliers of certain specialist items.

We are grateful to the authors for their co-operation and hard work in producing this book, and to the series editors and staff of IRL Press for their continued guidance throughout its incubation.

Glasgow A.J.D.
December 1992 R.M.E.

Contents

Contributors

KENNETH W. BUCK
Department of Biology, Imperial College of Science, Technology and Medicine, Prince Consort Road, London, SW7 2BB, UK.

SHIU-WAN CHAN
Department of Medical Microbiology, University of Edinburgh, Medical School Building, Teviot Place, Edinburgh, EH8 9AG, UK.

DAVID J. EVANS
Department of Microbiology, University of Reading, Whiteknights, Reading, RG6 2AJ, UK.

RONALD T. HAY
Department of Biochemistry and Microbiology, University of St Andrews, Irvine Building, North Street, St Andrews, Fife, KY16 9AL, UK.

ROBERT J. HAYES
Department of Biology, Imperial College of Science, Technology and Medicine, Prince Consort Road, London, SW7 2BB, UK.

ROBERT A. LAMB
Howard Hughes Medical Institute, Department of Biochemistry, Molecular Biology and Cell Biology, Northwestern University, 2153 Sheridan Road, Evanston, IL 60208-3500, USA.

JOHN McLAUCHLAN
MRC Virology Unit, Institute of Virology, Church Street, Glasgow, G11 5JR, UK.

REAY G. PATERSON
Department of Biochemistry, Molecular Biology and Cell Biology, Northwestern University, 2153 Sheridan Road, Evanston, IL 60208-3500, USA.

ROBERT D. POSSEE
NERC Institute of Virology and Environmental Microbiology, Mansfield Road, Oxford, OX1 3SR, UK.

FRAZER J. RIXON
MRC Virology Unit, Institute of Virology, Church Street, Glasgow, G11 5JR, UK.

THOMAS F. SCHULZ
Chester Beatty Laboratories, The Institute of Cancer Research, London, SW3 6JB, UK.

PETER SIMMONDS
Department of Medical Microbiology, University of Edinburgh, Medical School Building, Teviot Place, Edinburgh, EH8 9AG, UK.

GEOFFREY L. SMITH
Sir William Dunn School of Pathology, University of Oxford, South Parks Road, Oxford, OX1 3RE, UK.

BRUNO SPIRE
Laboratoire INSERM de Virologie, Faculté de Médecine Nord, Boulevarde Pierre Dramard, 13326 Marseille Cedex 13, France.

LORNA M. D. STEWART
Imperial Cancer Research Fund, St Bartholomew's Hospital, Dominion House, 59 Bartholomew Close, London, EC1A 7BE, UK.

ANDREW W. STOKER
Department of Human Anatomy, University of Oxford, South Parks Road, Oxford, OX1 3QX, UK.

NIGEL D. STOW
MRC Virology Unit, Institute of Virology, Church Street, Glasgow, G11 5JR, UK.

Abbreviations

3β-HSD	3β-hydroxysteroid dehydrogenase
ACDP	Advisory Committee on Dangerous Pathogens
ACGM	Advisory Committee on Genetic Modification
AcMNPV	*Autographa californica* nuclear polyhedrosis virus
Ad2	adenovirus serotype 2
Ad4	adenovirus serotype 4
AlMV	alfalfa mosaic virus
AMV	avian myeloblastosis virus
ALPA	antibody-linked polymerase assay
ATCC	American Type Culture Collection
ASLV	avian sarcoma and leukosis viruses
BCdR	5-bromo-deoxycytidine
BHK	baby hamster kidney
BMV	brome mosaic virus
BND	benzoylated naphthoylated DEAE
BPV	bovine papillomavirus
BSA	bovine serum albumin
BUdR	5-bromo-deoxyuridine
CAT	chloramphenicol acetyl transferase
CAV	cell-associated virus
CLB	cell lysis buffer
CMC	carboxymethyl cellulose
CMV	cucumber mosaic virus
CNS	central nervous system
c.p.e.	cytopathic effect
CRV	cell-released virus
CTL	cytotoxic T lymphocytes
CV	coxsackievirus
DBP	DNA-binding protein
ddNTP	dideoxynucleoside triphosphate
DEPC	diethyl pyrocarbonate
DME	Dulbecco's modified Eagle's medium
DMEM	Dulbecco's minimal essential medium
DMSO	dimethylsulphoxide
DSM	dried skimmed milk
DTT	dithiothreitol
E	early
Ecogpt	*E. coli* guanine phosphoribosyltransferase gene
EDTA	ethylenediaminetetraacetic acid

EEV	extracellular enveloped virus
EGTA	ethylene glycol-bis(β-aminoethyl ether)*N,N,N′,N′*-tetraacetic acid
EMEM	Eagle's minimal essential medium
endo H	endo β-*N*-acetylglucosaminidase H
ER	endoplasmic reticulum
F	fusion protein
FCS	fetal calf serum
FHV	flockhouse virus
GMEM	Glasgow minimal essential medium
GV	granulosis virus
HA	haemagglutinin
HBS	Hepes-buffered saline
HCMV	human cytomegalovirus
HCV	hepatitis C virus
HIV	human immunodeficiency virus
HN	haemagglutinin–neuraminidase
hprt	hypoxanthine phosphoribosyltransferase gene
HPV	human papillomavirus
HSV	herpes simplex virus
HTLV	human T-lymphotropic virus
IE	immediate early
INV	intracellular naked virus
IPTG	isopropyl β-D-thiogalactopyranoside
ITR	inverted terminal repeat
L	late
lacZ	*E. coli* β-galactosidase gene
LAT	latency-associated transcript
LCR	ligase chain reaction
LTR	long terminal repeat
M	matrix protein
mAb	monoclonal antibody
MDBK	Madin Darby bovine kidney (cell)
MDCK	Madin Darby canine kidney (cell)
MEM	minimal essential medium
MI	mock-infected
MLV	murine leukaemia viruses
MNPV	multiple nucleocapsid (nuclear) polyhedrosis virus
Mo-MLV	Moloney murine leukaemia virus
m.o.i.	multiplicity of infection
MPA	mycophenolic acid
M_r	relative molecular mass
NA	neuraminidase
NCS	newborn calf serum

NFI	nuclear factor I
NFI_{DBD}	DNA-binding domain of NFI
NFIII	nuclear factor III
NOV	non-occluded virus
NP	nucleoprotein
NPT	non-permissive temperature
NPV	nuclear polyhedrosis virus
NS	non-structural protein
OD_x	optical density at x nm
ONPG	*o*-nitrophenyl β-D-galactopyranoside
OPV	oral poliovirus vaccine
ORF	open reading frame
P	phosphoprotein
PBL	peripheral blood lymphocytes
PBS	phosphate-buffered saline
PCR	polymerase chain reaction
p.f.u.	plaque forming units
p.i.	post infection
PBMC	peripheral blood mononuclear cells
PCR	polymerase chain reaction
PMSF	phenylmethylsulphonyl fluoride
pol	DNA polymerase
PPO	2,5-diaphenyloxazole
PRV	pseudorabies virus
PT	permissive temperature
pTP	pre-terminal protein
rd	replication-defective
RdRp	RNA-dependent RNA polymerase
REV	reticuloendotheliosis viruses
Rf	relative gel migration
RFLP	restriction fragment length polymorphism
RI	replicative intermediate
RSB	reticulocyte standard buffer
RSV	Rous sarcoma virus
RT	reverse transcriptase
SDS	sodium dodecyl sulphate
SDS–PAGE	SDS–polyacrylamide gel electrophoresis
Sf	*Spodoptera frugiperda*
SIN	self-inactivation
SIV	simian immunodeficiency virus
SNPV	single nucleocapsid (nuclear) polyhedrosis virus
SV5	simian virus 5
SV40	simian virus 40
TBS	Tris-buffered saline

TCA	trichloroacetic acid
TEMED	*N,N,N′,N′*-tetramethylethylenediamine
tk	thymidine kinase gene
TM	transmembrane
TMV	tobacco mosaic virus
TP	terminal protein
ts	temperature sensitive
UV	ultraviolet
UWGCG	University of Wisconsin Genetics Computer Group
VSV	vesicular stomatitis virus
WT	wild type
X-gal	5-bromo-4-chloro-3-indolyl β-D-galactopyranoside

1

Analysis of replication complexes of positive strand RNA plant viruses

ROBERT J. HAYES and KENNETH W. BUCK

1. Introduction

Viruses with genomes of positive strand (messenger-sense) RNA are found in animals, plants, and bacteria. Examples include poliovirus, foot and mouth disease virus, flockhouse virus (FHV), tobacco mosaic virus (TMV), cucumber mosaic virus (CMV), and the bacteriophages MS2 and Qβ (1). However, they are most abundant in plants. About 77% (approximately 500) of known plant viruses have positive strand RNA genomes (2).

Enzymes that replicate RNA templates (RNA-dependent RNA polymerase, RdRp) apparently do not have proof-reading activities and have error rates of 10^{-4} to 10^{-5} (3). This may well have limited the size of viral RNA genomes, the largest of which among plant viruses are the closteroviruses (about 20 kb) and among animal viruses the coronaviruses (about 32 kb). Because of the high error rate, RNA of a given virus may be a population in dynamic equilibrium, with viable mutants arising at a high rate on the one hand and being selected against on the other.

Replication of a positive strand RNA virus is an apparently simple process (for a review see ref. 4). An RdRp recognizes and binds to a promoter sequence at the 3′ end of the genomic positive strand RNA. Using the four NTPs as substrates and genomic RNA as template, a complementary copy or negative strand is synthesized. The RdRp then recognizes and binds to a promoter sequence at the 3′ end of the negative strand and synthesizes a progeny positive strand using the negative strand as template. This process implies an exponential increase in RNA synthesis. *In vivo* RNA synthesis often starts off exponentially and then becomes linear, with synthesis becoming asymmetric, resulting in the production of a 50- to 100-fold excess of positive over negative strands.

Although in some RNA viruses there is complementarity, or partial complementarity, between sequences at the 5′ and 3′ ends of the RNA, it is likely that separate recognition of promoters at the 3′ ends of the positive and negative strands is required. Mutants of several viruses, such as brome mosaic virus (BMV), have been obtained that give altered ratios of the positive and negative strands (5). Hence the excess of positive strands could be explained, in part, by an intrinsically stronger promoter on the negative strand template. However, sequestration of some positive strands by ribosomes for translation and, late in the infection cycle, for encapsidation could also contribute to greater amounts of positive strands.

There are two possible methods by which this replication process could occur. In the first, the progeny negative strand remains hydrogen-bonded to the positive strand template to form a double-stranded RNA molecule. Production of progeny positive strand then occurs by strand-displacement synthesis. In the second model, the progeny negative strand is hydrogen-bonded to its template only in the region where RNA synthesis is taking place. Immediately afterwards, these regions are unwound so that the replicative intermediates are full-length positive strands with partial-length negative strand tails. The product of the first stage of replication is free positive and negative strand RNA. The negative strand can then act as a template to produce further free positive strands by a similar mechanism.

The following evidence favours the second mechanism for bacteriophage Qβ and CMV.

(a) Free negative strand was detected in reactions with both Qβ and CMV polymerases, whereas the first model implies that no free negative strand is produced.

(b) Qβ RNA replication intermediates were fully sensitive to RNase A at high salt (6).

(c) Double-stranded RNA was not a template for purified Qβ or CMV polymerases (6, 7).

Recently, however, evidence in favour of the first model was presented for alfalfa mosaic virus (AlMV) (8).

Both models indicate that at least two enzymes are required for replication of RNA: an RdRp to catalyse formation of phosphodiester bonds between adjacent nucleotides and a helicase to separate the two strands after synthesis. For a few viruses the protein subunit with polymerase activity has been isolated from infected cells or produced in *Escherichia coli* from cDNA clones. For other viruses, putative RdRps have been identified from conserved amino acid sequence motifs. Motifs characteristic of nucleic acid helicases have also been identified in proteins encoded by eukaryotic positive strand RNA viruses (9–13). However, only in the case of plum pox virus (14) has a protein with helicase activity been identified experimentally, but its role in replication was not reported.

It is likely that eukaryotic RNA viruses also encode enzymes for capping the 5′ end of the virus RNA, and amino acid sequence motifs characteristic of methyltransferases have been identified in a number of viral proteins (15).

Evidence for methyltransferase activity has been obtained for a Sindbis virus protein (16) and for guanyltransferase activity in a putative replication protein of TMV expressed in *E. coli* (17).

Proteins encoded by genes downstream from the 5′ proximal gene in some eukaryotic RNA viruses are expressed from subgenomic RNAs corresponding to a 3′ region of the viral RNA, such as the coat proteins of TMV, BMV, and CMV and the movement protein of TMV. For a few viruses it has been shown that such subgenomic RNAs are transcribed from subgenomic RNA promoters located internally in the negative strand RNA. Recognition of subgenomic promoters is apparently distinct from recognition of the 3′ terminal promoter on the negative strand RNA, since a crude BMV RdRp preparation was able to catalyse synthesis of a coat protein subgenomic RNA, but not the full-length genomic RNA, from a negative strand template (18).

Elucidation of the mechanism of virus RNA replication, the functions of virus- and host-encoded proteins, and their interactions with each other and with the viral RNA template requires the isolation and purification of enzyme complexes capable of complete replication of added viral RNA template. This has been achieved for the RdRp of bacteriophage Qβ, which has been purified to homogeneity and shown to consist of one virus-encoded protein and three host-encoded proteins (6). An additional host protein is needed for initiation of RNA synthesis on positive strand, but not negative strand, templates. Several template-dependent RdRps of eukaryotic viruses have been isolated that are able to catalyse the synthesis of full-length negative strands on positive strand templates and, where appropriate, subgenomic RNA on negative strand templates, such as poliovirus (19–21), black beetle virus (22), turnip yellow mosaic virus (23, 24), BMV (25–27), cowpea chlorotic mottle virus (28), and AlMV (29, 30). RdRp preparations from BMV-infected plants have been valuable for defining the promoter at the 3′ terminus of the positive strand and the internal subgenomic promoter in the negative strand of RNA 3 (18, 31). However, it is only recently that enzyme preparations capable of catalysing complete replication of the viral RNA have been obtained, namely for CMV (7, 32) and FHV (33).

CMV has a genome comprising three positive strand RNAs, RNA 1 (3.4 kb), RNA 2 (3.0 kb), and RNA 3 (2.2 kb), which encode non-structural proteins 1a, 2a, and 3a, respectively. Proteins 1a and 2a have sequence motifs characteristic of RNA replicases, that is helicase and methyltransferase (1a) and polymerase (2a) motifs. Protein 3a is believed to be required for virus cell-to-cell movement. The coat protein is encoded by RNA 4 (1.0 kb), a subgenomic RNA corresponding to the 3′ region of RNA 3 (for a review see ref. 34). A replicase complex has been isolated from tobacco plants infected with CMV and purified to near homogeneity (7). Using CMV RNAs 1, 2, or 3 as template, complete replication of each RNA was demonstrated. In addition, RNA 4 was produced with RNA 3 as template. No activity was observed with RNAs in different taxonomic groups (1), namely TMV, tomato bushy stunt virus, and red clover mosaic virus; that is, the replicase was template-specific. The purified replicase was shown to contain the virus-encoded proteins 1a and 2a and a major tobacco protein with a relative molecular mass (M_r) of about 50 000. Purified preparations contained small amounts of several other tobacco proteins, one or more of which might also be essential for replicase activity.

Tobacco and other plants contain an RdRp whose activity is stimulated after infection by CMV or some other viruses (35). This enzyme appears to be a single polypeptide with an M_r of about 130 000. Unlike the CMV replicase, it can use a variety of RNAs as template and the products of the reaction are short oligonucleotides. Although the cellular function of this enzyme is unknown, there is no evidence that it plays any role in the replication of CMV or any other positive strand RNA virus. It could not be detected in purified CMV replicase preparations, nor did it have any effect on the CMV replicase reaction when added to the purified CMV replicase. Host RdRp is likely to be a major contaminant during early stages of the purification of plant viral replicases and care is needed to remove this activity.

This chapter describes many of the techniques frequently used in the purification of a solubilized replication complex from CMV-infected plant material. Many of the techniques could also be applied to study the replication of other plant viruses and animal viruses, although empirical tests will be required to select procedures and conditions best suited for a particular virus. It is assumed that a ready supply of virus, viral RNA, and infected plant material is available. For details on the infection of plants and isolation and purification of plant viruses and viral RNA, the reader is directed towards ref. 36.

2. Production of infectious transcripts

2.1 Production of cDNA clones

The production of cDNA clones from which infectious viral RNA can be transcribed *in vitro* has become an essential tool for genetic studies of positive strand RNA viruses. It is useful in replication studies for producing wild type or modified templates for replication using isolated replication complexes, or for studying the effects of mutations in replicase genes or other regions of the genome on replication *in vivo*. The technique involves linking full-length cDNA copies of the genomic RNA to a bacteriophage promoter and transcribing *in vitro* with an RNA polymerase. Initially, the bacteriophage λ *PR* promoter was used together with *E. coli* RNA polymerase (37, 38), but more recently infectious transcripts of a number of viruses have been synthesized with the more reliable and efficient RNA polymerases from bacteriophages T3, T7, and SP6 (for example see ref. 39).

The most common procedure used to produce cDNA clones from which infectious transcripts can be obtained involves cloning full-length double-stranded cDNA into a vector containing the appropriate bacteriophage promoter. *In vitro* mutagenesis is then used to eliminate, or reduce, the number of nucleotides between the transcription initiation site and the 5′ end of the viral RNA, because the presence of more than one additional nucleotide can greatly reduce or even abolish the infectivity of the transcripts. Interestingly, a few extra nucleotides at the 3′ end of viral RNA transcripts appear to have little effect on infectivity.

cDNA copies of the viral RNA are synthesized by avian myeloblastosis virus or Molony murine leukaemia virus (Mo-MLV) reverse transcriptase (RT) using an oligonucleotide complementary to the 3′ end of the genomic RNA (*Protocol 1*). To aid ligation of the double-stranded cDNA product into the transcription vector, incorporation of a suitable unique restriction endonuclease cleavage site into the 3′ end of the oligonucleotide is advisable, preferably one that leaves a 5′ protruding end (see Section 2.2). The second strand cDNA is then produced using the Klenow fragment of DNA polymerase I and/or RT in conjunction with a second oligonucleotide complementary to the 5′ viral sequence (*Protocol 2*). Again, it is wise to incorporate a unique restriction site and, if possible, a site different from the 3′ oligonucleotide should be used to aid directional cloning into the vector. Of course, care should be taken when choosing the restriction sites so that cDNA ligated into the vector is in the desired orientation!

Protocol 1. First strand cDNA synthesis

Materials

- 10 × RT buffer: 500 mM Tris–HCl (pH 8.3 at 22°C), 750 mM KCl, 100 mM dithiothreitol (DTT), 30 mM $MgCl_2$ (supplied by Gibco-BRL)
- 5 mM dNTPs: 5 mM each of dATP, dTTP, dCTP, dGTP
- [^{32}P]dCTP (10 mCi/ml, 800 Ci/mmol)
- 3′ oligonucleotide
- Mo-MLV RT (Gibco-BRL)
- Sephacryl S-400

Method

1. Add the following to a 0.5 ml microcentrifuge tube, mix and incubate for 1 h at 37°C.

• 10 × RT buffer	5 µl
• 5 mM dNTPs	5 µl
• [^{32}P]dCTP	1 µl
• 3′ oligonucleotide	50 µg

Protocol 1. *Continued*

• viral RNA	10 μg
• Mo-MLV RT	5000 U
• water to a final volume of	50 μl

2. Stop the reaction and hydrolyse the RNA template by adding 1 μl of 0.5 M EDTA and 50 μl of 0.6 M NaOH. Incubate at 65°C for 1 h. Remove 10 μl to determine the length of first strand cDNA by alkaline-agarose gel electrophoresis (43).
3. Purify the remainder of the denatured cDNA by passing it through a 1 ml column of Sephacryl S-400 equilibrated in 0.3 M NaOH. Collect the void volume.
4. Add an equal volume of 0.3 M HCl and extract with an equal volume of 1:1 (v/v) phenol:chloroform.
5. Precipitate the DNA by adding 3 M sodium acetate (pH 5.2) to 0.3 M and 2.5 volumes of ethanol.

Protocol 2. Second strand cDNA synthesis with Klenow fragment

Materials

- buffer 1:200 mM Tris–HCl (pH 7.4), 20 mM $MgCl_2$, 500 mM NaCl
- buffer 2:100 mM Tris–HCl (pH 7.4), 50 mM $MgCl_2$, 20 mM DTT, 5 mM dNTPs
- 5′ oligonucleotide (1 mg/ml)
- Klenow fragment of *E. coli* DNA polymerase I

Method

1. Resuspend the first strand cDNA from *Protocol 1* in 10 μl of water. Add 1 μg of 5′ oligonucleotide and 1 μl of buffer 1. Overlay with mineral oil, heat to 70°C in a beaker of water for 2 min and then cool slowly to 30°C over a period of about 30 min.
2. Add 1 μl of buffer 2 and 10 U of Klenow fragment. Incubate at 37°C for 90 min.
3. Add 1 μl of 0.5 M EDTA to stop the reaction, extract with phenol: chloroform and ethanol precipitate the double-stranded DNA (see step 5 of *Protocol 1*).

Most of the commercially available, high copy number vectors include a number of unique restriction sites to aid cloning, together with immediately adjacent RNA polymerase promoter sites. A feature of some vectors which is crucial to success in producing infectious transcripts is a single-stranded phage origin of replication; as mentioned above, after the cDNA is cloned into the vector it is necessary to remove one or more nucleotides from the 5′ end of the cDNA by *in vitro* mutagenesis. The class III promoter sequence for T7 RNA polymerase is TAATACGACTCACTATAGGG, with transcription beginning at the underlined G. Ligation of cDNA immediately downstream of this sequence would result in transcripts with three extra G residues at the 5′ end. It is, however, possible to replace one or more of these G residues with viral sequences, resulting in a reduction in reaction yield. With CMV it was found that replacement of the underlined G residue with C results in a 10- to 15-fold reduction in yield, and replacement with A or U gives a 5-fold reduction. Any replacement of the second G results in a 2-fold reduction, while smaller effects are observed with replacement of the third G (R. J. Hayes, unpublished results). Therefore, it is usual to ligate the cDNA into the vector (for example pTZ18U (Pharmacia)), prepare single-stranded DNA, and then remove the undesired nucleotides. A compromise between yield and infectivity is often required.

The *in vitro* mutagenesis step is made unnecessary by incorporating a T7 RNA polymerase site within the 5′ oligonucleotide used during second strand cDNA synthesis. The cDNA is then ligated into one of the pUC or pEMBL plasmids (40).

The polymerase chain reaction (PCR) allows rapid amplification of rare cDNA molecules. This technique has been used to clone infectious transcripts of CMV and red clover necrotic mosaic virus (39, 41, 42). Given the high error rate of *Taq* polymerase (the number of mutations per base duplication is thought to be in the region of 2×10^{-4}), the T7 RNA transcripts would be expected to include a number of nucleotide substitutions resulting in reduced or abolished infectivity. However, while T7 RNA polymerase transcripts from some PCR products are not infectious, we have found that a number of these PCR T7 RNA polymerase transcripts are infectious at levels approaching that of the viral RNA genome. In infectivity tests some transcripts are shown to be phenotypically different from the viral RNA, thereby suggesting that PCR cloning could be used to introduce random mutations into the viral genome.

First strand cDNA synthesis is carried out as described in *Protocol 1* with oligonucleotides containing suitable restriction endonuclease sites. The first strand cDNA products are then fractionated on a Sephacryl S-400 column to remove small products that interfere with amplification, and are then used in a PCR reaction as described in *Protocol 3*. The 3′ oligonucleotide is the one used to prime first strand cDNA synthesis. The 5′ oligonucleotide also contains a unique restriction endonuclease site, and can contain a T7 RNA promoter sequence. After PCR, the products are subjected to restriction endonuclease cleavage and purified by agarose gel electrophoresis. The excised product is then ligated into a suitable vector and transformed into competent *E. coli*.

Protocol 3. Second strand cDNA synthesis by PCR

Materials

- thermal cycler (Perkin–Elmer Cetus)
- PCR buffer: 670 mM Tris–HCl (pH 8.8 at 25°C), 170 mM $(NH_4)_2SO_4$, 20 mM $MgCl_2$, 100 mM 2-mercaptoethanol, 2 mg/ml bovine serum albumin (BSA), 65 μM EDTA, 2 mM dNTPs
- 3′ and 5′ oligonucleotides
- *Taq* DNA polymerase (AmpliTaq; Cetus)

Method

1. In a 0.5 ml microcentrifuge tube set up a 100 μl reaction mixture containing the first strand cDNA from *Protocol 1*, 10 μl of PCR buffer and 1 μg each of the 5′ and 3′ oligonucleotides. Overlay with mineral oil.
2. Denature the cDNA by heating at 95°C for 5 min.
3. Add 2 U of *Taq* polymerase and subject the reaction to a 30 cycle programme of 92°C for 1 min, 40°C for 1 min, and 72°C for 3 min[a] using a thermal cycler. Increase the elongation step (72°C) by 10 sec at each step.
4. Remove the aqueous phase from underneath the mineral oil, extract it with an equal volume of chloroform, and ethanol precipitate the DNA.

[a] Use an extension time of approximately 1 min per 1000 nucleotides.

2.2 *In vitro* transcription reactions

The plasmid miniprep protocol combined with a proteinase K step (38) is usually sufficient to prepare DNA for *in vitro* production of short transcripts. However, we have found that for the production of long viral RNA transcripts (> 1 kb), DNA should be purified by banding on caesium chloride gradients, by polyethylene glycol precipitation, or by using a plasmid purification kit such as Magic MiniPreps DNA purification columns (Progema). After purification, completely linearize the vector with the restriction endonuclease corresponding to the 3′ end restriction site. Restriction enzymes that generate 3′ protruding ends, such as *Pst*I and *Kpn*I, should be avoided because extraneous transcripts with additional sequences complementary to the expected transcript, as well as sequences corresponding to the vector DNA, are synthesized in addition to the expected transcript. If such restriction endonuclease sites are unavoidable, the 3′ protruding end should be made blunt-ended using Klenow fragment (*Protocol 4*).

Protocol 4. Conversion of a 3′ overhang to a blunt end

Materials

- *in vitro* transcription reaction (see *Protocol 5*)

Method

1. Set up the *in vitro* transcription reaction as described in *Protocol 5*, omitting the 10 × NTP mixture and RNA polymerase.
2. Add 10 U of the Klenow fragment of DNA polymerase I and incubate for 30 min at 22°C.
3. Proceed with the transcription reaction (*Protocol 5*) by adding the 10 × NTP mixture and RNA polymerase.

Many RNA viruses have genomes that are capped at the 5′ end, and often this cap is essential for infectivity. Although *in vitro* synthesized RNAs can be capped enzymatically by guanylyl transferase (44), this is quite expensive. Alternatively, the cap analogue can be incorporated into the RNA during the transcription reaction (*Protocol 5*). By including a high concentration of the cap together with a low concentration of GTP, most of the synthesized molecules are capped. The reaction conditions described in *Protocol 5* work equally well for T7, T3, and SP6 RNA polymerases.

After *in vitro* transcription, the DNA template can be removed from the reaction by adding DNase I (*Protocol 5*). An aliquot of the reaction products should then be analysed to estimate the length of the reaction products. We do this routinely by electrophoresis in a formaldehyde–agarose gel and staining the gel with ethidium bromide.

Protocol 5. *In vitro* synthesis of capped RNA transcripts

Materials

- 10 × RNA transcription buffer: 400 mM Tris–HCl (pH 7.5), 60 mM $MgCl_2$, 20 mM spermidine, 100 mM DTT, 1 mg/ml BSA
- 10 × NTP mixture: 5 mM ATP, 5 mM CTP, 0.5 mM GTP, 5 mM UTP
- cap analogue: 5 mM m^7 GpppG (Pharmacia)
- RNase inhibitor (InhibitAce, Northumbria Biologicals)
- T7, T3, or SP6 RNA polymerase

Method

1. Add the following to a 1.5 ml microcentrifuge tube:
   - 10 × RNA transcription buffer — 10 μl
   - 10 × NTP mixture — 10 μl

Protocol 5. *Continued*

- 5 mM m^7 GpppG — 10 μl
- RNase inhibitor — 100 U
- DNA template[a] — 10 μg
- water to a final volume of — 100 μl
- RNA polymerase — 40 U

2. Incubate for 1 h at 37°C. For a higher yield of transcripts, add another 40 U of RNA polymerase and continue the incubation for a further 1 h.
3. Remove the DNA template by adding 10 U of RNase-free DNase I. Incubate at 37°C for 15 min.
4. Extract with an equal volume of phenol:chloroform and ethanol precipitate.

[a] Add the DNA just before the enzyme to prevent precipitation of the DNA by spermidine.

3. Preparation of CMV RdRp

The RdRp from a number of RNA viruses has been found to be membrane-bound. Electron micrographs of thin sections of CMV-infected plant cells showed that development of characteristic small vesicles is associated with the tonoplast, which was postulated to be the site of CMV RNA replication (45). Indeed, we have found that tonoplast vesicles isolated from CMV-infected tobacco leaves by the method described in ref. 46 have high levels of RdRp activity. However, membranes involved in replication of different RNA plant viruses could differ. Therefore, we describe a procedure involving isolation of membranes by differential centrifugation which works well for CMV and is likely to be applicable to a wide range of viruses (*Protocol 6*).

While differential centrifugation of membranes allows the rapid preparation of a crude fraction, further purification of the RdRp is complicated by the necessity to solubilize the proteins involved. The major problem in preparing RdRp is whether all the proteins or other factors essential for replication are present in the membrane or in the solubilized fraction. The inability of some RdRp preparations to complete the replication cycle may be due to the absence of one or more essential components or to the presence of high levels of inhibitors.

3.1 Preparation of crude membrane-bound CMV RdRp

Plant tissues contain a range of proteases, alkaloids, and phenolic compounds which may inhibit enzyme activity either directly or indirectly. The nature of these compounds varies from plant to plant, and therefore trial homogenizations should be performed in the presence of a range of protease inhibitors,

Table 1. Protease inhibitors, reductants, and antiphenolic compounds added to homogenization buffer

	Stock solution	Working solution	Comments
Reductants			
Dithiothreitol	1 M	10 mM	Prepare in water. Store at −20°C.
2-mercaptoethanol	14.4 M	14 mM	Dilute 1:1000 in water.
Protease inhibitors			
Phenylmethyl-sulphonyl fluoride (PMSF)	100 mM	1 mM	Prepare in 95% ethanol, 5% isopropanol. Store at −20°C for up to 6 months. Has a short half-life (20 min) in buffer. **Very toxic.**
Benzamide	100 mM	1 mM	Prepare in ethanol.
Benzamidine	100 mM	1 mM	Prepare in water.
Antipain	1 mg/ml	1 μg/ml	Prepare in water.
Leupeptin	1 mg/ml	1 μg/ml	Prepare in water.
Pepstatin	0.1 mg/ml	1 μg/ml	Prepare in methanol.
Aprotinin	1 mg/ml	5 μg/ml	Prepare in water. Can be inactivated by high levels of reductants. Use benzamide–benzamidine combination as an alternative.
EGTA	0.5 M	1 mM	Prepare in water. Will not dissolve until pH is approx. 8.0.
Antipolyphenol			
PVP (M_r 40 000)			Add directly to homogenization buffer to 2%. Allow to hydrate overnight.
PVPP			Add directly to homogenization buffer to 1.5%. Allow to hydrate overnight.

reductants, and antiphenolic compounds by grinding 1–5 g of material with a pestle and mortar. The homogenized plant material should be assayed for RdRp activity immediately (see *Protocol 7*), and also after 30 min and 1 h at 4°C. The stability of the RdRp in the homogenization buffer can then be determined. The most commonly used inhibitors are shown in *Table 1*, together with suggested concentrations and preparations. **Note** that because the reductants and most of the protease inhibitors are labile, they should be added directly before homogenization.

As CMV has a wide host range, we isolate the RdRp from infected *Nicotiana tabacum* plants, these are easy to grow, are readily infected, and have a large amount of leaf tissue. Leaves of young *N. tabacum* cv. Samsun plants are inoculated with a 1 mg/ml solution of CMV (Q strain). The infected leaves are harvested three days later, before symptoms are detectable, and deribbed. The leaf tissue is homogenized in buffer A (*Protocol 6*), originally described for the purification of BMV RdRp (25) and very similar to that used for FHV (33), filtered through muslin, and centrifuged to remove intact cells, large debris, and nuclei. The supernatant is carefully removed, in order not to disturb the loose pellet, and then centrifuged at high speed. The crude membrane-bound RdRp is resuspended in buffer B (*Protocol 6*), and frozen in small aliquots at −70°C. The aliquots can be thawed and frozen at least three times without significant loss of activity.

Protocol 6. Preparation of crude membrane-bound CMV RdRp

Materials

- buffer A: 50 mM Tris–HCl (pH 8.0), 15 mM $MgCl_2$, 10 mM KCl, 20% (v/v) glycerol, 10 mM DTT, 1 μg/ml leupeptin, 1 μg/ml pepstatin, 1 mM benzamide, 1 mM benzamidine
- buffer B: 50 mM Tris–HCl (pH 8.2), 15 mM $MgCl_2$, 5% (v/v) glycerol, 1 mM DTT, 1 μg/ml leupeptin, 1 μg/ml pepstatin, 5 μg/ml aprotinin

Method

1. Harvest the infected leaves. Derib, weigh, and homogenize them in two volumes of buffer A at 4°C.
2. Filter the homogenate through six layers of muslin.
3. Centrifuge at 500 *g* for 10 min at 4°C.
4. Carefully remove the supernatant and centrifuge it at 30 000 *g* for 30 min at 4°C.
5. Resuspend the pellet at 5 mg protein/ml (measured by Bradford assay) in buffer B, dispense into small aliquots and store at −70°C.

At this stage the RdRp is highly active, but in our hands is capable of producing only double-stranded viral RNA. However, a similar fraction from FHV-infected insect cells is capable of producing both single-stranded and double-stranded RNA. As it is not necessary at this stage to add CMV viral RNA to the RdRp assay (*Protocol 7*), we conclude that a substantial fraction of the RdRp is attached to viral RNA template, and that the double-stranded RNA products result from RdRp completing previously initiated complementary strand synthesis. Indeed, the addition of large amounts of viral RNA to this fraction often reduces incorporation, probably because high concentrations of viral RNA inhibit RdRp activity. Endogenous RNA can be removed from the RdRp template by treating with micrococcal nuclease (27). This enzyme, often used to prepare RNA-dependent rabbit reticulocyte lysate, is capable of degrading single-stranded RNA. It has an absolute requirement for Ca^{2+} ions, which are usually supplied in the form of calcium acetate. After micrococcal nuclease digestion, the Ca^{2+} ions are removed by chelating with

EGTA, and viral RNA or T7 RNA polymerase transcripts are then added to the RdRp (*Protocol 8*).

Protocol 7. Assay of RdRp activity

Materials

- DE81 discs (Whatman)
- 2 × RdRp assay buffer: 100 mM Tris–HCl (pH 8.2), 8% (v/v) glycerol, 20 mM $MgCl_2$, 2 mM ATP, 2 mM GTP, 2 mM CTP, 50 μM UTP, 20 mM DTT
- Bentonite (prepared as described in *Protocol 9*)
- [^{32}P]UT (10 mCi/ml, 800 Ci/mmol)
- TE: 10 mM Tris–HCl (pH 8.0), 1 mM EDTA

Method

1. Add 10 μl of RdRp fraction to 12.5 μl of 2 × RdRp assay buffer. Add 0.5 μl of bentonite.
2. To test for endogenous RNA, add 1 μl (10 μCi) of [^{32}P]UTP. To test activity with exogenous RNA, add 5 μg of template (viral RNA or T7 RNA polymerase transcript) and 1 μl of [^{32}P]UTP.
3. Incubate both reactions for 1 h at 30 °C.
4. Spot 10 μl on to DE81 discs, dry the discs, and wash them four times for 15 min with 0.4 M Na_2HPO_4. Then wash the discs twice with water, dry them, and determine radioisotopic incorporation by scintillation counting.
   Alternatively, add 25 μl of phenol (equilibrated with water) to the reactions, vortex, and centrifuge. Transfer the supernatants to fresh tubes. Add 25 μl of TE to the phenol phase, vortex, and re-centrifuge. Pool the superantants from each reaction and extract twice with 24:1 (v/v) chloroform:isoamyl alcohol. Add 0.2 volumes of 10 M ammonium acetate and 2.5 volumes of ethanol. Place on dry ice for 15 min, thaw at room temperature with occasional inversion, and centrifuge for 15 min at 10 000 *g* at 4 °C. Approximately 90% of the unincorporated radioactive nucleotides should be in the supernatant. Dispose of the supernatant, wash the pellet with 80% (v/v) ethanol and air dry. Analyse the products by urea–polyacrylamide gel electrophoresis (*Protocol 10*).

Protocol 8. Micrococcal nuclease treatment of RdRp fractions

1. Add 10 μl of RdRp fraction to 12.5 μl of 2 × RdRp assay buffer. Add 1 μl of 50 mM calcium acetate and 100 U of microccocal nuclease. Incubate at 30 °C for 30 min.

Protocol 8. *Continued*

2. Add 0.5 M EGTA (see *Table 1*) to a final concentration of 5 mM.
3. Add template RNA and [^{32}P]UTP, and continue the RdRp assay as described in *Protocol 7*.

3.2 RdRp assays

Plant cells contain high levels of RNase, and as the amount of contaminating RNase present at the crude membrane bound stage varies considerably from preparation to preparation, it is advisable to include an RNase inhibitor in the assay (*Protocol 7*). We have tested a number of commercially available inhibitors, but none was found to be more efficient than bentonite (*Protocol 9*), a traditional inhibitor of RNase (47) which has the advantage of being very cheap.

Protocol 9. Preparation of bentonite

Materials

- Bentonite powder (BDH)

Method

1. Add 20 g of bentonite powder to 400 ml of 10 mM Tris–HCl (pH 7.6) and vortex extensively.
2. Centrifuge at approximately 3000 *g* for 1 min. Collect the supernatant and discard the pellet.
3. Repeat step 2 twice.
4. Centrifuge the supernatant at 6000 *g* for 10 min.
5. Resuspend the pellet in 100 ml of 10 mM Tris–HCl (pH 7.6). The concentration should be approximately 40 mg/ml.

Incorporation of labelled nucleotides into trichloroacetic acid-precipitable radioactivity is often used to measure RdRp activity. We use DE81 discs (see *Protocol 7*). However, while this is a quick method for measuring the activity and stability of RdRp fractions, care should be used when analysing the results. Sometimes incorporation measured by this method does not correspond to the amount of viral RNA synthesized, perhaps due to an endogenous host RdRp (see Section 1), and therefore the products of the RdRp assay should be analysed by gel electrophoresis and autoradiography.

The products of the assay can be resolved by electrophoresis through a standard agarose gel, followed by drying down and autoradiographing the gel. Single-stranded RNA migrates through the gel faster than double-stranded RNA. An obvious problem is the presence of contaminating RNase which can digest the single-stranded products during electrophoresis. Ideally, the gel tank used for the gel should be used only for RNA work, and should be treated with diethyl pyrocarbonate after use. Using urea–polyacrylamide gel (*Protocol 10*) overcomes this problem. The 4% polyacrylamide gel is prepared containing 8 M urea. Samples from the RdRp assay are resuspended in 1% (w/v) SDS and heated at 60°C just prior to electrophoresis. The double-stranded viral RNA is not denatured either at this temperature in 1% (w/v) SDS or by the 8 M urea in the gel at room temperature, but any secondary structure in the single-stranded RNA is removed, thus giving good resolution. After removal of the urea and staining with ethidium bromide, the double- and single-stranded viral RNA can often be visualized. The gel is then dried down and the labelled RdRp products are detected by autoradiography.

Protocol 10. Urea–polyacrylamide gel electrophoresis

Materials

- gel electrophoresis apparatus (BioRad Mini-Protean II; gel size approximately 80 × 70 × 0.8 mm)
- acrylamide solution: 40% (w/v) acrylamide, 0.4% (w/v) *N,N′*-methylene-bisacrylamide
- urea/SDS: 10 M urea containing 1.25% (w/v) sodium dodecyl sulphate (SDS)
- 50 × TAE buffer: 2 M Tris–acetate, 50 mM EDTA (pH 8.2)
- loading buffer: 0.05% (w/v) bromophenol blue, 0.05% (w/v) xylene cyanol, 25% (v/v) glycerol

Method

1. For each gel of 5 ml, combine 0.5 ml of acrylamide solution, 4 ml of urea/SDS, 0.1 ml of 50 × TAE buffer, 0.05 ml of 10% (w/v) ammonium persulphate, 0.3 ml of water, and 5 μl of *N,N,N′,N′*-tetramethyethylenediamine (TEMED). Insert a ten-well comb and allow the gel to polymerize for approximately 1 h.
2. Remove the comb and pre-electrophorese the gel for 1 h with 1 × TAE, 0.1% (w/v) SDS in the upper electrode chamber and 1 × TAE, 0.1% SDS, 8 M urea in the lower electrode chamber.
3. Dissolve the samples from *Protocol 7* in 9 μl of water. Add 1 μl of 10% (w/v) SDS and heat for 20 minutes at 60°C.
4. Add 2–3 μl of loading buffer to the samples. Flush out the wells with a syringe and needle containing upper electrode buffer and load the samples immediately. Electrophorese at 100 V for 4 h or 30 V overnight at room temperature.

Protocol 10. *Continued*

5. Remove the gel from the tank and carefully part the glass plates. The gel may stick to one of the plates. Immerse it in water and gently ease the gel off the plate by rubbing the edge with a finger. The gel is 'sloppy' but not too fragile. Wash it three times in fresh water (100 ml) for 15 min to remove urea. During the last wash, add 5 μl of 10 mg/ml ethidium bromide to stain the RNA.
6. Place on a UV transilluminator to observe bands. Single-stranded RNA migrates more slowly than double-stranded RNA in this gel system. Dry the gel and expose it to X-ray film.

3.3 Solubilization of CMV RdRp

There are a number of commercially available detergents which are capable of solubilizing proteins present in the crude fraction. The main criterion for successful solubilization of the RdRp is, of course, activity in the solubilized fraction (i.e. what proportion of RdRp is solubilized in the active form). We define a solubilized RdRp as one which is found in the supernatant after centrifugation at 100 000 *g* for one hour. This rather arbitrary standard must be applied with care; in high-density media non-solubilized RdRp may also be found in the supernatant. For a detailed discussion of this problem see ref. 48.

Tests should be performed to ascertain the usefulness of a number of detergents at a range of concentrations. The choice of detergents is likely to be influenced by both cost and whether the detergent is compatible with the chromatography media used in the later stages of purification. Stock solutions of 10% are usually prepared in homogenization buffer. Protease inhibitors and reducing reagents should be present at twice their final concentration in the solubilization medium. Initially, try a number of different detergents at approximately 1%. We perform these initial tests on ice in Beckman 11 × 34 mm tubes. Add 50–100 μl of detergent to an equal volume of crude membrane fraction (5 mg/ml), mix gently, and leave on ice for 1 h. Centrifuge the tubes in a Beckman TL 100.2 rotor at 100 000 *g* for 1 h at 2°C. Remove the supernatant and resuspend the pellet in an equal volume of solubilization buffer. Assay both the supernatant and the resuspended pellet immediately and 1, 2, and 8 h afterwards. This time course will provide some indication of the stability of RdRp in the detergent. After initial tests, a few successful detergents should be tested for concentration effects; repeat the tests using a range of final concentrations from 0.1% to 5%.

We have tested a number of different detergents, and tried combinations of most. Solubilization ranged from approximately zero with all concentrations of *n*-dodecyl β-D-maltoside to approximately 70% with 2% (v/v) *n*-octyl β-D-glucopyranoside. 3-[3-cholamidopropyl)-dimethyl-ammonio]-1-propane sulphonate (CHAPS) at 5% (v/v) produced 40% solubilization, while Triton

X-100 or Nondiet P-40 (NP40) at 1% (w/v) approach 55–65% efficiency. Because of the cost, we routinely use the non-ionic detergent Triton X-100.

A combination of high concentrations of detergent and salt can also be effective in solubilizing RdRp. Quadt and Jaspers (26) used 10% (w/v) NP40 and 750 mM KCl for 30 min at 4°C to solubilize the RdRp of BMV and AlMV (8). This treatment also works well for CMV (30), and may be useful for a number of other viruses (*Protocol 11*). This method has the advantage of completely removing the RNA template during solubilization, thereby avoiding the micrococcal nuclease step. Interestingly, the relatively crude CMV RdRp prepared by this method is capable of completely replicating CMV satellite RNA. However, as high salt concentrations inhibit RdRp activity, it is necessary to remove the salt by dialysis in order to determine the efficiency of solubilization. Alternatively, salt removal can be combined with a subsequent purification step such as glycerol gradient centrifugation or size exclusion chromatography.

Protocol 11. NP40/KCl solubilization of RdRp

Materials

- crude RdRp preparation (*Protocol 6*)
- NP40/KCl: 25% (v/v) NP40, 1.875 M KCl
- Sephacryl S-400

Method

1. To the crude RdRp fraction add NP40 and KCl to final concentrations of 10% and 750 mM, respectively (i.e. 3 ml of RdRp plus 2 ml of NP40/KCl).
2. Stir for 30 min and centrifuge at 4°C for 1 h at 100 000 *g*.
3. Discard the pellet, add a half volume of buffer B (*Protocol 6*) to the supernatant and load the mixture on to a Sephacryl S-400 column (see Section 4.1) or a glycerol gradient (see Section 4.2).

4. Further purification of RdRp

Having solubilized the RdRp, the next stage of purification must be considered. Possibilities include fractionation according to size by size-exclusion chromatography, centrifugation through glycerol gradients, or fractionation according to electrical charge on ion-exchange media. The main advantage of ion exchange is that it allows the sample to be concentrated; the main disadvantage being that it relies upon changes in salt concentration and so each fraction needs to be dialysed to negate the effect of salt on RdRp activity. If high salt levels have been used for solubilization (*Protocol 11*), the solubilized fraction will have to be dialysed extensively before loading on to the ion-exchange media. As solubilizations are usually performed in small volumes, we routinely use size fractionation as the next step in the purification of CMV RdRp. The dilute RdRp is then concentrated by ion-exchange chromatography.

4.1 Size-exclusion chromatography

As detergent–protein complexes may exhibit twice the apparent M_r expected of the protein complexes alone, and as non-denaturing detergents such as Triton X-100 may not fully denature protein complexes, media with large pore sizes are required for fractionation of RdRp. It is usual to reduce the amount of detergent (and protease inhibitors) in the column buffer, but reducing the concentration too far can also reduce the level of resolution and the degree of purification obtained. For maximum resolution, a range of detergents in the column buffer should be tested. The optimal level of detergent in the buffer (always greater than the critical micelle concentration) will be a compromise between maximal resolution and high RdRp recovery.

A Pharmacia C26/70 column is packed to a bed height of approximately 65 cm with Sephacryl S-400 equilibrated in column buffer (50 mM Tris–HCl (pH 8.2), 15 mM $MgCl_2$, 5% (v/v) glycerol, 1 mM DTT, 0.1 μg/ml leupeptin, 0.1 μg/ml pepstatin, 5 μg/ml aprotinin, 0.1% (v/v) Triton X-100) at 4°C. For CMV we use Triton X-100, but this may differ for other virus RdRps. Approximately 5–7 ml of solubilized RdRp is loaded on to the top of the column and passed through the column in buffer at 1 ml/min with the aid of a peristaltic pump. Fractions of 2 ml are collected, and 12.5 μl of each is tested for RdRp activity.

Figure 1 shows the elution profile of CMV RdRp. The active peak elutes in or close to the void volume as a clear solution before the green chlorophyll-containing major peak. It can be stored at −70°C for at least 3 months without loss of double-stranded RNA synthesizing activity.

CMV RdRp solubilized by the standard method of using Triton X-100 is not completely template-free at this stage, although the addition of CMV RNA significantly increases incorporation of labelled UTP into double-stranded RNA products. However, if the NP40/KCl solubilization method is used (*Protocol 11*) the RdRp is essentially template-dependent even without micrococcal nuclease digestion. This indicates that during the solubilization process the high concentrations of detergent and salt have the effect of removing template from the RdRp. RdRp fractions no longer contain the high levels of KCl used for solubilization and, therefore, do not need to be dialysed prior to assay.

4.2 Glycerol gradients

Although we rarely use glycerol gradients, they have been used successfully as a first stage in the purification process for solubilized BMV and AlMV

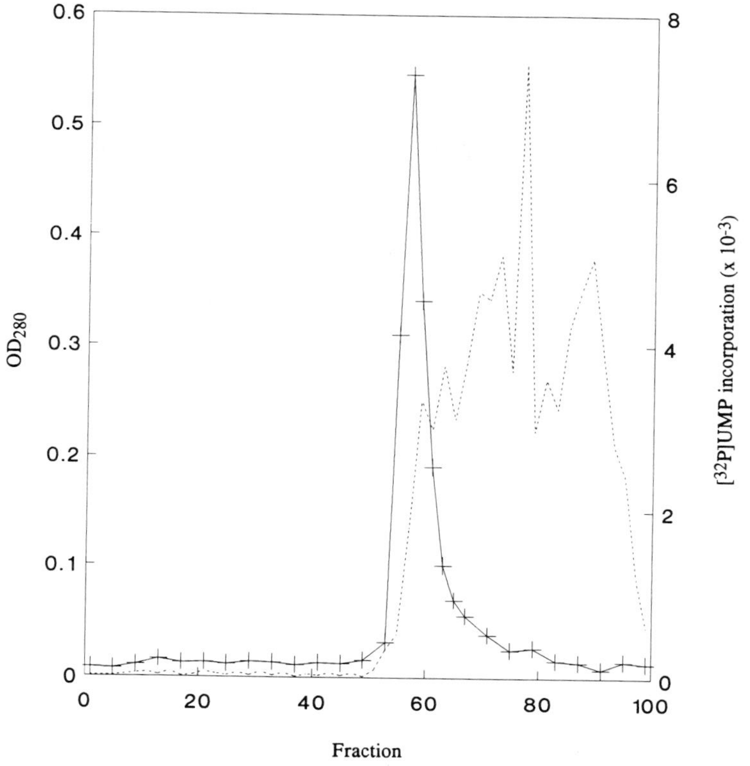

Figure 1. Sephacryl S-400 chromatography of solubilized CMV RdRp. Approximately 5 ml of solubilized crude membrane fraction was applied to the column. Fractions of 2 ml were collected and assayed for RdRp activity (continuous line). The protein content of each fraction (OD_{280}) is represented by the broken line.

RdRps (8, 26). They are useful when high concentrations of detergent and salt have been used in solubilization. The solubilized RdRp is layered on to a 22–44% (v/v) glycerol gradient in buffer B (*Protocol 6*) and centrifuged in a Beckman SW28 rotor at 80 000 *g* for 20 h at 2°C. When 2 ml fractions are assayed, the main peak of activity is found approximately halfway down the gradient. The fractions do not need to be dialysed prior to assay, and the RdRp is template-dependent.

4.3 Ion-exchange chromatography

Chromatography on anion-exchange columns has been a useful next step in purifying RdRp. Media used successfully include Sepharose Q, Mono Q, and DEAE Biogel A. As mentioned above, one of the main advantages of ion exchange is the ability to concentrate dilute protein samples. The active fractions from the Sephacryl S-400 column are applied to a 1 × 10 cm

Sepharose Q column equilibrated with column buffer at 4°C. After washing with 50 ml of column buffer at approximately 1 ml/min, proteins bound to the column are eluted by a 0–1 M KCl gradient in column buffer at 10 mM/ml. Fractions of 3 ml are collected.

For rapid dialysis, 100 μl of each fraction is dialysed against column buffer using a dialysis manifold (Gibco-BRL). After dialysing for 1 h the KCl concentration in each fraction is approximately the same, and low enough to allow RdRp activity to be assayed. Once the active fractions are identified, the remaining 2.9 ml of each fraction are dialysed extensively and stored in aliquots at −70°C.

Figure 2 shows the Sepharose Q Fast Flow elution profile of CMV RdRp. The RdRp in peak 1 (centred on fraction 20) is completely template-dependent (i.e. addition of CMV RNA is required for incorporation of labelled nucleotides). In contrast, there is significant incorporation with the

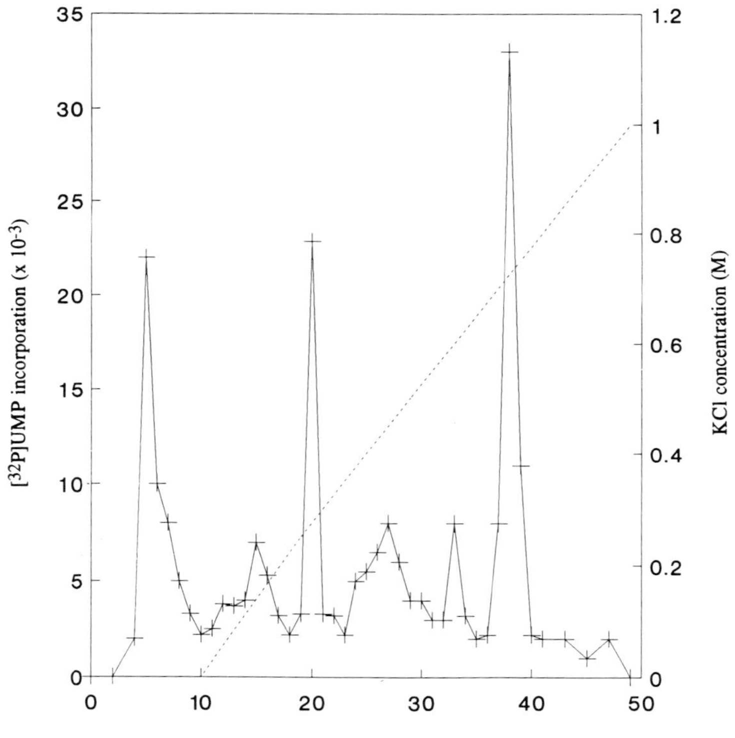

Figure 2. Sepharose Q Fast Flow chromatography of CMV RdRp. The active fractions from the Sephacryl S-400 column were applied to the Sepharose column, and proteins were eluted with a linear gradient of KCl (broken line). After dialysis each fraction was assayed for RdRp activity (continuous line).

fractions in peak 2 (centred on fraction 38) even without addition of viral RNA to the RdRp assays. This incorporation, stimulated at least six-fold by the addition of viral RNA, probably occurs because peak 2 contains RdRp still complexed to template. Not surprisingly, this peak is not observed if the RdRp sample applied to the column has been solubilized with NP40/KCl.

4.4 Further purification of RdRp

The RdRp can be purified further by additional rounds of higher resolution size-exclusion and ion-exchange chromatography, essentially as described above. We use Superose 6 and Mono Q columns attached to an FPLC apparatus. The glycerol present in the column buffer creates substantial back pressure, and so we run both columns at approximately 0.2 ml/min.

Alternative chromatography media may allow further purification of the RdRp or speed up the whole process. Affinity chromatography has been the most successfully applied method for the purification of integral membrane proteins, including multisubunit receptor complexes, with routine purifications of between 1000- and 10 000-fold. Three types of affinity chromatography for RdRp purification are currently under investigation:

(a) *Binding of RdRp to immobilized viral RNA*. There are promoter regions on the viral RNA which are recognized by RdRp at the beginning of positive and negative strand synthesis (see Section 1). It may be possible to immobilize RNA containing these regions to the column resin via hydrophilic spacer arm. Such columns would allow very rapid purification, provided RNase activity in the sample could be inhibited.

(b) *Lectins*. Although there is no information regarding glycosylation of one or more polypeptides present in the RdRp, these carbohydrate-binding compounds offer a rapid and mild method by which to purify plasma membrane proteins.

(c) *Antibodies prepared against virally-encoded components of the RdRp*. Although this technique is rapid, yielding very pure proteins, the elution conditions are usually harsh owing to the high-binding affinity of antibodies for antigens. It may be possible to use antibodies with lower affinities or polyclonal antibodies raised to peptides. Bound proteins can be eluted under mild conditions with excess peptide.

5. Analysis of RdRp products

Figure 3 shows the products of CMV RdRp after two different stages of purification. Each fraction had been treated with micrococcal nuclease to remove any residual endogenous RNA. Addition of viral RNA typically results in some partial products; clearer autoradiographs are often obtained using synthetic T7 RNA transcripts as templates. In the early stages of

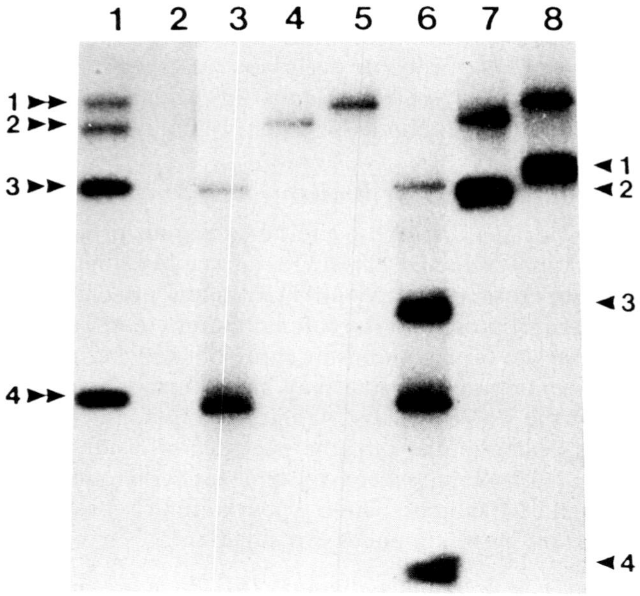

Figure 3. Non-denaturing agarose gel electrophoresis of products of RdRp reactions. Lanes 1 to 5, products of RdRp after Sepharose Q Fast Flow chromatography and micrococcal nuclease treatment; lanes 6 to 8, products of RdRp after Mono Q chromatography. Reactions were programmed with: lane 1, CMV RNA; lane 2, no added RNA; lanes 3 and 6, RNA 3; lanes 4 and 7, RNA 2; lanes 5 and 8, RNA 1. Double arrowheads indicate double-stranded RNA bands. Single arrowheads indicate single-stranded RNA. Bands were detected by autoradiography. (Reproduced from reference 6 with kind permission from *Cell Press.*)

purification only double-stranded products are detected. However, the purer RdRp also produces significant amounts of single-stranded RNA (*Figure 3*, lanes 6, 7, and 8). This is in contrast to the FHV RdRp which is capable of single-stranded RNA synthesis at the earliest stages of purification. Solubilization of FHV RdRp abolishes single-stranded RNA production, and it has been proposed that the membrane environment is essential for full FHV RdRp activity (33). As Triton X-100 is a mild detergent, it is possible that the CMV RdRp is still associated with a number of membrane components.

5.1 RNase A digestion

Nuclease tests using RNase A can be used to distinguish between single- and double-stranded RNA. Resuspend the products of the RdRp assay in 10 μl of 0.1 × SSC or 2 × SSC (see Chapter 5, *Protocol 2* for composition). Add 0.05 μg of RNase A and incubate at 37°C for 30 min. Precipitate the products and analyse by one of the methods in *Protocol 7*. If the labelled RNA is digested in 0.1 × SSC and 2 × SSC it is single-stranded, but if digested only in 0.1 × SSC it is double-stranded.

5.2 RNase H analysis

Detection of labelled single-stranded RNA by gel electrophoresis does not prove that the RdRp is capable of completing the replication cycle. It is theoretically possible that in some cases the RdRp might produce labelled negative strand RNA molecules which are then released from the replication complex. Although no positive genomic sense RNA synthesis would have occurred, labelled single-stranded RNA would be detected. Another possible source of labelled RNA is contaminating terminal transferase activity in the RdRp preparation. Therefore, any labelled single-stranded RNA needs to be analysed rigorously to ensure that the RNA is indeed newly synthesized genomic RNA, and that it is evenly labelled along its length.

To address these considerations, a reasonable amount of labelled RNA is required. After detection of single-stranded RNA production by RdRp using the standard assay (*Protocol 7*), we repeat the assay with a five-fold greater amount of labelled UTP. The newly synthesized RNA is then extracted from the gel and analysed by RNase H digestion. This enzyme digests the RNA of a RNA:DNA hybrid, but RNA not hybridized to DNA is left intact. By annealing a short oligonucleotide to an RNA, it is possible to cut the RNA molecule at a defined sequence. As shown in *Figure 4*, this technique has been used to analyse the products of CMV RdRp. Single-stranded RNA 1 shown in *Figure 3*, lane 8 was purified from a low-melting point agarose gel. Oligonucleotides, P1 and P2 (*Figure 4b*), with sequences complementary to internal sequences of RNA 1, were then hybridized to the extracted RNA. Digestion with RNase H hydrolysed the RNA specifically at the sites of hybridization with P1 and P2 to produce fragments of the calculated sizes (1.0 kb, 1.5 kb, and 0.8 kb) (*Figure 4a*, lane 3) with labelled nucleotide incorporation in proportion to their length. Hence, the single-stranded RNA 1 consists mainly of positive strand RNA uniformly labelled along its length.

The small amount of RNA not digested by RNase H (*Figure 4a*, lane 3) was shown to be the RNA 1 negative strand. In similar experiments using oligonucleotides P3 and P4 with sequences complementary to the internal sequences of the RNA 1 negative strand (*Figure 4b*) and RNase H digestion, most of the RNA (i.e. positive strand) remain undigested, but products of the sizes (0.9 kb, 1.8 kb, and 0.6 kb) calculated for specific cleavage of RNA 1 negative strand at the sites of hybridization with P3 and P4 were formed (*Figure 4a*, lane 4).

The results in *Figure 4a* show that the ratio of positive to negative strand RNA 1 synthesis with CMV RdRp is approximately 7:1. Similar results were obtained with RNA 2 and RNA 3. These RNase H experiments show conclusively whether the isolated RdRp is capable of synthesizing both positive and negative sense RNA, and provide a method for differentiating between

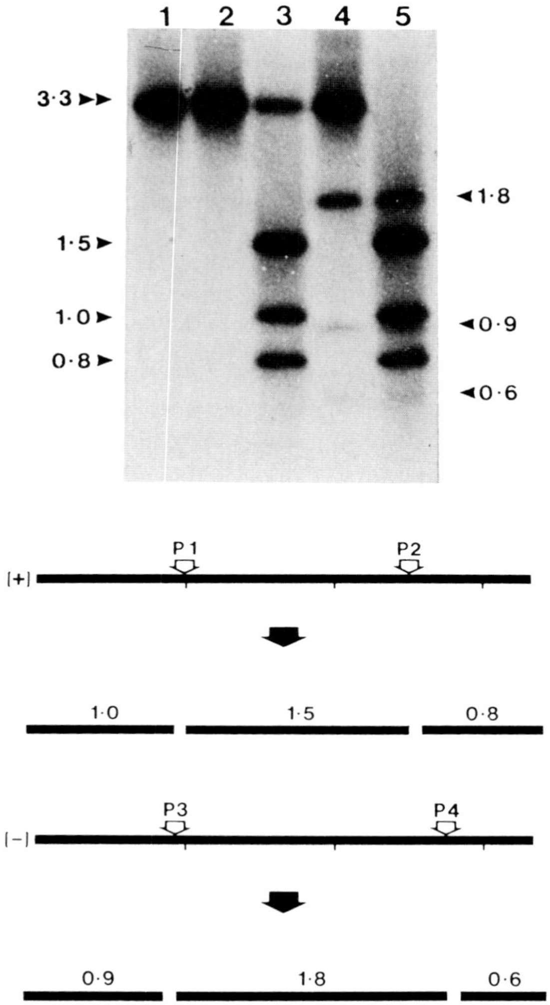

synthesized RNA and RNA which has become terminally labelled by contaminating enzymes.

6. Virus-encoded and host polypeptides in RdRp fractions

6.1 Western blotting

Analysis of the protein content of the crude-membrane RdRp fraction by SDS–PAGE often fails to reveal the presence of the viral polypeptides due to

Figure 4. Analysis of single-stranded RNA 1 synthesized in an RdRp reaction. (a) The band corresponding to RNA 1 (*Figure 3*, lane 8) was extracted from the gel, annealed to oligonucleotides and treated with RNase H. The products were then electrophoresed through 1.2% agarose–formaldehyde gels and detected by autoradiography. Lane 1, single-stranded RNA 1 reaction product; lane 2, single-stranded RNA 1 reaction product + RNase H; lane 3, single-stranded RNA 1 reaction product + RNase H + P1 + P2; lane 4, single-stranded RNA 1 reaction product + RNase H + P3 + P4; lane 5, single-stranded RNA 1 reaction product + RNase H + P1 + P2 + P3 + P4. Sizes in kb are shown on the sides of the gel. The genomic RNA is indicated by the double arrowhead and the RNase H products by single arrowheads. (b) Diagram showing the expected products from RNase H digestion of positive strand RNA 1 annealed to P1 and P2 or negative strand annealed to P3 and P4. The sequences of the oligonucleotides are : P1,5′-ATAGGTCATACCATTG-3′ (nucleotides 988–1003); P2, 5′-GACAGCATGAAGTTTC-3′ (nucleotides 2488–2503); P3, 5′-GTCTTATGTTCACGAT-3′ (nucleotides 928–948); P4, 5′-AAGTGAGGAAGTCTGT-3′ (nucleotides 2717–2733). (Reproduced from reference 6 with kind permission from *Cell Press*.)

the high levels of contaminating host polypeptides. However, as the RdRp is purified further and most of the host polypeptides are removed, virally-encoded polypeptides can often be detected by standard silver-staining. To prove that these polypeptides are indeed virally-encoded, Western blot analysis should be performed. *Figure 5*, lanes 3 and 4 show polypeptides present in partially purified RdRp and an equivalent fraction from uninfected plants. Antibodies were raised against the virally-encoded proteins 1a and 2a produced in *E. coli* from an expression vector. In Western blots, the 1a antibodies reacted specifically with the 1a protein expressed in *E. coli* and with a protein of the same electrophoretic mobility in the RdRp fraction (*Figure 5*, lanes 9 and 11). Similarly, the 2a antibodies reacted specifically with the 2a protein expressed in *E. coli* and with a protein of the same mobility in the RdRp fraction (*Figure 5*, lanes 12 and 15).

SDS–PAGE may require some modifications when used with membrane proteins. SDS may not denature proteins fully, and in such cases 8 M urea should be included in the sample and gel buffers. Also, some membrane proteins such as H^+-ATPase are aggregated by boiling. An alternative is to incubate samples at 25–60°C for 20–90 min.

6.2 Antibody-linked polymerase assay

Evidence that the viral proteins are subunits of the RdRp and have not fortuitously co-purified with it can be obtained from an antibody-linked polymerase assay (ALPA; *Protocol 12*). This technique, first described by Van der Meer *et al.* (49), is shown diagrammatically in *Figure 6*. Polypeptides in the RdRp are fractionated by SDS–PAGE, transferred to nitrocellulose, and incubated with the viral antisera exactly as described for Western blotting. The isolated RdRp is then added and incubated with the immobilized polypeptide–antibody complexes. As antibodies have more than one antigen binding site, a proportion will be capable of binding the corresponding

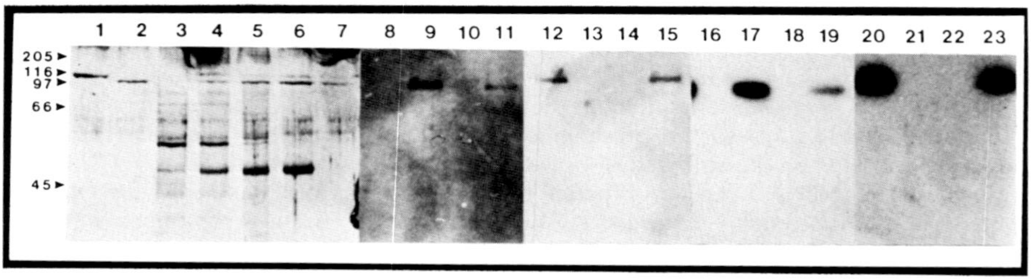

Figure 5. Gel electrophoretic analysis of proteins of RdRp fractions. Proteins were subjected to SDS–PAGE and detected by silver staining (lanes 1–7), by Western blotting and probing with antiserum to CMV protein 1a (lanes 8–11) or antiserum to CMV protein 2a (lanes 12–15), or by blotting followed by ALPA with antiserum to 1a (lanes 16–19) or with antiserum to protein 2a (lanes 20–23). Lanes 1, 8, 12, 16, and 20, protein 2a expressed in *E. coli*; lanes 2, 9, 13, 17, and 21, protein 1a expressed in *E. coli*; lanes 3, 10, 14, 18, and 22, Sephacryl S-400 chromatography fraction from healthy plant; lanes 4, 11, 15, 19, and 23, Sephacryl S-400 chromatography fraction from CMV-infected plant; lane 5, Sepharose Q Fast Flow fraction from CMV-infected plant; lane 6, Mono Q fraction from CMV-infected plant; lane 7, inactive fraction from Mono Q column. The M_rs of marker proteins are shown to the left of the gel. (Reproduced from reference 6 with kind permission from *Cell Press*.)

polypeptide in the RdRp, so forming a 'bridge' between the immobilized and RdRp polypeptides. Therefore, binding of RdRp to the immobilized polypeptide shows that the polypeptide is a component of RdRp. The immobilized polypeptide–antibody–RdRp complex is then detected by autoradiography after allowing the RdRp to perform RNA synthesis in the presence of labelled nucleotides.

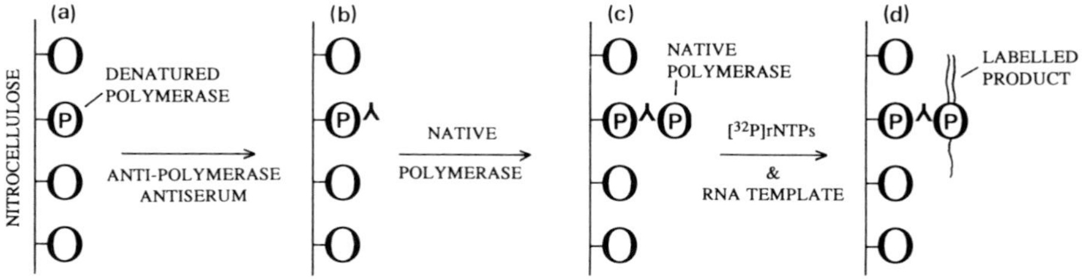

Figure 6. Diagram of antibody-linked polymerase assay (ALPA) (44). (a) Proteins are separated by SDS–PAGE and transferred to nitrocellulose; (b) the immobilized proteins are incubated with antiserum raised against a viral protein; (c) the immobilized proteins are then incubated with active RdRp; (d) the denatured polymerase–antibody–RdRp complex is then incubated with labelled nucleotides and RNA template. The RdRp synthesizes labelled double-stranded RNA which can be detected by autoradiography.

Protocol 12. Antibody-linked polymerase assay

Materials

- TBS buffer: 50 mM Tris–HCL (pH 7.4), 200 mM NaCl
- TBS/TB buffer: TBS buffer containing 0.2% (v/v) Tween 20 and 3% (w/v) BSA
- TBS/TS buffer: TBS buffer containing 0.2% (v/v) Triton X-100 and 0.2% (w/v) SDS
- polymerase buffer: 50 mM Tris–HCl (pH 8.2), 4% (v/v) glycerol, 10 mM $MgCl_2$
- PDT buffer: polymerase buffer containing 10 mM DTT, 0.1% (v/v) Triton X-100
- assay buffer: polymerase buffer containing 10 mM DTT, 1 mM GTP, 1 mM ATP, 1 mM CTP, 0.15 μM [^{32}P]UTP (10 mCi/ml, 800 Ci/mmol)
- CMV viral RNA
- TCA solution: 10% (w/v) trichloroacetic acid (TCA) containing 1% (w/v) sodium pyrophosphate
- antiserum against RdRp
- pre-immune serum

Method

1. Separate proteins by SDS–PAGE and transfer them to nitrocellulose sheets. Treat the sheets as follows.
2. Soak for 1 h in TBS/TB buffer at room temperature with gentle agitation.
3. Incubate overnight at room temperature in TBS/TB buffer containing pre-immune serum or RdRp antiserum.
4. Wash for 1 h in four changes of TBS/TS buffer.
5. Wash for 30 min in three changes of TBS buffer.
6. Equilibrate by washing for 30 min in three changes of polymerase buffer.
7. Incubate for 4 h at 4°C in PDT buffer containing partially purified RdRp.
8. Wash at 4°C for 2–3 h with five changes of PDT buffer.
9. Incubate at 30°C for 30 min in assay buffer containing 10 μg CMV viral RNA.
10. Quickly wash in polymerase buffer and transfer to ice-cold TCA solution. Wash for 2 h with five changes of ice-cold TCA solution.
11. Dry and expose to X-ray film at −80°C with intensifying screens.

Lanes 16 to 23 of *Figure 5* show the results of such an assay carried out using CMV RdRp and antibodies directed against the two virally-encoded polypeptides 1a and 2a (see Section 6.1). A partially purified RdRp fraction was subjected to electrophoresis, blotted on to nitrocellulose and incubated with either 1a or 2a antibodies. The membranes were then incubated with RdRp to allow the RdRp to bind to the second antigen-binding site on the IgG antibodies. After further incubation with assay buffer, the labelled products were detected by autoradiography. Bands in the positions of proteins 1a or 2a were detected using 1a antibodies (lanes 17 and 19) or 2a antibodies (lanes 20 and 23), respectively, but not with pre-immune serum. Therefore, by using only a fairly crude RdRp preparation, the ALPA provides strong evidence that polypeptides 1a and 2a are indeed components of the CMV RdRp.

There are two potential problems with the ALPA. The first is that many antisera contain significant levels of RNase, and therefore in a number of cases the antibodies need to be purified beforehand on protein A affinity columns (BioRad). Another potential problem is the variable ability of an antiserum to inhibit RdRp activity. Antibodies directed against polypeptide components of the RdRp have very different effects on enzymatic activity; if the antibodies bind at or near regions involved in the control of RdRp activity or areas directly involved in the enzymatic process, the RdRp activity might be greatly diminished or even abolished. If such an antiserum were used in the ALPA, a negligible or very weak signal would result, even though the antigen is part of the RdRp. An example of such a polyclonal antiserum is one we produced using a peptide corresponding to a putative enzymatic site in CMV polypeptide 2a. It is excellent in Western blotting experiments, is highly efficient in inhibiting the RdRp, but has never worked in ALPA.

Other antibodies may bind to different regions equally well but without such an effect on RdRp activity. These differing effects are most readily observed with monoclonal antibodies (mAbs). Two mAbs might be equally effective in Western blotting analysis, but differ dramatically in their effectiveness in an ALPA. Such differences can also be observed with polyclonal antisera, but the differences are usually less dramatic. By raising polyclonal antisera to different regions of the virally encoded protein, the proportion of antibodies which inhibit the RdRp—and therefore adversely effect the ALPA—can be varied.

6.3 Inhibition studies

Antibodies which inhibit RdRp activity are useful in confirming that a polypeptide is a subunit of the RdRp instead of merely a contaminant of the preparation. Although they might not be suitable for an ALPA, the ability of an antibody to specifically inhibit the RdRp provides good evidence that the immunizing antigen forms part of the RdRp complex. It may be necessary to purify the antibodies prior to use and, as with the ALPA, antisera can vary significantly in their ability to inhibit the RdRp. Ten microlitres of RdRp are incubated with 1–5 μl of antiserum or pre-immune serum for 1 h on ice and then used in a polymerase assay (*Protocol 7*). Some polyclonal antibodies are highly effective, such as those raised against the whole RdRp or against a region thought to be involved in the polymerization step of RNA synthesis. Others are far less impressive, including polyclonals against the whole virally-encoded polypeptides 1a or 2a. This shows that if the antiserum fails to be useful in an ALPA, it may often excel in inhibition studies and vice versa.

7. RNA–protein interactions

The RdRp binds to the 3′ end of genomic RNA to initiate negative strand RNA synthesis and to the 3′ end of the negative strand RNA to initiate genomic RNA synthesis. The replication strategy of some viruses also includes subgenomic RNA synthesis which involves binding of the RdRp internally on negative strand RNA (see Section 1). These interactions are thought to be specific because the RdRp from one virus will not replicate host mRNA or viral RNA from another virus (except when the viruses are very closely related).

Interaction between RNA and proteins is an area of research which is expanding rapidly. Two techniques are commonly used. The first is RNA mobility shift analysis. The principle behind this technique is simple—a labelled RNA transcript complexed to a protein migrates through a non-denaturing polyacrylamide gel more slowly than an unbound transcript. This method should be particularly useful in studying RNA sequences involved in the interaction between RdRp and viral RNA. For example, short labelled viral RNA transcripts could be added to the RdRp. Those that are retarded during electrophoresis contain the site of RdRp binding. *In vitro* mutagenesis could then be used to identify the bases which are directly involved. The technique of RNA mobility shift also allows proteins involved in the interaction to be identified, as the addition of antibodies against one of the polypeptide components of an RNA–protein complex further reduces mobility on a gel.

Protocol 13. Preparation of RNA transcripts for analysis of RNA–protein interactions

Materials

- vertical gel electrophoresis apparatus (for example BioRad Sequi-Gen Nucleic Acid Sequencing System)
- 10 × RNA transcription buffer: 400 mM Tris–HCl (pH 7.5), 60 mM $MgCl_2$, 20 mM spermidine, 100 mM DTT, 1 mg/ml BSA

Protocol 13. *Continued*

- 10 × NTPs: 5 mM ATP, 5 mM CTP, 5 mM GTP
- linearized plasmid DNA (0.2 mg/ml)
- T7 RNA polymerase (10 U/μl)
- [α-^{32}P]UTP (10 Ci/mmol, 800 Ci/mmol)
- gel loading buffer: 95% (v/v) formamide, 20 mM EDTA, 0.05% (w/v) bromophenol blue, 0.05% (w/v) xylene cyanol (or USB stop solution supplied with Sequenase kit)
- 10 × TBE: 0.9 M Tris–borate (pH 8.0), 0.02 M EDTA
- solution 1: 1 × TBE, 15% (w/v) acrylamide, 0.75% (w/v) *N,N′*-methylenebisacrylamide, 7 M urea
- solution 2: 1 × TBE, 7 M urea

Method

Caution: RNA transcripts prepared by this protocol require large amounts of radiolabel. Extreme care should be exercised during all stages.

1. Add the following to a microcentrifuge tube and incubate at 37°C for 1 h.
   - 10 × RNA transcription buffer 2 μl
   - 10 × NTPs 2 μl
   - RNase inhibitor (InhibitAce) 1 μl
   - DNA template 2 μl
   - [α-^{32}P]UTP 11 μl
   - T7 RNA polymerase 2 μl
2. During incubation, prepare a denaturing polyacrylamide gel. Mix 12 ml of solution 1, 24 ml of solution 2, 165 μl of 10% (w/v) ammonium persulphate, and 24 μl of TEMED. Quickly pour the gel between the plates assembled with 0.4 mm spacers. One of the glass plates should be siliconized prior to use (for example using Repelcote; BDH). The gel should polymerize within 20 min, but allow it to stand for 1 h.
3. Add 15 μl of 5 M ammonium acetate and 100 μl of ethanol to the T7 polymerase reaction from step 1. Incubate at −20°C for 30 min and centrifuge for 15 min in a microcentrifuge.
4. Remove the supernatant, dry the RNA, and resuspend it in 7 μl of gel loading buffer. (Samples can be stored at −20°C.)
5. Pre-electrophorese the gel in 1 × TBE at 1500 V until the temperature of the gel reaches 50°C.
6. Denature the RNA samples by heating to 100°C for 3 min. Immediately load them on to the gel. Electrophorese the gel at 1500 V until the bromophenol blue dye reaches the bottom.

7. Carefully remove the gel from the electrophoresis tank. **Caution**: the lower chamber buffer contains substantial amounts of radioactivity.
8. Separate the glass plates; the gel should stick to the non-siliconized plate. Cover the gel with Saran Wrap or Clingfilm.
9. In the darkroom, cut a piece of X-ray film slightly smaller than the gel and lay it on top of the wrapped gel; mark its position with a pen. A 20 sec exposure should be sufficient if transcript synthesis has been successful.
10. Develop the X-ray film and, by aligning the marker pen marks, determine the location of the labelled RNA. Cut out the appropriate gel slice and place it in a 1.5 ml microcentrifuge tube.
11. Add the following to the tube and incubate at 65°C for 2 h.

• water	180 μl
• 3 M sodium acetate (pH 5.5)	20 μl
• 0.25 M EDTA	1 μl
• 20% (w/v) SDS	1 μl

12. Remove the buffer to a fresh microcentrifuge tube and add 500 μl of ethanol. Incubate at −20°C for at least 1 h.
13. Pellet the RNA by centrifugation for 15 min. Dry the pellet, resuspend in 50 μl of water and measure the radioactivity (Cerenkov) of 1 μl in a scintillation counter. We usually obtain 5×10^5 to 1×10^6 c.p.m./μl, but this depends on the sequence and length of the transcript. Adjust the volume to obtain 1×10^5 c.p.m./μl and store at −20°C.

Protocol 14. RNA mobility shift assay

Materials

- gel electrophoresis apparatus (for example BioRad Protean II)
- Hepes buffer: 10 mM Hepes–NaOH (pH 7.5), 40 mM KCl, 10 mM $MgCl_2$, 5 mM DTT, 8.5% (v/v) glycerol
- acrylamide solution: 39% (w/v) acrylamide, 1% (w/v) *N,N'*-methylene-bisacrylamide
- 10 × Tris–glycine: 500 mM Tris base, 500 mM glycine (pH 8.8)
- ^{32}P-labelled RNA from *Protocol 13*
- purified RdRp
- bromophenol blue/glycerol: 0.02% (w/v) bromophenol blue in 10% (v/v) glycerol 10 mg/ml wheatgerm tRNA

Protocol 14. *Continued*

Method

1. Dialyse the RdRp fraction against Hepes buffer.
2. Prepare a native polyacryalmide gel at room temperature containing 4% (w/v) polyacryalmide, 1 × Tris–glycine, 0.1% (w/v) ammonium persulphate, 0.1% (v/v) TEMED. Pre-electrophorese the gel at 13 V/cm for 30 min in 1 × Tris–glycine running buffer.
3. Set up a binding reaction containing 10 μl of RdRp, 3 μl of labelled RNA, and 2 μl of tRNA. Incubate for 15 min at room temperature.
4. Add 1 μl of 0.5 mg/ml RNase A and incubate for 10 min at room temperature.
5. Add 2 μl of glycerol to the binding reaction and load on to the gel. In a separate well, load a small amount of glycerol/bromophenol blue. Electrophorese at 13 V/cm until the bromophenol blue reaches the bottom of gel. Dry the gel and autoradiograph it.

Protocol 15. UV cross-linking

Materials

- UV cross-linker (Stratalinker 1800; Stratagene)
- loading buffer: 8% (w/v) SDS, 0.05% (w/v) bromophenol blue, 3 M 2-mercaptoethanol, 12% (v/v) glycerol

Method

1. Carry out step 1 of *Protocol 14*.
2. Add 14 μl of RdRp and 1 μl of RNA to a 1.5 ml microcentrifuge tube. Incubate for 10 min at room temperature.
3. Open the microcentrifuge tube and irradiate with 1.8 J of UV light.
4. Digest unprotected RNA for 20 min at 37°C with 1 μl of 0.5 mg/ml RNase A.
5. Add 0.3 volumes of loading buffer and boil for 5 min. Electrophorese through a 10% (w/v) SDS–polyacrylamide gel until the bromophenol blue reaches the bottom. Stain, dry, and autoradiograph the gel.

There are two important points which should be considered. The first is that the RNA transcript should be 100–500 nucleotides in length. As the viral RNAs of CMV range from 2100 to 3300 nucleotides, we subcloned the desired regions into another transcription vector. Transcripts of 150 nucleotides were then synthesized essentially as described in *Protocol 13*. Longer transcripts may require the addition of some unlabelled UTP (25 μM). We find that the best results are obtained with RNA transcripts purified on a denaturing 6% polyacrylamide gel (described in *Protocol 13*).

The second consideration is that solubilized RdRp may contain a number of proteins which bind non-specifically to the RNA transcript. A polyanion, such as heparin, can be included in the binding assay to suppress these interactions. A commonly used alternative is to include excess tRNA (*Protocol 14*).

The second major technique used in the study of RNA binding proteins is UV cross-linking (*Protocol 15*). As with RNA mobility shift, labelled transcript is hybridized to RdRp. The complex is then subjected to UV irradiation, which covalently links the protein to the RNA. After RNase treatment to digest the unbound portion of the transcript, the short RNA–protein complex is analysed by SDS–PAGE. Because the portion of RNA complexed is only a few bases long, the size of the protein is increased only by 1 to 2 kDa. Autoradiography of the gel allows the size of the RNA-binding protein to be estimated with some degree of accuracy.

Acknowledgements

We are grateful to BP Nutrition, ICI Seeds, and the Agricultural and Food Research Council for financial support, and to the AFRC for a fellowship to R.J.H. We would also like to thank Professor Wilhelm Griissem, Dr Gadi Schuster, and Dr Susan Abrahamson for their kind help and advice to R.J.H.

References

1. Francki, R. I. B., Fauquet, C. M., Knudson, D. L., and Brown, F. (ed.) (1991). *Classification and nomenclature of viruses. Fifth report of the international committee on taxonomy of viruses. Arch. Virol. Supplementum 2.* Springer-Verlag, Wien, New York.
2. Steinhauer, D. A. and Holland, J. J. (1987). *Annu. Rev. Microb.*, **41,** 409.
3. Zaitlin, M. and Hull, R. (1987). *Annu. Rev. Plant Physiol.*, **38,** 291.
4. David, C., Gargouridi-Bouzid, R., and Haenni, A.-L. (1992). *Prog. Nucl. Acid Res. Mol. Biol.*, **41,** 157.
5. Kroner, P., Richards, D., Traynor, P., and Ahlquist, P. (1989). *J. Virol.*, **63,** 5302.
6. Blumenthal, T. and Carmichael, G. G. (1979). *Annu. Rev. Biochem.*, **48,** 525.
7. Hayes, R. J. and Buck, K. W. (1990). *Cell*, **63,** 363.
8. Quadt, R., Rosdorff, H. J. M., Hunt, T. W., and Jaspers, E. M. J. (1991). *Virology*, **182,** 309.
9. Argos, P. (1988). *Nucl. Acids Res.*, **16,** 9909.
10. Habili, N. and Symons, R. H. (1989). *Nucl. Acids Res.*, **17,** 9543.
11. Bruenn, J. A. (1991). *Nucl. Acids Res.*, **19,** 217.

12. Koonin, E. V. (1991). *J. Gen. Virol.,* **72,** 2197.
13. Gorbalenya, A. E. and Koonin, E. V. (1989). *Nucl. Acids Res.,* **17,** 8413.
14. Lain, S., Riechmann, J. L., and Garcia, J. A. (1990). *Nucl. Acids Res.,* **18,** 7003.
15. Goldbach, R. and Wellink, J. (1988). *Intervirology,* **29,** 260.
16. Mi, S., Durbin, R., Huong, H. V., Rice, C. M., and Stoller, V. (1989). *Virology,* **170,** 385.
17. Dunigan, D. D., Lomonossoff, G. P., and Zaitlin, M. (1990). *Abstr. VIIIth International Congress of Virology, Berlin,* p. 477.
18. Marsh, L. E., Dreher, T. W., and Hall, T. C. (1988). *Nucl. Acids Res.,* **16,** 981.
19. Van Dyke, T. A. and Flanegan, J. B. (1980). *J. Virol.,* **35,** 732.
20. Baron, M. H. and Baltimore, D. (1982). *J. Biol. Chem.,* **257,** 12359.
21. Hey, T. D., Richards, O. C., and Ehlenfeld, E. (1987). *J. Virol.,* **61,** 802.
22. Saunders, K. and Kaesberg, P. (1985). *Virology,* **144,** 373.
23. Mouches, C., Bove, C., Barreau, C., and Bove, J. M. (1975). *Ann. Microbiol. (Inst. Pasteur),* **127A,** 75.
24. Candresse, T., Mouches, C., and Bove, J. M. (1986). *Virology,* **152,** 322.
25. Quadt, R., Verbeck, H. J. M., and Jaspers, E. M. J. (1988). *Virology,* **165,** 256.
26. Quadt, R. and Jaspers, E. M. J. (1990). *Virology,* **178,** 189.
27. Miller, W. A. and Hall, T. C. (1983). *Virology,* **125,** 236.
28. Miller, W. A. and Hall, T. C. (1984). *Virology,* **132,** 53.
29. Houwing, C. J. and Jaspers, E. M. J. (1986). *FEBS Lett.,* **209,** 284.
30. Quadt, R. and Jaspers, E. M. J. (1991). *FEBS Lett.,* **279,** 273.
31. Dreher, T. W. and Hall, T. C. (1988). *J. Mol. Biol.,* **201,** 31.
32. Wu, G., Kaper, J. M., and Jaspers, E. M. J. (1991). *FEBS Lett.,* **292,** 213.
33. Wu, S.-X. and Kaesberg, P. (1991). *Virology,* **183,** 392.
34. Palukaitis, P., Roossinck, M. J., Dietzgen, R. G., and Francki, R. I. B. (1992). *Adv. Virus Res.,* **41,** 281.
35. Frankel-Conrat, H. (1986). *CRC Crit. Rev. Plant Sci.,* **4,** 213.
36. Matthews, R. E. F. (1991). *Plant virology* (3rd edn). Academic Press, London.
37. Ahlquist, P. and Janda, M. (1984). *Mol. Cell. Biol.,* **4,** 2876.
38. Ahlquist, P., French, R., Janda, M., and Loesch-Fries, L. S. (1984). *Proc. Natl Acad. Sci. USA,* **81,** 7066.
39. Hayes, R. J. and Buck, K. W. (1990). *J. Gen. Virol.,* **71,** 2503.
40. Weiland, J. J. and Dreher, T. W. (1989). *Nucl. Acids Res.,* **17,** 4675.
41. Boccard, F. and Baulcombe, D. C. (1992). *Gene,* **114,** 223.
42. Osman, T. A. M., Ingles, P. J., Miller, S. J., and Buck, K. W. (1991). *J. Gen. Virol.,* **72,** 1793.
43. Sambrook, J., Fritsch, E. F., and Maniatis, T. (1989). *Molecular cloning: a laboratory manual* (2nd edn). Cold Spring Harbor Press, New York.
44. Green, M. R., Maniatis, T., and Melton, D. A. (1983). *Cell,* **32,** 681.
45. Hatta, T. and Francki, R. I. B. (1981). *J. Gen. Virol.,* **53,** 343.
46. Bremberger, C., Haschke, H.-P., and Lüttge, V. (1988). *Planta,* **175,** 465.
47. Singer, B. and Frankel-Conrat, H. (1961). *Virology,* **14,** 59.
48. Findlay, J. B. C. (1990). In *Protein purification methods: a practical approach* (ed. E. L. V. Harris and S. Angal), pp. 59–83. Oxford University Press.
49. Van der Meer, J., Dorssers, L., and Zabel, P. (1983). *EMBO J.,* **2,** 233.

2

The molecular biology of influenza viruses and paramyxoviruses

REAY G. PATERSON and ROBERT A. LAMB

1. Introduction

Orthomyxoviruses and paramyxoviruses are important aetiological agents of disease in man and animals. Each year disease outbreaks caused by these viruses result in extensive morbidity and mortality.

In recent years major advances have been made in our knowledge and understanding of the structure and function of orthomyxovirus and paramyxovirus genes and the proteins that they encode. In particular, the complete genome sequences of many strains of influenza A, B, and C viruses and paramyxoviruses have been obtained. The expression of cDNA copies of individual genes has led to the elucidation of coding strategies used by these viruses and has also permitted study of the synthesis and maturation of many virally-encoded proteins in the absence of other virus gene products.

Influenza viruses and paramyxoviruses are enveloped viruses with RNA genomes of negative sense polarity. The genome in the case of influenza viruses (totally approximately 13 000 nucleotides) is segmented, consisting of eight RNA segments in the influenza A and B viruses and seven segments in influenza C virus, whereas the genome of paramyxoviruses is a continuous RNA chain of approximately 15 000 nucleotides. The discussion of the virus proteins in this chapter is not intended to be a comprehensive review of the subject (for reviews, see refs 1 and 2), but will be confined to those of the influenza A and B viruses and the prototype paramyxovirus simian virus 5 (SV5).

The influenza virus and SV5 genomes do not consist of naked RNA but are found to be tightly associated with the nucleoprotein (NP), which is a major structural component of the virus particle. Associated with the RNA–NP complex is the RNA-dependent RNA polymerase, which in the influenza viruses is formed of three polymerase proteins, PA, PB1, and PB2 and in SV5 is made up of the large protein (L), the phosphoprotein (P), and possibly a third small protein, V. Underlying the virus envelope which surrounds the ribonucleoprotein complex is the matrix protein (M). Inserted into the virus envelope, which is derived from the host-cell plasma membrane, are the major virus-encoded integral membrane proteins, the haemagglutinin (HA) and neuraminidase (NA) proteins of influenza virus and the fusion protein (F) and haemagglutinin–neuraminidase (HN) proteins of SV5. A third small integral membrane protein, M_2, which has recently been found to be an ion channel, is also found in small amounts as a structural component of influenza A virus particles. Influenza B virus and SV5 both encode an additional integral membrane protein (NB) and small hydrophobic (SH) proteins, respectively; however, at present it is not known whether these proteins are structural components of the virion. Finally, influenza viruses are known to encode two non-structural proteins, NS_1 and NS_2, and recently an additional influenza B virus-encoded protein, BM_2, has been identified in influenza B virus-infected cells.

The integral membrane proteins encoded by influenza viruses and paramyxoviruses are some of the better studied proteins in the field of virology, in part because they are expressed at fairly high levels in virus-infected cells and both polyclonal and monoclonal antibodies (mAbs) specific for most of these proteins are available. In this chapter we will describe the methods we have used to examine the synthesis, maturation, and biological activities of influenza virus- and SV5-encoded proteins.

2. Virus growth

2.1 Growth of SV5 and influenza viruses in tissue culture

In our laboratory, the cell type of choice for *growth* of the paramyxovirus SV5 is the Madin Darby bovine kidney (MDBK) cell (*Protocol 1*). This cell line is resistant to SV5-mediated cell fusion and exhibits minimal cytopathic effect (c.p.e.) upon infection by this virus. Resistance to fusion is an important characteristic with respect to the ability of the cell type to support growth of SV5 to high titre, as it has been found that cells that exhibit massive fusion when infected by SV5 do not produce high yields of virus. The cell type of choice for *growth* of all strains of influenza A virus and influenza B virus is the Madin Darby canine kidney (MDCK) cell, with the exception of influenza A/WSN/33 virus which grows to higher titre in MDBK cells (*Protocol 2*). For growth of influenza viruses, trypsin (plasminogen present in serum will substitute) has to be included in the medium to cleave HA to its biologically active form. However, we routinely propagate influenza viruses in embryonated chicken eggs, which gives virus with a cleaved HA (*Protocol 3*). We prefer CV-1 cells for many biochemical analyses of macromolecular synthesis during a single cycle of viral growth. However, for many avian strains of influenza virus MDCK cells have to be used because of the strict requirement by HA for the appropriate linkage within sialic acid for attachment to the target cell.

2.2 Maintenance of cell lines

Monolayer cultures of MDBK, MDCK, and CV-1 cells are grown in Dulbecco's modified Eagle's (DME) medium supplemented with 10% (v/v) fetal calf serum (FCS). Monolayer cultures of BHK-21F cells are maintained in DME supplemented with 10% (v/v) newborn calf serum and 10% (v/v) tryptose phosphate broth.

Protocol 1. Growth of SV5 in tissue culture

Materials

- monolayers of MDBK cells in 100 mm plates
- phosphate-buffered saline containing divalent cations (PBS^+): 140 mM NaCl, 2.7 mM KCl, 8 mM Na_2HPO_4, 1.4 mM KH_2PO_4, 0.7 mM $CaCl_2$, 0.5 mM $MgCl_2$ (pH 7.2)
- BSA/MEM: minimal essential medium (with Earle's salts) supplemented with 1% (w/v) bovine serum albumin (BSA)
- replacement medium: DME supplemented with 2% (v/v) FCS

Method

1. Wash the cell monolayers once with 5 ml PBS^+.
2. Inoculate the monolayers with 2 ml of the appropriate dilution of virus stock in BSA/MEM to give a multiplicity of infection (m.o.i.) of 0.2 plaque-forming units (p.f.u.)/cell.
3. Incubate at 37°C for 1.5 h, tilting the plates every 15–20 min to ensure that the monolayers remains moist.
4. Add 13 ml of replacement medium.
5. Incubate at 37°C for 4 days. Assay the growth of the virus every day by performing an HA assay (*Protocol 4*).
6. Harvest the virus once the HA titre has stopped increasing (a typical HA titre for SV5 is 512–1024). Remove the medium and transfer it to centrifuge tubes. Centrifuge at 2250 *g* at 4°C for 15 min (e.g. 3000 r.p.m. in an IEC DPR-6000 centrifuge using a 253 12-place rotor). Carefully remove the supernatant to a new tube without disturbing the pellet of cellular debris. Add BSA to a final concentration of 1% (w/v), aliquot into vials, quick-freeze using dry ice/methanol and store at −70°C.

Protocol 2. Growth of influenza A and B viruses in tissue culture

Materials

- monolayers of MDCK cells in 100 mm plates

Protocol 2. *Continued*

- 1 mg/ml *N*-acetyl trypsin (Sigma); store at −20 °C
- harvesting medium: DME containing 2 μg/ml *N*-acetyl trypsin

Method

1. Follow steps 1–3 of *Protocol 1* for both influenza A and B viruses.
2. Add 13 ml of harvesting medium.
3. Incubate for 2–3 days at 37 °C. Inspect the cells under the microscope for c.p.e. and perform an HA assay (*Protocol 4*) each day.
4. Harvest the virus when the HA titre has stopped increasing. HA titres vary greatly depending on the strain of virus. Influenza A/WSN/33 virus can grow to a titre of 1024, whereas some avian strains grow to only 128.

Protocol 3. Growth of influenza A and B viruses in embryonated eggs

Materials

- embryonated chicken eggs
- egg candling device
- humidified egg incubator, preferably with an automatic turning system for the egg trays
- low speed dentist's drill or jeweller's engraver
- histology embedding wax (any brand)
- BSA/DME: DME containing 1% (w/v) BSA

Method

1. Incubate embryonated chicken eggs at 37 °C in a humidified egg incubator for 11 days.
2. Candle the eggs to check that the embryos are alive and, using a soft pencil, mark a position on the egg shell that is devoid of underlying blood vessels. A dark red band around the egg indicates that the embryo has died.
3. Lay the eggs on their side in egg cartons and swab them with ethanol. Endeavour to keep the egg surface around the pencil mark aseptic.
4. Using a drill or engraver, grind away the egg shell at the pencil mark but do not penetrate through the shell. Some investigators also drill a second hole at the airsac end of the egg to permit easy uptake of inoculum, but we do not find it necessary.

5. Inoculate the eggs through the weakened shell with 0.1 ml of an appropriate dilution[a] of influenza virus in BSA/DME using a 1 ml tuberculin syringe and a 23 gauge needle. It helps to hold the syringe at an oblique angle and to have the bevel of the syringe needle facing down towards the egg. Slowly inoculate the virus so that fluid is not forced out of the egg. If this does occur, the fluid will be re-absorbed rapidly.
6. Cover the hole with molten wax using a swab applicator.
7. Turn the eggs so that the air-sac ends are uppermost and incubate at 35 °C for 2–3 days, depending on the strain of influenza virus. Make sure that the humidifier contains water.
8. After an appropriate period of virus growth,[b] chill the eggs to 4 °C for 1–2 h.
9. Swab the eggs with ethanol.
10. Crack open the eggs with forceps (using ethanol-flamed forceps if sterile virus is required) or a commercial egg opener (in the manner of opening a boiled egg for breakfast) and remove the egg shell top.
11. With one hand, use (sterile) forceps to enter the egg away from the yolk and hold down the embryo, and, with the other hand, suck out allantoic fluid (5–10 ml) using a pipette and bulb.[c]
12. Place the allantoic fluid in a (sterile) centrifuge tube[d] and centrifuge at 2250 *g* for 20 min to remove debris (see step 6, *Protocol 1*). Quick-freeze the allantoic fluid in aliquots and carry out HA and plaque assays.
13. Dispose of the embryonated eggs in a manner that satisfies local rules for infectious material and the use of animals. Autoclaving is the method preferred at many institutions.

[a] The dilution depends somewhat on the strain of influenza virus, but we aim for 10^4 p.f.u./egg.
[b] This can be determined by harvesting allantoic fluid from 1–2 eggs and performing an HA assay.
[c] Look for viral c.p.e. (for example translucent membrane). Any eggs that are black or green are likely to be bacterially contaminated and should not be used.
[d] If sterile virus is required, it is sensible to harvest virus in carefully marked batches and to test the sterility of allantoic fluid on blood agar plates.

2.3 Quantitating virus yield

The HA protein of influenza virus and the HN protein of SV5 possess receptor-binding activity, and function to attach virus particles to target cells by binding to sialic acid residues on the surface of the plasma membrane. As a consequence of sialic acid-binding activity, influenza virus and SV5 virus particles have the ability to cause agglutination of erythrocytes, and this can be used as a way of assaying the growth of virus (*Protocol 4*). The NA and HN proteins of influenza virus and SV5, respectively, have receptor-destroying or neuraminidase activity, and therefore cause elution of the red blood cells from the virus particles. The HA and HN proteins can bind sialic acid at 4°C, a temperature at which the neuraminidase activity of NA and HN is reduced; therefore, HA assays are incubated at 4°C.

Protocol 4. HA assay

Materials

- polyvinyl chloride 96-well plates (U-bottom)
- 0.5% suspension of red blood cells, prepared as in *Protocol 5*
- phosphate-buffered saline (PBS): 170 mM NaCl, 3.4 mM KCl, 10 mM Na_2HPO_4, 1.8 mM KH_2PO_4 (pH 7.2)

Method

1. Aliquot 100 μl of ice-cold PBS into each well of a 96-well plate. Twelve wells per sample is more than sufficient and allows some wells to be used for non-agglutinated controls.
2. Add 100 μl of virus stock to the first well. Mix by pipetting up and down with an adjustable pipette and transfer 100 μl to the next well.
3. Using a fresh tip, repeat the mixing and transfer as in step 2. Continue making two-fold dilutions until the last well is reached. After mixing the virus and PBS into the final well, discard 100 μl.
4. Add 100 μl of chicken red blood cells to the 100 μl of diluted virus in each well.
5. Incubate for 1–2 h at 4°C.
6. The HA titre is read as the reciprocal of the highest dilution at which agglutination is seen, with the dilution in the first well being 1:4.

Protocol 5. Preparation of chicken blood for HA assays

Materials

- chicken blood (Environmental Diagnostics)
- Alsever's solution: 4.2 g/l NaCl, 8.0 g/l trisodium citrate (dihydrate), 20.5 g/l glucose, bring pH to 6.1 with 10% (w/v) citric acid

Method

1. Transfer blood to a 50 ml Corning tube and centrifuge at 1000 *g* (for example 2000 r.p.m. in an IEC DPR-6000 centrifuge using the 253 12-place rotor) for 10 min at 4°C.

2. Remove the supernatant and buffy coat carefully and resuspend the pellet in 50 ml of Alsever's solution. Centrifuge at 1000 *g* as in step 1.
3. Remove the supernatant and resuspend the pellet in 50 ml of Alsever's solution. Centrifuge at 2250 *g* for 10 min at 4°C as in step 1.
4. Remove the supernatant and measure the volume of the pellet carefully.
5. Dilute the packed red cells with four volumes of Alsever's solution to give a concentration of 20% red cells. Store the suspension at 4°C.
6. Use the blood at a concentration of 0.5% in PBS for HA assays (*Protocol 4*).

2.4 Quantitating virus infectivity

After harvesting the virus the titre is determined by performing a plaque assay (*Protocols 6–8*). SV5 plaque assays are usually carried out on CV-1 or BHK-21F cells. Influenza virus plaque assays are carried out using MDCK cells for most strains except for influenza A/WSN/33 virus, where MDBK cells are used.

Protocol 6. SV5 plaque assay

Materials

- monolayers of CV-1 or BHK-21F cells in 60 mm plates
- 2% (w/v) agarose
- FCS/DME: 4% (v/v) FCS in 2 × DME
- 1% (w/v) neutral red stain

Method

1. Prepare ten-fold dilutions of virus stock (e.g. 10^{-5} to 10^{-8}) in MEM/BSA.
2. Wash the confluent monolayers once with PBS.
3. Inoculate the monolayers in duplicate with 0.5 ml of the virus dilutions and incubate them at 37°C for 1–2 h, tilting the plates periodically.
4. While the plates are incubating, melt the agarose by autoclaving; 3 ml agarose/plate is needed.
5. Cool the agarose to 45°C in a water bath.
6. Remove the virus inoculum and wash the monolayers with PBS.
7. Mix the agarose with an equal volume of FCS/DME which has been warmed at 37°C[a].
8. Add 6 ml of agarose–FCS/DME to each plate. Leave at room temperature until the agarose solidifies.

Protocol 6. *Continued*

9. Incubate the cultures at 37°C in inverted orientation for 4 days, at which time plaques are usually visible.
10. Stain the monolayers by adding a further 4 ml of agarose–FCS/DME containing 0.01% (w/v) neutral red stain. Plaques can normally be visualized 7–12 h later.

[a] Ensure that the autoclaved agarose is at 45°C and that the FCS/DME is at 37°C before mixing. Do not allow them to cool, and dispense them as quickly as possible. Autoclave sterile solidifed agarose; do not microwave it, since this may not destroy contaminants.

Protocol 7. Influenza A virus plaque assay

Materials

- monolayers of MDCK cells[a] in 60 mm dishes
- influenza A overlay medium: mix equal volumes of 2% agarose and 2 × DME containing 200 U/ml penicillin, 200 μg/ml streptomycin, 2 μg/ml *N*-acetyl trypsin[b] (see step 7, *Protocol 6*)
- staining medium: mix equal volumes of 2% agarose containing 0.01% (w/v) neutral red stain and 2 × DME containing 2 μg/ml *N*-acetyl trypsin

Method

1. Dilute the virus serially in BSA/MEM, using appropriate dilutions (usually 10^{-6}, 10^{-7}, 10^{-8}).
2. Infect MDCK cells as in steps 2 and 3 of *Protocol 6*, incubating 0.5 ml virus/plate for 2 h.
3. While the plates are incubating, autoclave 2% agarose.
4. Wash the cells with warm PBS^+ to remove the virus inoculum.
5. Add 8 ml/plate of influenza A overlay medium and leave it to set at room temperature for 10 min. Incubate the plates in inverted orientation at 37°C for 2–3 days.
6. Add 3 ml staining medium/plate and count the plaques within a few hours.

[a] Use MDCK cells for most strains of influenza A virus. For A/WSN/33, use MDBK or CEF cells. For strains that have a cleaved HA (for example A/Cornell(H7N7)), omit trypsin from the overlay as the HA is acid-labile and trypsin-sensitive.

[b] Trypsin is necessary to cleave HA to HA1 and HA2 to make the virus infectious. *N*-acetyl trypsin is more resistant to auto-digestion than other forms of trypsin.

Protocol 8. Influenza B virus plaque assay

Materials

- freshly confluent MDCK cells in 60 mm dishes[a]
- influenza B overlay medium: mix equal volumes of 2% agarose and 2 × DME containing 200 U/ml penicillin, 200 μg/ml streptomycin, 2 μg/ml *N*-acetyl trypsin,[b] 40 mM Hepes–NaOH (pH 7.1), 1% (w/v) BSA (see step 7, *Protocol 6*)

Method

1. Dilute the virus serially, as in step 1, *Protocol 6*.
2. Wash the cells thoroughly twice with PBS^+. This is to ensure removal of FCS, so that serum inhibitors will not block virus infection by binding to sialic acid.
3. Add 0.5 ml of the virus dilutions to the cells and allow the virus to adsorb by rocking at 37°C for 2 h.
4. While the plates are incubating, autoclave 2% agarose.
5. Remove the inoculum by aspiration. Wash the monolayers once with PBS^+.
6. Overlay the plates as in step 5, *Protocol 7*, using 7 ml influenza B overlay medium/plate.
7. Plaques should be visible without staining by 3 days. If necessary, carry out step 6 of *Protocol 7*.

[a] The lineage of MDCK cells used to plaque influenza B virus is quite important. We find that we have to keep the cells within 30 passages of those obtained from the American Type Culture Collection.

[b] *N*-acetyl trypsin is kept as a 1 mg/ml stock solution in water and stored at −20°C.

2.5 Purification of virions

For experiments in which cells are to be infected with virus, the clarified tissue culture supernatant stock is of sufficiently high titre (usually about 10^8 p.f.u./ml). However, there are many experiments, for example injecting animals to raise antibodies or determining whether virus-encoded proteins are structural components of the virus particle, where a preparation of purified virus is required. Purification of SV5 and influenza virus is a relatively quick procedure (*Protocol 9*), and results in reasonably pure preparations of virus particles. The purity of the virus preparation can be checked by SDS–PAGE (Section 3.3).

Protocol 9. Purification of SV5 and influenza virus

Materials

- NTE: 100 mM NaCl, 10 mM Tris–HCl, 1 mM EDTA (pH 7.4)
- 15% and 60% (w/v) sucrose in NTE
- 15% and 60% (w/v) sodium–potassium tartrate in PBS
- TE: 10 mM Tris–HCl (pH 7.4), 1 mM EDTA

Method

1. Grow and harvest virus as described in *Protocols 1–3*.
2. After clarifying the culture medium (or allantoic fluid) by centrifugation, omit the addition of BSA and pellet the virus by centrifuging at 118 000 *g* for 45 min at 4°C (40 000 r.p.m. in a Beckman Type 70 Ti rotor). Resuspend the pellets in NTE, using 100 μl/plate.
3. Layer the virus suspension on top of linear 15–60% (w/v) sucrose gradients. Use either Beckman SW41 or SW28 tubes, depending on the amount of pelleted virus. As a guide, SW28 gradients are used when harvesting more than 20 plates or allantoic fluid from more than 36 eggs. Centrifuge at 71 000 *g* for 1 h at 4°C (24 000 r.p.m. in the SW41 rotor, 23 000 r.p.m. in the SW28 rotor).
4. Collect the broad band of virus from the side of the tube using a syringe and needle or from the top of the tube using a syringe and bent needle inserted under the band.
5. Dilute the virus in NTE and pellet it by centrifuging as described in step 2.
6. Resuspend the virus in PBS and layer it on top of 15–60% linear sodium–potassium tartrate gradients in SW41 tubes. Centrifuge at 97 000 *g* (28 000 r.p.m.) for 2 h at 20°C.
7. Collect the band, which is well defined and flocculent in nature, as described in step 4 and dialyse it overnight against 4 litres of PBS at 4°C.
8. Pellet the virus by centrifuging at 118 000 *g* (40 000 r.p.m. in a Beckman Type 70.1 Ti rotor) for 45 min at 4°C and dissolve the pellet at the desired concentration in TE.

3. Biochemical analysis of virus proteins

For experiments in which the aim is to carry out a biochemical analysis of virus-encoded proteins, the strategy for virus infection differs from that used for the growth of virus stocks. Whereas the infection for growth of virus is performed at a low m.o.i. in order to minimize the production of defective interfering particles, to look at virus-encoded proteins synthesized in infected cells the infection is carried out at a high m.o.i. so that every cell is infected. Infection is carried out at an m.o.i. of approximately 10–30 p.f.u./cell for SV5 and influenza viruses in CV-1 cells or for avian strains of influenza virus in MDCK cells (*Protocol 10*).

Unlike many strains of influenza A virus, paramyxoviruses are generally less efficient at inhibiting host cell macromolecular synthesis. Since there is no known involvement of cellular DNA-dependent RNA polymerase II in the paramyxovirus replicative cycle (unlike that of influenza viruses), infected cells are incubated in the presence of actinomycin D. Thus, at the optimal time for harvesting SV5-infected cell poly-A(+) RNA or for labelling infected cell proteins (16–18 h post-infection (p.i.)), in the presence of actinomycin D the poly-A(+) RNA is enriched for SV5-specific mRNA species and background host protein synthesis is much reduced. The optimal time for examining viral protein synthesis depends on the virus being studied and, therefore, has to be determined by labelling infected cells at different times after infection.

Protocol 10. Infection of cells with SV5 and influenza viruses for analysis of virus-specific protein synthesis

Materials

- 500 μg/ml actinomycin D in ethanol; store at −20°C

Method

1. Wash monolayers of CV-1 cells with PBS.
2. Add the virus inoculum at a dilution such that the m.o.i. is 10–30 p.f.u./cell.
3. Incubate for 30–60 min, depending on the balance between having a fairly synchronous infection and maximizing the input m.o.i.
4. Replace the virus inoculum with DME. For SV5 infections, supplement the DME with 2% (v/v) FCS and 5 μg/ml actinomycin D.
5. Incubate until 16–18 h p.i. for SV5, 3–6 h p.i. for influenza A virus and 6–10 h p.i. for influenza B virus.

At this point the methodology diverges, depending on the type of analysis to be carried out. If the proteins are to be analysed by staining a gel following SDS–PAGE or by Western blot analysis, remove the medium and lyse the cell monolayers using SDS–PAGE sample buffer. For components of the SDS–PAGE sample buffer, see *Protocol 13*.

3.1 Radiolabelling infected cell proteins

For procedures where the proteins are to be metabolically labelled prior to further analysis, the medium is replaced at the appropriate time with medium deficient in the radioactive precursor that is to be incorporated into the proteins, and the cultures are incubated for 15–30 min; this is termed 'starving'. The radioactive amino acid precursors most commonly used in our laboratory are [^{35}S]methionine and [^{35}S]cysteine. An economical source of radioactive label is derived from a [^{35}S]*Escherichia coli* hydrolysate containing 70% L-methionine and 20% L-cysteine, and is available from several vendors (for example Tran[^{35}S]-label from ICN-Flow). If DME deficient in methionine and cysteine is used, optimal labelling will be obtained using this reagent. We have determined that medium deficient in both methionine and cysteine can also be used to label proteins efficiently with either [^{35}S]methionine or [^{35}S]cysteine. Medium deficient in methionine and cysteine can be custom ordered from Gibco-BRL or can be prepared by the investigator using a kit such as the MEM Select-Amine Kit also sold by Gibco-BRL. For proteins that are methionine- and cysteine-poor, there are many ^{3}H-labelled amino acids that can be utilized as radioactive precursors.

At the conclusion of the starvation period, the medium is removed and replaced with medium deficient in methionine and cysteine but containing 20–100 μCi/ml of Tran[^{35}S]-label or [^{35}S]methionine or 20–150 μCi/ml of [^{35}S]cysteine, the amount of radioisotope used depending on economic considerations and the period required for the radioactive signal to be detected. The volume of medium used for labelling is the same as that used for virus infection (see *Table 1*). A labelling period of 1–2 h is sufficient to give good labelling of all the virus-encoded proteins and is also long enough to allow a reasonable amount of cleavage of paramyxovirus F_0 into F_1 and F_2. At the end of the labelling period, the medium is removed, the monolayers are washed with PBS, and if the proteins are to be analysed directly by SDS–

Table 1. Volumes of inoculum, replacement medium, and lysis buffer

	Diameter of well or plate (mm)			
	17.6	**35**	**60**	**100**
Inoculum (ml)	0.2	0.3	0.5	2
Replacement medium (ml)	1	2	5	10
SDS–PAGE sample buffer (ml)[a]	0.3	0.5	1	2
RIPA buffer[b] or 1% (w/v) SDS (ml)	0.5	0.75	1	2

[a] See *Protocol 13* for composition; omit the DTT if the samples are to be analysed under non-reducing conditions.
[b] See *Protocol 11* for composition.

PAGE without prior immunoprecipitation or other treatment, the monolayers are lysed in SDS–PAGE sample buffer (see *Table 1*). Samples that are prepared by lysing monolayers directly in SDS–PAGE sample buffer may be very viscous; if this is the case they should be sonicated using a probe sonicator to shear the DNA. In addition to making the samples easier to pipette, this will also improve the appearance of the autoradiographs.

More often than not, the labelled proteins are destined for further analysis prior to separation by SDS–PAGE. Many of these analyses require the proteins to be immunoprecipitated. In this laboratory, cell lysates are most commonly prepared using RIPA buffer (*Protocol 11*).

Protocol 11. Preparation of clarified cell lysates for immunoprecipitation

Materials

- RIPA buffer: 0.15 M NaCl, 1% (w/v) sodium deoxycholate, 1% (v/v) Triton X-100, 0.1% (w/v) sodium dodecyl sulphate (SDS), 0.1 M Tris–HCl (pH 7.4), 1 mM phenylmethylsulphonyl fluoride (PMSF), 210 ng/ml (0.24 trypsin inhibiting U/ml) aprotinin, 10 mM iodoacetimide

Method

1. At the conclusion of the labelling period, remove the medium.
2. Wash the cell monolayer with PBS.
3. Add RIPA buffer (see *Table 1* for volume) and scrape the lysed cells into ultracentrifuge tubes.
4. Clarify the lysate by centrifuging at 62 000 *g* (30 000 r.p.m. in a Beckman Ti 70.1 rotor) for 30 min at 4°C.
5. Store the supernatant at −20°C.

Lysates prepared using this method usually immunoprecipitate well with the majority of antibodies; however, there are some antibodies that give rise to non-specific precipitation of some host or viral proteins. In such cases the result may be 'cleaner' if the cells are lysed using a solution of 1% (w/v) SDS (ref. 3 and see *Table 1*). When making the decision as to which method to use for preparing cell lysates for use in immunoprecipitation, it should be borne in mind that some antibodies will not recognize proteins that have been denatured in SDS.

In the methods described above, infected cells are labelled for a relatively long period of time (1–2 h), which permits the detection of proteins synthesized in their final mature form and also of proteins synthesized as precursors, in addition to products arising from their processing during the labelling period. If one wishes to study the rate of processing or transport, or even to examine the possibility that a protein may arise from processing of a precursor, a pulse–chase experiment has to be carried out (*Protocol 12*). In this type of experiment, a very short labelling period is used, typically 3–10 min. The length of the labelling period, or pulse, is determined by the rate of the intracellular process that is being examined; the more rapid the process, the shorter the pulse required. Because of the short labelling times used, a higher concentration of radioactive precursor has to be used in order to achieve sufficient incorporation. Not surprisingly, the shorter the pulse, the higher the concentration of label that has to be used; for a 3 min pulse, it may be necessary to use 500 μCi/ml on a 35 mm plate. Ideally, experiments utilizing very short labelling times are best carried out in a 37°C warm room, since a proportionally large amount of time can be spent removing the dishes from incubators in order to add or remove label. If a warm room is available, the starvation period is carried out in a 37°C incubator and the labelling and collection of the first few time points carried out in the warm room. The dishes for later time points can be maintained in the incubator for the chase period. However, because of the relative lack of CO_2 in a warm room, the medium should be supplemented with 20 mM Hepes–NaOH (pH 7.1) to maintain pH.

Protocol 12. Pulse–chase labelling of infected cells

Materials

- starvation medium: DME deficient in methionine and cysteine
- labelling medium: starvation medium supplemented with 500 μCi/ml [^{35}S]methionine
- chasing medium: DME supplemented with 2 mM methionine

Method

1. Infect cells as described in *Protocol 10*, and at the desired time replace the medium with starvation medium. Use the same volume as that for replacement medium in the virus infection (see *Table 1*). Incubate for 30 min in 37°C incubator.
2. Transfer the dishes to a 37°C warm room,[a] and replace the starvation medium with an equal volume of labelling medium. Incubate for 5 min.
3. Replace the labelling medium with chasing medium and incubate for the desired times (for example 5, 10, 15, 20, 30, 40, 60 min).
4. Lyse the cells as described in Section 3.1 or *Protocol 11*.

[a] If a warm room is unavailable, the dishes may be transported to and from the incubator on a polystyrene tray kept in the incubator. This helps minimize the temperature loss.

3.2 Immunoprecipitation of viral proteins

Most analyses performed on proteins synthesized in SV5- or influenza virus-infected cells require that the virus protein under examination be immunoprecipitated. Two types of infected cell lysate are utilized for immunoprecipitation in our laboratory, as described in *Protocol 13* (SDS lysates) and *Protocol 14* (RIPA lysates). Several variables have to be determined for the particular system being studied:

(a) How much lysate is required for the immunoprecipitation? This will depend on the level of expression and labelling efficiency of the protein of interest.

(b) How much antibody is required? This will depend on the titre and affinity of the antibody for the antigen.

(c) For most biochemical experiments, it is important that the amount of antibody should be in excess of that required to quantitatively immunoprecipitate the protein of interest. This is determined by immunoprecipitating as described below, using increasing quantities of antibody and determining a plateau of precipitable material.

Protocol 13. SDS immunoprecipitation

Materials

- SDS immunoprecipitation dilution buffer: 1.25% (v/v) Triton X-100, 190 mM NaCl, 60 mM Tris–HCl (pH 7.4), 6 mM EDTA, 210 ng/ml (0.24 trypsin-inhibiting U/ml) aprotinin
- SDS wash solution I: 0.1% (v/v) Triton X-100, 0.02% (w/v) SDS, 150 mM NaCl, 50 mM Tris–HCl (pH 7.4), 5 mM EDTA
- SDS wash solution II: 150 mM NaCl, 50 mM Tris–HCl (pH 7.4), 2.5 mM EDTA
- protein A–Sepharose (Pharmacia)
- SDS–PAGE sample buffer: 4% (w/v) SDS, 40% (v/v) glycerol, 3% (w/v) dithiothreitol (DTT),[a] 60 mM Tris–HCl (pH 6.8), a few grains of bromophenol blue
- an appropriate antibody

Method

1. Lyse labelled infected cell monolayers using an appropriate volume (*Table 1*) of 1% (w/v) SDS.
2. Aliquot the amount of lysate to be used into a 1.5 ml microcentrifuge tube. For example, for one well of a 24-well plate (17.6 mm diameter) infected with SV5 and lysed in 500 μl, use 200 μl of lysate for immunoprecipitation. Boil for 5 min.

Protocol 13. *Continued*

3. Add four volumes of SDS immunoprecipitation dilution buffer and centrifuge for 5 min at 16 000 *g* in a microcentrifuge.
4. Transfer the supernatant to a new tube and add antibody (usually 5–15 μl of polyclonal antibody). Incubate at 4°C for 3 h.
5. Add 30 μl of protein A–Sepharose and incubate at 4°C for 30 min with rocking to ensure good mixing of the beads.
6. Pellet the beads by centrifuging at 16 000 *g* for 30 sec, and aspirate the supernatant.
7. Wash the beads by resuspending them in 1 ml of SDS wash solution I and centrifuging as described in step 6, and aspirate the supernatant. Repeat this step three times.
8. Wash the beads by resuspending them in 1 ml of SDS wash solution II using the procedure described in step 7. Remove as much of the supernatant as possible.
9. Resuspend the beads in 50 μl of SDS–PAGE sample buffer by vortexing briefly. Centrifuge briefly (2 sec) at 16 000 *g* to retrieve beads from the wall of the tube. Boil for 3 min prior to SDS–PAGE.

[a] If the samples are to be run under non-reducing conditions, omit the DTT.

Protocol 14. Immunoprecipitation of proteins solubilized in RIPA buffer

Materials

- RIPA wash buffer: 0.3 M NaCl, 1% (w/v) sodium deoxycholate, 1% (v/v) Triton X-100, 0.1% (w/v) SDS, 0.1 M Tris–HCl (pH 7.4)

Method

1. Aliquot 200 μl of the RIPA lysate, prepared as described in *Protocol 11*, into a 1.5 ml microcentrifuge tube.
2. Incubate with antibody as described in step 4, *Protocol 13*. If a mAb is being used, the volume will usually be less than that used in the case of a polyclonal antibody. Incubate at 4°C for 3 h.
3. Carry out steps 5 and 6 of *Protocol 13*.[a]
4. Wash the beads twice in 1 ml of RIPA wash buffer by resuspending them, centrifuging at 16 000 *g* for 30 sec, and aspirating the supernatant.
5. Wash the beads twice in RIPA buffer (see *Protocol 11*) lacking protease inhibitors, using the procedure described in step 4.

6. Wash the beads once with SDS wash solution II as described in step 8, *Protocol 13*.
7. Resuspend the beads in SDS–PAGE sample buffer as described in step 9, *Protocol 13*.

[a] Not all mAb isotypes can bind to protein A. If a mAb isotype of the human IgG3, mouse IgG1, or any of the rat subclasses except IgG2C is used, an additional step is required consisting of a 30 min incubation at 4°C in the presence of a rabbit anti-species-specific secondary antibody prior to the addition of the protein A–Sepharose beads.

3.3 Analysis of polypeptides

The most common method of analysing polypeptides synthesized in infected cells is by one-dimensional SDS–PAGE. In this procedure, the proteins are solubilized by boiling in the detergent SDS. SDS is negatively charged at neutral pH and binds to proteins to give them an overall negative charge. Thus, proteins migrate towards the anode and are separated according to molecular mass (M_r). Normally, a reducing agent such as DTT or 2-mercaptoethanol is included in the SDS sample buffer to reduce disulphide bonds, such that the polypeptides in a disulphide-linked protein will migrate according to their individual subunit M_rs. *Table 2* lists the reagents required for SDS–PAGE. The volumes are for use with gels set up using 20 × 20 cm gel plates separated by 1.5 mm spacers and containing a 3 cm deep stacking gel. If plates of different dimensions are used the volumes will have to be adjusted accordingly. There is no need to de-gas the gel solutions, but it is necessary to overlay the separating gel in order for it to polymerize. In our laboratory, we overlay with water-saturated *n*-butanol.

Once the separating gel has polymerized, the butanol is washed off and the stacking gel is layered on top. For general analysis of SV5-infected cell polypeptides, a 15% gel is ideal since it resolves all the virus-encoded proteins while retaining F_2 on the gel. For analysis of immunoprecipitated samples, the protein A–Sepharose beads are also loaded on the gel: cutting 2 mm off the end of the pipette tip makes pipetting the beads easier.

If the gel contains radiolabelled samples, after electrophoresis it is treated for fluorography by soaking for 30 min in glacial acetic acid followed by 30 min in a 22% (w/v) solution of 2,5-diaphenyloxazole (PPO) in glacial acetic acid, and the PPO precipitated by immersing the gel in cold water for 30 min. The gel is then dried under vacuum and exposed to X-ray film at −70°C.

If there is a low abundance of the polypeptide in infected cells or virions, radiolabelling of samples may not be sufficiently sensitive to allow its detection. In such cases, analysis of the sample using a Western blot procedure may prove to be more useful. In a Western blot assay, proteins are separated by SDS–PAGE and are then electrophoretically transferred to a solid support

Table 2. SDS–PAGE recipes (volumes in ml)

Separating gels	**Concentration**			
	10%	**15%**	**17.5%**	**20%**
30% (w/v) acrylamide	15	22.5	26.24	30
2.5% (w/v) *N,N′*-methylenebisacrylamide	2.34	1.56	1.35	1.17
10% (w/v) SDS	0.45	0.45	0.45	0.45
1 M Tris–HCl (pH 8.8)	16.8	16.8	16.8	—
2 M Tris–HCl (pH 8.8)	—	—	—	8.4
water	10.41	3.69	0.15	4.9
TEMED	0.015	0.015	0.015	0.015
10% (w/v) ammonium persulphate	0.25	0.25	0.25	0.25

3.2% stacking gel

30% (w/v) acrylamide: 0.8% (w/v) bisacrylamide	2.25
1 M Tris–HCl (pH 6.8)	2.5
10% (w/v) SDS	0.2
glycerol	1.0
water	14.3
TEMED	0.04
1.5 % (w/v) ammonium persulphate	1.0

10 × Tris–glycine (Laemmli) gel running buffer

Tris base	30.28 g/l
glycine	144.13 g/l
SDS	10 g/l

such as nitrocellulose. The proteins are detected by incubation of the membrane first with a primary antibody which has the potential to recognize the protein of interest and then with either ^{125}I-labelled secondary antibody or ^{125}I-labelled protein A. Alternatively, several antibody–enzyme conjugate-based detection methods are available. *Protocol 15* is adapted from ref. 4.

Protocol 15. Western blot analysis of proteins

Materials

- transfer apparatus (for example BioRad or Hoeffer)
- nitrocellulose filters: 0.45 μm pore size
- 3MM chromatography paper (Whatman) or equivalent
- Tris–glycine buffer: 75.7 g/l Tris base, 13.5 g/l glycine

- Western buffer: 80 ml/l Tris–glycine buffer, 200 ml/l methanol (pH >9.0)
- blocking solution: 5% (w/v) non-fat powdered milk, 0.01% (v/v) anti-foam A (Sigma), 0.02% (w/v) sodium azide in PBS
- antibody buffer: 2.5% (w/v) BSA, 0.02% (w/v) sodium azide in PBS

Method

1. Subject the samples to SDS–PAGE (see *Table 2*).
2. Cut away the stacking gel and any excess separating gel on either side of the samples, and cut nitrocellulose sheet and six pieces of 3MM paper to the same size as the gel.
3. Wet the gel and nitrocellulose in Western buffer. Assemble the transfer stack in the following order:
   - plastic cassette
   - porous sponge
   - three pieces of 3MM paper (wetted in Western buffer)
   - gel
   - nitrocellulose
   - three pieces of 3MM paper (wetted in Western buffer)
   - porous sponge
   - plastic cassette

   Take care to ensure that there are no air bubbles in the stack. It is easiest to assemble it in a dish containing Western buffer. Place the stack in the apparatus with the nitrocellulose towards the anode. Fill the apparatus with Western buffer. Transfer at 0.15 A for 12–15 h while cooling using a cooling coil and cold water.
4. Remove the stack from the apparatus and blot the nitrocellulose dry by placing it on 3MM paper. If the nitrocellulose is to be probed using more than one primary antibody, it should be cut into strips.
5. Mark the nitrocellulose with a permanent marker for orientation and block it with blocking solution for at least 1 h at room temperature.
6. Incubate with the primary antibody diluted in antibody buffer for 1 h at room temperature with agitation. The antibody dilution used at this step will depend on antibody titre.[a]
7. Wash the nitrocellulose once for 10 min in PBS, twice for 10 min in 0.05% (v/v) NP40 in PBS, and once for 10 min in PBS.
8. Incubate with ^{125}I-labelled secondary antibody diluted in blocking solution (0.1 μCi/ml) for 1 h at room temperature with agitation,[b] and wash as in step 7.

Protocol 15. *Continued*

9. Air dry the nitrocellulose and mark it with ^{14}C-labelled ink to permit alignment with the X-ray film.

10. Expose to X-ray film at −70°C using an intensifying screen.

[a] A 1:500 dilution of a reasonably high titre polyclonal antibody works well in our hands, while a 1:4000 dilution of a high titre mAb has been used successfully.
[b] Incubation with the labelled second antibody is best carried out in a heat-sealed plastic pouch.

Detection of proteins using a radioactively labelled secondary antibody has an advantage over enzyme-linked antibodies, in that quantitation of proteins in different samples can be readily carried out either by cutting the band out of the nitrocellulose and measuring the radioactivity of the bound antibody by scintillation counting or by densitometric analysis of the X-ray film. An example of a Western blot performed using this *Protocol 15* to detect the SV5 V protein in purified SV5 virions is shown in *Figure 1*.

A variation of the electrophoretic transfer procedure described above can

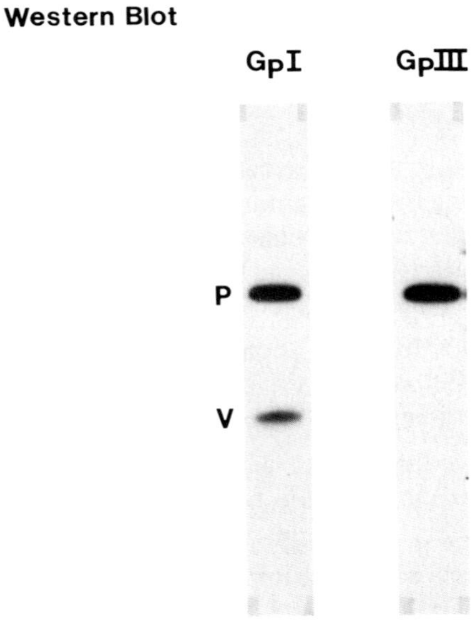

Figure 1. Analysis of polypeptides by Western blotting. Polypeptides synthesized in SV5-infected cells were analysed as described in *Protocol 15*. Two mouse mAbs that recognize the SV5 P protein were used as the primary antibodies, the group I (GpI) antibodies recognize the V protein in addition to P, whereas the group III (GpIII) antibodies recognize only the P protein. The secondary antibody used was a ^{125}I-labelled goat anti-mouse antibody.

be used to analyse proteins for their ability to bind RNA (*Protocol 16*). This assay is termed a Northwestern blot and utilizes ^{32}P-labelled RNA as a probe rather than an antibody. If the labelled RNA binds to a protein it is a good indicator that the protein binds RNA. However, if the probe fails to bind to a protein the result is inconclusive, as some proteins known to bind RNA from other types of assay do not bind RNA in a Northwestern blot assay.

Protocol 16. Analysis of protein–RNA interactions by a Northwestern blot assay

Materials

- 50 × Denhardt's solution (10 g/l Ficoll, 10 g/l BSA, 10 g/l polyvinylpyrolidone; filter through filter paper and add 70 mg sodium azide
- Northwestern buffer: 10 mM Tris–HCl (pH 7.5), 50 mM NaCl, 1 mM EDTA, 1 × Denhardt's solution
- a ^{32}P-labelled RNA probe[a]

Method

1. Carry out steps 1–4, *Protocol 15*.
2. Block the nitrocellulose for 1 h at room temperature by agitating in Northwestern buffer.
3. Heat-seal the blot into a pouch and probe it in Northwestern buffer. If using virion RNA or *in vitro* transcribed RNA as the probe, include 5 μg/ml tRNA in the hybridization solution. Incubate for 1 h at room temperature.
4. Wash the nitrocellulose five times for 10 min at room temperature in Northwestern buffer.[b]
5. Air-dry the nitrocellulose, mark it with ^{14}C-labelled ink to allow alignment and expose it to X-ray film.

[a] The probe can be viral RNA extracted from purified virus grown in the presence of [^{32}P]orthophosphate, total cytoplasmic RNA extracted using the guanidinium thiocyanate method (13), or ^{32}P-labelled RNA transcribed *in vitro* using SP6 or T7 DNA-dependent RNA polymerase.

[b] The first two washes should be carried out in the pouch. For the remaining washes, the blot can be removed from the pouch and washed in a dish.

---

4. Analysis of integral membrane glycoproteins

Paramyxoviruses and orthomyxoviruses encode integral membrane glycoproteins that function in attachment and penetration of virus particles into the target host-cell cytoplasm. In paramyxoviruses these are F and HN, and in influenza A and B virus it is HA and NA that carry out these functions. Much has been learned recently about the rate of transport of these proteins from their site of synthesis on the rough endoplasmic reticulum (ER) membrane to their final destination in the cell, the plasma membrane. A large amount of information has been gained from studying the carbohydrate moieties of these proteins. Influenza virus NA and HA and paramyxovirus F and HN all possess N-linked carbohydrate. A protein can be shown to be modified by the addition of N-linked carbohydrates in several different ways, the easiest of which is to synthesize the protein in the presence of the drug tunicamycin (*Protocol 17*). Tunicamycin inhibits formation of the dolicholphosphate-*N*-acetylglucosamine complex, which functions as the donor for transfer of *N*-acetylglucosamine to the appropriate asparagine residues in the polypeptide backbone, and thus blocks formation of N-linked carbohydrate.

Protocol 17. Labelling virus-infected cell polypeptides in the presence of tunicamycin

Materials

- 1 mg/ml tunicamycin (Calbiochem) in dimethylsulphoxide (DMSO)

Method

1. Infect cells as described in *Protocol 10*, but omit the actinomycin D mentioned in step 4.
2. At 14 h p.i. for SV5, 5–6 h p.i. for influenza A virus, and 6–8 h p.i. for influenza B virus, wash the infected cell monolayers with DME and then replace the medium with DME supplemented with 1 μg/ml tunicamycin[a] and incubate for 2 h.
3. Replace the medium with starvation medium supplemented with 1 μg/ml tunicamycin. Incubate for 30 min at 37°C.
4. Replace the starvation medium with labelling medium supplemented with 1 μg/ml tunicamycin. Incubate for 1–2 h at 37°C.
5. Lyse the cells as described in *Protocol 11*.

[a] For some cell types (for example MDCK) it is necessary to use 2 μg/ml tunicamycin.

When a polypeptide labelled in the presence of tunicamycin is immunoprecipitated, analysed by SDS–PAGE, and compared with a sample labelled in the absence of tunicamycin, it will exhibit a relatively greater electrophoretic mobility if it normally possesses N-linked carbohydrate. *Figure 2* shows an example of proteins synthesized in the presence and absence of tunicamycin. Some mAbs fail to precipitate unglycosylated proteins, not because the antibody recognizes a carbohydrate determinant, but because the polypeptide does not fold into a conformation recognizable by the mAb (i.e. for some

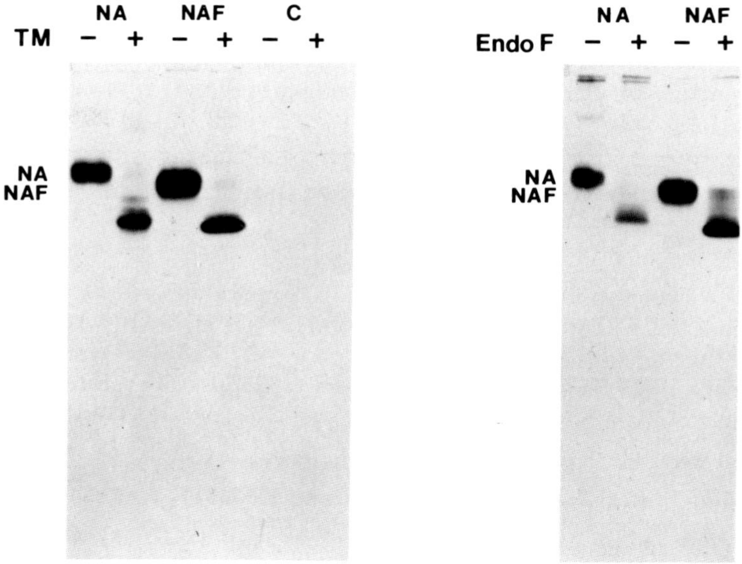

Figure 2. Synthesis of a glycoprotein in the presence of tunicamycin, an inhibitor of N-linked glycosylation. Monolayers of CV-1 cells infected with SV40 recombinants expressing influenza virus NA, a secreted chimeric glycoprotein (NAF), or an unrelated protein (C) were labelled with Tran[^{35}S]-label in the presence (+) or absence (−) of tunicamycin as described in *Protocol 17*. Cell lysates were prepared and immunoprecipitation was achieved using a polyclonal antibody specific for the influenza NA protein as described in *Protocols 11* and *14*, the proteins were analysed by SDS–PAGE. Tm, tunicamycin; EndoF, *N*-glycosidase F. (This figure is taken from ref. 5 with permission.)

proteins the presence of carbohydrate chains influences the ability of the polypeptide chain to fold into a native form). Thus, in the first instance it is often better to use a polyclonal antibody, as such antibodies may contain a subset of IgG species which recognize determinants from the denatured protein.

4.1 Endoglycosidase treatment

There are two endoglycosidases that have proven to be very useful in the study of proteins that possess N-linked carbohydrate. These are *N*-glycosidase F (sometimes referred to as peptide *N*-glycosidase F) and endo β-*N*-acetylglucosaminidase H (endo H). *N*-glycosidase F removes all forms of N-linked carbohydrate from proteins, whereas endo H removes only carbohydrate that is still in the high mannose form. The conversion of carbohydrate chains from the high mannose form to the complex form that is resistant to endo H digestion occurs in the *medial* Golgi compartment. Therefore, resistance of carbohydrate chains to endo H can be utilized as a measure of whether a protein has reached the medial Golgi. *N*-glycosidase F and endo H digestions are carried out on proteins that have been immunoprecipitated and eluted from protein A–Sepharose beads by boiling in SDS (*Protocols 18* and *19*).

Protocol 18. *N*-glycosidase F digestion of glycoproteins

Materials

- 250 × protease inhibitor solution: 50 μl of pepstatin A (500 μg/ml in 75% ethanol), 25 μl of chymostatin (1 mg/ml in 75% DMSO), 50 μl of leupeptin (500 μg/ml in water), 50 μl of antipain (500 μg/ml in water), 250 μl of aprotinin (24.36 trypsin inhibiting U/ml), 575 μl water; store at −20°C (all reagents from Sigma)
- *N*-glycosidase F (Boehringer–Mannheim)
- boiling buffer: 0.4% (w/v) SDS, 20 mM Na_2HPO_4 (pH 8.0)
- protease inhibitor buffer: 40 mM Na_2HPO_4 (pH 8.0), 20 mM EDTA, 1% (w/v) NP40

Method

1. Perform immunoprecipitation of glycoproteins (*Protocol 14*) and, following the last wash of the protein A–Sepharose beads, remove as much of the wash solution as possible.
2. Add 22 μl of boiling buffer and boil for 4 min. Centrifuge for 40 sec in a microcentrifuge. Remove the supernatant (approx. 25 μl) to a new tube. Add an equal volume of freshly prepared 1 × protease inhibitor solution in protease inhibitor buffer. Add 0.2 U of *N*-glycosidase F. Do not add the enzyme to control samples. Digest overnight at 37°C.
3. Add 35 μl of SDS–PAGE sample buffer (*Protocol 13*), boil for 3 min, and analyse by SDS–PAGE.

Protocol 19. Endo H digestion of glycoproteins

Materials

- endoglycosidase H (ICN Biomedicals)
- 0.1 M sodium citrate (pH 5.3): 25.5 ml 0.1 M citric acid plus 74.5 ml 0.1 M sodium citrate
- extraction buffer: 50 mM Tris–HCl (pH 7.4), 0.5% (w/v) SDS

Method

1. Carry out step 1, *Protocol 18*.
2. Add 20 μl of extraction buffer. Mix, centrifuge for 2 sec in a microcentrifuge, boil for 4 min, spin for 40 sec, and transfer 22 μl to a new tube.
3. Add 22 μl of 0.1 M sodium citrate (pH 5.3) containing 1 mM PMSF and 2 mU of endo H to each sample. Do not add the enzyme to control samples. Digest for 18–24 h at 37°C.
4. Carry out step 3, *Protocol 18*.

4.2 Analysis of protein oligomerization

The majority of viral integral membrane proteins that have been studied exist as oligomers in their mature biologically active state. As a result, much effort has been spent in trying to determine the oligomeric structure of many virus-encoded integral membrane proteins. In addition, as more becomes known about the processes involved in maturation of secreted and integral membrane proteins, attention has also been turned to the process of protein folding in the ER. A number of assays have been found to be useful in determining the kinetics of folding of proteins in the ER and in determining their oligomeric nature. Three assays are described below.

Chemical cross-linking (*Protocol 20*, adapted from ref. 6) has proven to be a very powerful technique with which to examine protein–protein interactions such as those that occur in oligomeric proteins. There is a degree of luck involved in cross-linking, as the different subunits have to be an appropriate distance apart in order for reactive amino acids to be cross-linked to each other. However, there are many bifunctional reagents possessing differing lengths of alkyl spacers that are available commercially, hence it is likely that one can be found that will work for a protein of interest. The Pierce Chemical Company is a major supplier of these reagents, and their catalogue contains useful information concerning their use. Because of the serendipitous nature of cross-linking it is worthwhile investing in a variety of reagents.

Sucrose density-gradient centrifugation (*Protocol 21*) is another method that can be used to make a crude estimate of the oligomeric nature of a protein. This method has an advantage over cross-linking in that it results in a quantitative analysis of the fraction of the protein of interest that exists as oligomers at any one time, since oligomeric proteins are usually only partially cross-linked. A disadvantage of this method is that some oligomeric proteins are not stable under the conditions used for gradient analysis and dissociate upon exposure to the forces of centrifugation. In addition, sucrose density-gradient analysis alone does not permit a true determination of the number of subunits in the oligomer. Because sedimentation is dependent on the shape of molecules in addition to their M_rs, the sedimentation pattern of known oligomers cannot be extrapolated to an unknown molecule.

The third method is the use of conformation-specific antibodies to probe the folding and oligomerization of proteins. This method is obviously applicable only if a bank of antibodies that recognize different forms of the protein is available.

Protocol 20. Chemical cross-linking of proteins

Materials

- cross-linking reagents[a]

Method

1. Infect cells in 60 mm dishes and radioactively label the infected cell proteins as described in *Protocols 10* and *12*.
2. Wash the cells with PBS and incubate in 50 mM EDTA in PBS at 37°C to remove them from the dish.
3. Pellet the cells by centrifuging for 5 min at 560 *g* (1500 r.p.m. in an IEC DPR-6000 centrifuge using the 253 12-place rotor).
4. Resuspend the cells in 500 μl PBS (pH 8.5)/plate.
5. Aliquot 100 μl of cells into tubes for cross-linking, add 2 μl of 10% (v/v) NP40, and 1 μl of cross-linking reagent in DMSO. Incubate for 30 min at 4°C.
6. Quench by adding glycine to a concentration of 50 mM. Add 500 μl of RIPA buffer, centrifuge for 3 min at 16 000 *g* in a microcentrifuge, and immunoprecipitate as described in *Protocol 14*.
7. Analyse by SDS–PAGE, using non-reducing conditions.

[a] Useful reagents to start with are dithio-bis(succinimidylpropionate), 3,3′-dithio-bis (sulphosuccinimidylpropionate), bis[2-(succinimidooxycarbonyloxy)ethyl]sulphone, and ethylene glycol-bis(succinimidylsuccinate). These reagents are all available from Pierce. Try these reagents at concentrations of 1 mM, 200 μM, and 40 μM. We routinely store the reagents as 100 mM stocks in DMSO at −20°C and make five-fold dilutions in DMSO prior to use, such that 1 μl of these dilutions added to 100 μl samples of cell lysate gives the required concentrations.

Protocol 21. Sucrose density gradient centrifugation

Materials

- 5%, 25%, and 60% (w/v) sucrose solutions in 10 mM sodium phosphate (pH 7.2), 100 mM NaCl, 0.1% (v/v) Triton X-100
- lysis buffer: 10 mM sodium phosphate (pH 7.2), 100 mM NaCl, 1% (v/v) Triton X-100

Method

1. Infect 100 mm plates of cells as described in *Protocol 10*.
2. Label the infected cells for 1 h with 100 μCi/ml Tran[^{35}S]-label as described in Section 3.1. Replace the labelling medium with chasing medium and incubate for 1–2 h.
3. Wash the cells once with PBS and lyse them using 1 ml lysis buffer/plate.
4. Centrifuge the lysate for 5 min in a microcentrifuge to pellet nuclei and cellular debris.
5. Layer the lysate (0.5 ml) on top of a 10.5 ml 5–25% sucrose gradient over a 0.75 ml 60% sucrose cushion.
6. Centrifuge the gradients for 16.75 h at 180 000 *g* (38 000 r.p.m. in a Beckman SW41 rotor) at 20°C.
7. Fractionate the gradients from the bottom of the tube into 16 fractions of 0.75 ml each.
8. For immunoprecipitation of the gradient fractions, add an equal volume of 2 × RIPA buffer (see *Protocol 11*) and immunoprecipitate as described in *Protocol 14*.

We have successfully used *Protocol 21* to analyse the oligomeric structure of a hybrid protein consisting of influenza virus NA with the fusion peptide of the SV5 F protein in place of the extended signal-anchor domain, as shown in *Figure 3*. Depending on the size of the protein, the gradient conditions should be optimized to achieve good separation of the different oligomeric forms.

As mentioned above, if a collection of mAbs specific for the protein of interest is available, it is possible that some may recognize different forms of the protein. To determine whether this is the case, carry out a pulse–chase experiment (*Protocol 12*) using as short a pulse as possible, immunoprecipitate the samples with the various mAbs, and compare their reactivities to that of a polyclonal antibody. If the reactivity of a particular mAb is delayed after the pulse-label in comparison to reactivity with a polyclonal antibody that recognizes all forms of the protein, then it may be a useful tool for experiments in conjunction with sucrose gradient analysis and cross-linking experiments. For an example of this type of study the reader is referred to ref. 7.

4.3 Association of proteins with the ER

In some instances it is necessary to determine whether a viral protein has been successfully inserted into the ER membrane, what orientation it is in with respect to the membrane, and how it is associated with the membrane. The first step in such experiments involves the preparation of microsomal membranes from infected cells (*Protocol 22*).

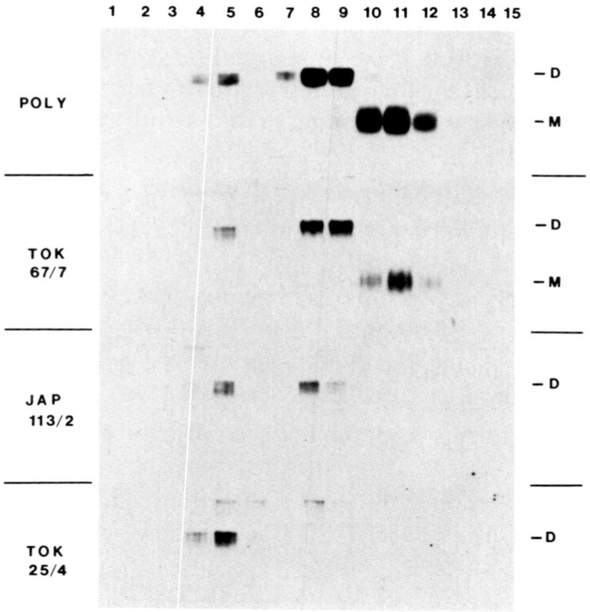

Figure 3. Analysis of the oligomeric form of a soluble glycoprotein on sucrose density gradients and immunoprecipitation of different oligomeric species using conformation-specific mAbs. Monolayers of CV-1 cells infected with the SV40 recombinant expressing the soluble and secreted hybrid protein NAF were labelled with Tran[^{35}S]-label for 1 h and incubated in chase medium for 2 h as described in *Protocol 12*. NAF secreted into the medium was fractionated on sucrose gradients as described in *Protocol 21*. Samples collected from the bottom of the gradient were immunoprecipitated with polyclonal antisera (poly) or three conformation-specific mAbs (Tok 67/7, Jap 113/2, or Tok 25/4) as described in *Protocol 14*. The polypeptides were analysed by SDS–PAGE under non-reducing conditions. Fraction 1, bottom of the gradient; fraction 15, top of the gradient; D, dimer; M, monomer. (Adapted from ref. 5 with permission.)

Protocol 22. Preparation of microsomal membranes

Materials

- Dounce homogenizer (Wheaton, with a type A pestle)
- DHB: 10 mM Tris–HCl (pH 7.5), 1 mM $MgCl_2$
- 50 mM triethanolamine acetate (TEA): prepared from triethanolamine by adjusting the pH with acetic acid

Method

1. Infect 100 mm dishes of cells and metabolically label the infected cell proteins as described in *Protocol 10* and Section 3.1, respectively.

2. Remove the medium and wash the monolayers twice with 5 ml PBS.
3. Add 2 ml of PBS to each dish and scrape off the cells. Pellet the cells by centrifuging at 560 *g* (see step 3, *Protocol 20*) for 5 min at 4°C.
4. Resuspend the pellets in 3 ml of DHB. Allow the cells to swell on ice for 5 min.
5. Disrupt the cells using 15 strokes in a Dounce homogenizer.
6. Layer 1.5 ml of the extract on top of a 10 ml cushion of 10% (w/w) sucrose in DHB and centrifuge for 4 h at 180 000 *g* (38 000 r.p.m. in a Beckman SW41 rotor) at 4°C.
7. Resuspend the microsomal pellets in 300 μl of 50 mM TEA (pH 7.5).
8. Immunoprecipitate according to *Protocol 13* or, after the addition of an equal volume of 2 × RIPA buffer, according to *Protocol 14*.

Once a protein has been found to be associated with microsomal membranes it may be necessary to determine whether it is associated peripherally or is an integral membrane protein. One can differentiate between these possibilities by treating microsomal membranes with alkali (*Protocol 23*). Such treatment has been shown to result in the extraction of peripheral proteins while leaving integral membrane proteins associated with the lipid bilayer (8).

Protocol 23. Alkali treatment of microsomal membranes

Materials

- 1 M Na_2CO_3 (adjusted to pH 11 with NaOH)
- cushion solution: 0.2 M sucrose, 30 mM Hepes–NaOH (pH 11), 150 mM potassium acetate, 2.5 mM magnesium acetate

Method

1. Prepare microsomes according to steps 1–7, *Protocol 22*.
2. Adjust the microsome suspension to pH 11 by adding of one-tenth of a volume of 1 M Na_2CO_3. Incubate on ice for 20 min.
3. Layer 310 μl of the suspension on top of 680 μl of the cushion solution and centrifuge at 135 000 *g* (45 000 r.p.m. in a Beckman TL-100 Tabletop ultracentrifuge using the TLS-55 swinging-bucket rotor) for 20 min at 4°C.
4. Collect the supernatant, resuspend the pellet in 310 μl of TEA (pH 11) (*Protocol 22*) and add 680 μl of cushion solution (this makes the pellet sample equivalent to the supernatant).
5. Add SDS to 1%, boil for 5 min, neutralize by adding 1 M HCl, dilute by adding four volumes of SDS immunoprecipitation dilution buffer, and immunoprecipitate as described in *Protocol 13*.

In some instances it may be necessary to determine the membrane topology of a protein, or it may be important to find out whether a protein has been translocated across the ER membrane. These questions can be addressed by treating microsomal membranes with a protease (*Protocol 24*). For these experiments to be feasible when analysing the membrane topology of a protein, it is essential that the domain of the protein that may be exposed to the action of exogenously added protease is of sufficient size, so that its removal can be detected by an increased mobility on SDS–PAGE. In addition, it is also essential that the protein is not resistant to protease digestion.

Protocol 24. Trypsin treatment of microsomes

Materials

- TPCK-treated trypsin[a] (Worthington)

Method

1. Carry out steps 1–5, *Protocol 22*.
2. Divide each sample into four 750 μl aliquots and add the following:
   (a) an equal volume of 2 × RIPA buffer
   (b) 250 μl of DHB (*Protocol 22*) (control microsomes)
   (c) 250 μl of DHB containing TPCK-treated trypsin, such that the final concentration of trypsin in the sample is 250 μg/ml
   (d) 250 μl of DHB containing TPCK-treated trypsin and NP40, such that the final concentrations in the sample are 250 μg/ml and 1% (v/v), respectively.
3. Incubate samples (b), (c), and (d) at 37 °C for 1 h and then add PMSF and aprotinin to final concentrations of 100 μM and 1.2 trypsin-inhibiting U/ml, respectively.
4. Layer the samples on top of 10 ml cushions of 10% (w/w) sucrose in DHB and centrifuge them for 4 h at 180 000 *g* (38 000 r.p.m. in a Beckman SW41 rotor) at 4 °C.
5. Dissolve the pellets in 0.5 ml of RIPA buffer (*Protocol 11*) and immunoprecipitate as described in *Protocol 14*.

[a] Trypsin that has been treated with tosyl-phenylalanine chloromethyl ketone (TPCK), which is an inhibitor of chymotrypsin, a potentially contaminating protease.

4.4 Immunofluorescent staining

Another method for analysing the intracellular distribution of proteins is by immunofluorescent staining. There are two basic types of procedure that can be used:

- Analysis of the distribution of proteins on the surface of cells (*Protocol 25*).
- Examination of the intracellular localization of proteins (*Protocol 26*). More specialized permeabilization techniques permit the visualization of molecules expressed at the cell surface and intracellularly.

In both procedures, an antibody specific for the protein of interest is added to the cells (the primary antibody), followed by a fluorescently-labelled secondary antibody that recognizes the primary antibody.

For immunofluorescence, the cells to be infected should be plated on glass coverslips (18 mm square is a convenient size) which have been placed in the appropriately sized tissue culture dish. The coverslips should be kept in a humid chamber to prevent the cells from drying out when they are incubated with the blocking solution or the antibodies. A large Petri dish with damp filter paper in the bottom works well, as does placing the coverslip on top of pieces of glass capillary tube in individual dishes and putting a few drops of PBS in the bottom of the dish.

Protocol 25. Cell surface immunofluorescent staining

Materials

- Coplin staining jars
- self-closing forceps
- 0.5% formaldehyde solution[a]
- mounting medium[b]
- primary antibody
- fluorescently-labelled secondary antibody (Organon Teknika-Cappel)
- BSA/PBS: 1% (w/v) BSA in PBS

Method

1. Infect monolayers of cells plated on coverslips as described in *Protocol 10*, but omit the addition of actinomycin D in the case of SV5.
2. Wash the cells by dipping the coverslips in a beaker of PBS several times while holding them with self-closing forceps.
3. Fix the cells by incubating the coverslips in 0.5% formaldehyde solution for 5 min at room temperature in a Coplin staining jar.
4. Wash the coverslips as described in step 2.
5. Incubate the coverslips in a 1% (w/v) solution of BSA/PBS for 30 min at room temperature to block non-specific binding of primary or secondary antibodies (150 μl added dropwise to an 18 mm square coverslip is sufficient to cover the monolayer).
6. Wash the coverslips as described in step 2.

Protocol 25. *Continued*

7. Incubate the coverslips for 30 min at room temperature with 150 μl of the primary antibody diluted in BSA/PBS.[c]
8. Wash the coverslips as described in step 2.
9. Incubate the coverslips for 30 min at room temperature with 150 μl of secondary antibody[d] diluted in BSA/PBS according to the manufacturer's instructions.
10. Wash the coverslips as described in step 2.
11. Incubate the coverslips for 5 min at room temperature in a Coplin staining jar containing PBS.
12. Place a drop of mounting medium on a glass slide. Drain excess liquid from the coverslip and invert it (cell side down) on to the mounting medium.
13. Place a pad of absorbent paper on top of the coverslip and press down gently to remove excess liquid. Take care not to let the coverslip slide.
14. Fix the coverslip in place on the slide by applying nail polish along the edges of the coverslip.

[a] Dissolve paraformaldehyde in a small volume of water by heating it in a boiling water bath. The formaldehyde is then made up to volume with cold PBS. Alternatively, methanol-free formaldehyde is available commercially.

[b] This is available commercially or can be prepared as follows. Mix 1 volume of 100 mM Tris–HCl (pH 8.0) with 3 volumes of glycerol. Adjust to pH 8.0 with NaOH or Tris and check several times over a 1 h period, since the pH usually falls to 7.0. Add *p*-phenylenediamine (a reagent which helps prevent photobleaching; Sigma) to 1% (w/v). Wrap the vessel in aluminium foil and rock it overnight at 4°C. Aliquot the medium into brown glass bottles and store at −20°C.

[c] A series of dilutions of the primary antibody should be made in BSA/PBS and tested to determine the optimal dilution, such that specific staining is intense and background staining is low. In some cases, owing to cross reactivity with a cellular antigen, it may be necessary to pre-absorb the antibody on formaldehyde-fixed uninfected cells prior to use.

[d] The fluorescently-labelled secondary antibody should be centrifuged prior to use to pellet insoluble fluorescent material. We centrifuge the antibody at 128 000 *g* (60 000 r.p.m. in a Beckman TL100 Tabletop ultracentrifuge using the TLA 100.2 rotor).

Protocol 26. Immunofluorescent staining of permeabilized cells

Method

1. Carry out steps 1–4, *Protocol 25*.
2. Permeabilize[a] the cells by incubating the coverslips for 5 min at −20°C in a Coplin staining jar containing acetone that has been pre-chilled to −20°C.
3. Wash the coverslips as described in step 2, *Protocol 21*.
4. Carry out steps 5–14, *Protocol 25*.

[a] There are many ways of permeabilizing cells. These include the use of methanol, Triton X-100 and saponin. The reader is referred to a text on fluorescent microscopy for details of these methods.

5. Expression of proteins from cDNA

Recently, a major emphasis in the molecular virology of myxoviruses and paramyxoviruses has been the cloning and sequencing of virus genes, with the ultimate goal for paramyxoviruses of constructing biologically active full-length cDNA copies of the virus genomes. The obvious step following the completion of the sequence of a full-length cDNA copy of a gene is to express the protein, in the absence of any of the other virus proteins, in order to determine whether a protein of the correct size is made, whether it is localized to the correct subcellular compartment, and whether an assay is available to determine biological activity.

5.1 Viral expression systems

There are many different eukaryotic viral vector systems available that can be used to study the protein of interest; two that we have found useful for biochemical and functional analysis of both influenza virus- and SV5-encoded proteins are the simian virus 40 (SV40) expression vector system (*Protocol 27*) and the system whereby the bacteriophage T7 DNA-dependent RNA polymerase is expressed in mammalian cells using a recombinant vaccinia virus (vTF7-3) (*Protocol 28*; ref. 9; see Chapter 9). In the latter system, the T7 RNA polymerase transcribes mRNA from transfected plasmids in which the gene being studied is under the control of the T7 promoter. The T7 RNA polymerase/vaccinia virus recombinant expression method is efficient and gives rise to high levels of expressed protein a few hours after transfection, and can be used in a broad range of host-cell types. However, this method involves carrying out transfections for each experiment, and vaccinia virus has a severe c.p.e. in most cell types which may not be ideal for certain experiments.

The SV40 expression system has certain advantages in that it is easy to make recombinant virus stocks through the use of a defective helper virus, and the c.p.e. is less severe. However, it is slower than some eukaryotic viral vector systems, as it takes about two weeks to prepare virus stocks and then most experiments are carried out 48–72 h p.i. In addition, the SV40 expression system is restricted to cells of monkey origin, the permissive host for SV40. The SV40 vector that we employ is a late region replacement vector (pSV103) that was adapted from the vector pSV2330 (10). Foreign cDNA is inserted into this vector such that it is under the control of the SV40 late promoter and polyadenylation signals. To enable propagation of the plasmid, the SV40 sequences are joined to pBR322 at an *Sst*I site which replaces the region of pBR322 from the *Eco*RI to the *Bam*HI sites. The helper virus used in this system is an early region deletion mutant (dl1055) which provides SV40 structural proteins (11). The SV40 plasmid DNAs have to be digested prior to transfection with *Sst*I to liberate the SV40 sequences from pBR322 sequences, and are then ligated at low concentrations so that the SV40 DNA recircularizes.

Protocol 27. Expression of heterologous proteins using recombinant SV40 virus vectors

Materials

- subconfluent monolayers of cells of monkey origin (for example CV-1) in 60 mm plates
- SV40 recombinant and helper DNAs recircularized and dissolved at 0.1 μg/μl
- DEAE–dextran solution: 200 μg/ml DEAE–dextran in DME containing 50 mM Tris–HCl (pH 7.28)[a]
- DMSO solution: 10% (v/v) DMSO in PBS containing 20 mM Hepes–NaOH (pH 7.3)
- transfection buffer: DME supplemented with 10% (v/v) FCS, 20 mM Hepes–NaOH (pH 7.3), 2 mM L-glutamine, 0.1 M sodium pyruvate

Method

1. Wash the monolayers three times with 5 ml of PBS and add 3 ml of DEAE–dextran solution to the monolayers. Then add 4 μg each of the helper and recombinant DNAs and mix by swirling. Incubate at 37°C for 3–5 h until about 30% of the cells exhibit a c.p.e.
2. Wash the monolayers three times with 5 ml of DME.
3. Add 3 ml of DMSO solution and incubate at room temperature for 2 min.
4. Wash the monolayers as described in step 2.
5. Add 10 ml of transfection buffer and incubate overnight at 37°C.
6. Replace the medium with 10 ml of fresh transfection buffer. Incubate for three days at 37°C and replace the medium again.
7. Incubate the monolayers for a further 7–9 days or until the c.p.e. is extensive. Collect the supernatant medium every three days during this period for virus stock. For the final virus harvest, freeze and thaw the cells three times. All the virus harvests are then clarified to remove cell debris.

[a] Mix equal volumes of 2 × DME and 100 mM Tris–HCl (pH 7.28) and dissolve DEAE–dextran to 200 μg/ml. Adjust the pH to 7.28 if necessary and filter-sterilize the solution. The pH is critical for the success of the transfection.

The virus stocks prepared according to *Protocol 27* can be titrated by immunofluorescence to determine to what extent they can be diluted. To infect cells with the SV40 recombinants, wash the cells once with PBS, add the inoculum, absorb for 1.5 h, and replace it with DME supplemented with 2% (v/v) FCS. The proteins synthesized in cells infected with these virus stocks are routinely labelled between 48 and 72 h p.i.

Protocol 28. Protein expression using a recombinant vaccinia virus which expresses bacteriophage T7 RNA polymerase

Materials

- monolayers of cells permissive for vaccinia virus[a] in 60 mm dishes
- Hepes-buffered saline (HBS): 0.14 M NaCl, 5 mM KCl, 1 mM $Na_2HPO_4{\cdot}2H_2O$, 0.1% (w/v) dextrose, 20 mM Hepes–NaOH (pH 7.05); filter-sterilize
- 2.5 M $CaCl_2$; filter-sterilize, store at −20°C
- recombinant vaccinia virus vTF7-3
- recombinant plasmid with target gene cloned under control of the T7 promoter

Method

1. Wash the cell monolayers with 5 ml of PBS.
2. Infect the cells with 10 p.f.u./cell vTF7-3 in an inoculum of 0.5 ml of BSA/PBS. Incubate for 1 h at 37°C.
3. Prepare calcium phosphate-precipitated DNA. Add 250 μl of HBS to 10 μg of plasmid in 20 μl water and vortex. Place the pipette tip on the surface of the solution and slowly add 12.5 μl of 2.5 M $CaCl_2$ down the side of the tube. Tap the tube with a finger to mix. Incubate at room temperature for 30 min. A suitable calcium phosphate precipitate is only slightly cloudy.
4. Remove the virus inoculum, add 3 ml of DME (pH 7.05). Add the calcium phosphate-precipitated DNA dropwise, mix and incubate at 37°C for 3 h.[b]

[a] Vaccinia virus has a wide host range, so that many different cell types can be used with this expression system. In most cell types, however, vaccinia virus exhibits a severe c.p.e. A cell line that is resistant to c.p.e. and thus is useful for biological experiments is the HeLa-T4 cell line which expresses the CD4 molecule (12).

[b] After incubating for 3 h there is sufficient protein synthesis to enable metabolic labelling of expressed proteins and biochemical experiments. However, the cells do remain viable for longer periods if a more extended time course is required.

5.2 Expression by translation of mRNA *in vitro*

It is unnecessary these days for the researcher to make a reticulocyte lysate, since there are several vendors. We use Promega reticulocyte lysate, which contains an energy-generating system consisting of phosphocreatine kinase and phosphocreatine, a mixture of tRNAs, haemin (to prevent inhibition of initiation), potassium acetate, and magnesium acetate. All that is required is to add mRNA (synthesized according to *Protocol 29*) and a mixture of amino acids lacking the amino acid to be added subsequently in radiolabelled form (*Protocol 30*).

Protocol 29. *In vitro* transcription of RNA using bacteriophage T7 RNA polymerase

Materials

- DE-81 filters (2.4 cm diameter; Whatman)
- 0.5 mM ^{7}mGpppG (Pharmacia)
- [^{3}H]GTP
- 5 × reaction buffer: 200 mM Tris–HCl (pH 8.0), 125 mM NaCl, 40 mM $MgCl_2$, 10 mM spermidine
- RNasin (Promega)
- 1 U/μl RNase-free DNase (RQ1 DNase; Promega)
- 50 U/μl T7 RNA polymerase (Gibco–BRL)
- 1 μg/μl linearized template DNA
- 1 mg/ml yeast tRNA

Method

1. Dry down 10 μCi of [^{3}H]GTP in a sterile microcentrifuge tube.
2. Prepare the reaction mixture by adding the following to a second tube.[a]

• 5 × reaction buffer	20 μl
• 10 mM ATP	5 μl
• 10 mM CTP	5 μl
• 10 mM UTP	5 μl
• 0.5 mM GTP	30 μl
• 0.5 M DTT	1 μl
• 0.5 mM ^{7}mGpppG	31 μl
• RNasin	1 μl

3. Add the reaction mixture (98 μl) to the tube containing the dried [^{3}H]GTP and add 2 μl of linearized template DNA.[b] Vortex, add 2 μl of T7 RNA polymerase, vortex again, and incubate at 37°C for 1 h.
4. Add a further 0.2–1 μl of T7 RNA polymerase and incubate for a further 1 h.
5. Add 10 μl of 0.1 mM EDTA and 2 μl of DNase and incubate at 37°C for 10 min.
6. Add 4 μl of 0.25 M EDTA and spot 2 μl on to each of two DE-81 filters to assess the efficiency of the reaction.[c]

7. Add 20 μl of yeast tRNA. Extract with phenol then chloroform, ethanol precipitate, and desalt on a 1 ml Sephadex G50 spin column.[d] Ethanol precipitate the labelled RNA and resuspend it in sterile water at a concentration of 0.2 μg/μl. For long-term storage, store the RNA under ethanol at −20°C.
8. To check the integrity and size of the RNA transcripts, analyse 0.5 μg on a polyacrylamide/urea gel.

[a] The reaction mixture should be kept at room temperature while it is being prepared, as DNA can precipitate in the presence of spermidine at 4°C.
[b] If possible, the DNA template should be linearized using a restriction endonuclease that leaves a 5′ overhang. Longer transcripts than expected can result from templates possessing 3′ protruding ends. It is also important that the restriction digest goes to completion.
[c] The DE-81 filter assay is carried out as described in Appendix E.19 of ref. 13. One sample is washed, and the unwashed filter acts as the control for the number of c.p.m. that are put into the reaction.
[d] The Sephadex G50 spin column used for desalting the RNA is used as described in Appendix E.37–E.38 of ref. 13.

Protocol 30. *In vitro* translation using nuclease-treated rabbit reticulocyte lysate

Materials

- rabbit reticulocyte lysate (Promega)
- radiolabelled amino acid
- mRNA: poly-A(+) RNA prepared from a 60 mm plate of infected cells according to ref. 13 or *in vitro* transcribed RNA (*Protocol 29*)
- TCA/casamino acids: 10% (w/v) TCA, 3% (w/v) casamino acids

Method

1. Dissolve the poly-A(+) RNA in 4.5 μl of sterile water, or take 4.5 μl of *in vitro* transcribed RNA, heat at 67°C for 10 min, and chill on ice.
2. Add the following to 17.5 μl of reticulocyte lysate[a] in a sterile 500 μl microcentrifuge tube and incubate at 30°C for 1 h:
   - 1 μl of a 1 mM mixture of 19 amino acids (i.e. lacking the amino acid to be added in radioactive form)
   - 2 μl (20 μCi) of [^{35}S]methionine or [^{35}S]cysteine, or a [^{3}H]amino acid
   - 4.5 μl of RNA or 4.5 μl of sterile water (i.e. control lacking RNA)
3. Add 2 μl of the translation mixture to 2 ml of TCA/casamino acids in a 13 mm diameter disposable borosilicate glass tube. Mix, incubate on ice for 5 min, dilute with 2 ml of water, mix again, and boil for 10–15 min.

Protocol 30. *Continued*

4. Using a vacuum manifold with a glass fibre filter in place, pour the precipitated samples through the apparatus. Rinse the tube with 10% (w/v) TCA and pour it through the filter. Repeat this washing procedure four more times.
5. Analyse 5 μl of the translation mixture by SDS–PAGE.

[a] As the lysate is provided in 1 ml aliquots, to avoid repeated freezing and thawing, it should be divided into 18 μl or 36 μl aliquots when first thawed.

6. Conclusions

In this chapter we have described the basic methods that can be used to grow and characterize paramyxoviruses and influenza viruses and their encoded proteins. We have not discussed the techniques involved in cloning or oligonucleotide-directed mutagenesis, as these have become routine for many investigators. The reader is referred to ref. 14 for descriptions of relevant techniques.

Acknowledgements

This work was supported by Public Health Service research grants AI-20201 and AI-23173 from the National Institute of Allergy and Infectious Diseases. R.A.L. is an Investigator of the Howard Hughes Medical Institute.

References

1. Lamb, R. A. (1989). In *The influenza viruses* (ed. R. M. Krug), pp. 1–87. Plenum Publishing Corp., New York.
2. Paterson, R. G. and Lamb, R. A. (1991). In *The paramyxoviruses* (ed. D. W. Kingsbury), pp. 181–214. Plenum Publishing Corp., New York.
3. Erickson, A. H. and Blobel, G. (1979). *J. Biol. Chem.,* **254,** 11771.
4. Burnette, W. N. (1981). *Anal. Biochem.,* **112,** 195.
5. Paterson, R. G. and Lamb, R. A. (1990). *J. Cell Biol.,* **110,** 999.
6. Whitt, M. A., Buonocore, L., Prehaud, C., and Rose, J. K. (1991). *Virology,* **185,** 681.
7. Ng, D. T. W., Randall, R. E., and Lamb, R. A. (1989). *J. Cell Biol.,* **109,** 3273.
8. Steck, T. L. and Yu, J. (1973). *J. Supramol. Struct.,* **1,** 220.
9. Fuerst, T. R., Niles, E. G., Studier, F. W., and Moss, B. (1986). *Proc. Natl Acad. Sci. USA,* **83,** 8122.
10. Markoff, L., Lin, B.-C., Sveda, M. M., and Lai, C.-J. (1984). *Mol. Cell Biol.,* **4,** 8.
11. Pipas, J. M., Peden, K. W. C., and Nathans, D. (1983). *Mol. Cell Biol.,* **3,** 203.

12. Maddon, P. J., Dalgleish, A. G., McDougal, J. S., Clapham, P. R., Weiss, R. A., and Axel, R. (1986). *Cell,* **47,** 333.
13. Sambrook, J., Fritsch, E. F., and Maniatis, T. (ed.) (1989). *Molecular cloning, a laboratory manual*, (2nd edn). Cold Spring Harbor Press, Cold Spring Harbor, NY.
14. Ausubel, F. M., Brent, R., Kingston, R. E., Moore, D. D., Seidman, J. G., Smith, J. A., and Struhl, K. (ed.) (1989). *Current protocols in molecular biology.* Greene Publishing Associates and Wiley-Interstate, New York.

3

Viral DNA replication

NIGEL D. STOW and RONALD T. HAY

1. Introduction

During the last decade considerable interest has focused upon the mechanisms by which the genomes of DNA animal viruses are replicated. These studies provide information relevant not only to our understanding of viral growth and potential strategies for intervention, but also to processes occurring during the replication of eukaryotic chromosomes. The genomes of DNA viruses provide attractive models for the study of cellular DNA synthesis because they are relatively simple, readily manipulated, accumulate to high levels during lytic infection, and, in some cases, can be faithfully replicated in cell-free systems. Some viruses (for example Epstein–Barr virus and bovine papilloma virus (BPV)) additionally establish latent infections in dividing cells during which genome replication is likely to be regulated by mechanisms similar to those operating upon the host chromosomes.

The DNA genomes of animal viruses are diverse with respect to their size (ranging from less than 4 to over 200 kbp), configuration (single- or double-stranded, linear or circular), and number of proteins with direct roles in replication which they encode. Nevertheless, studies on the replication of viral DNAs attempt to answer experimentally a number of common questions:

- At which sites is DNA synthesis initiated?
- Which viral and host proteins are required for genome replication?
- What enzymatic activities do these proteins have?
- How is DNA synthesis initiated?
- What are the mechanisms of strand elongation?
- How are the products of DNA replication matured and packaged into virus particles?

There are several major experimental approaches to these questions in which viral DNA synthesis is examined in permissive cells (here referred to as *in vivo*) or in cell-free extracts (*in vitro*). Numerous biochemical activities which are known to be involved in the synthesis of DNA have also been assayed individually. These include origin-binding proteins, DNA helicases,

DNA polymerases, DNA primases, single-stranded DNA-binding proteins, RNase H activities, DNA ligases, and DNA toposiomerases. Given the wide range of virus systems and the extensive list of topics studied, a complete description of experimental approaches to the investigation of viral DNA replication is clearly beyond the scope of this chapter. We have, therefore, confined our discussions to two viruses which have double-stranded DNA genomes, and have selected representative assays which illustrate some of the important strategies used to investigate DNA replication *in vivo* and *in vitro* and to assay for activities involved in DNA synthesis.

The viruses which will be discussed are human adenovirus, particularly serotypes 2 and 4 (Ad2 and Ad4), and herpes simplex virus type 1 (HSV-1). Several comprehensive reviews describing the replication of their DNAs are available (1–5), in addition to texts covering more general aspects of DNA synthesis (6, 7). The following sections briefly outline features of the viral genomes and their modes of replication relevant to this article. References to specific facets can be found in the above reviews.

1.1 Human adenoviruses Ad2 and Ad4

The replication of adenovirus DNA is an excellent system in which to study viral DNA replication as the viral genome can be replicated *in vitro* by the action of three viral proteins, DNA-binding protein (DBP), pre-terminal protein (pTP), and DNA polymerase (pol), and two cellular proteins, nuclear factors I (NFI) and III (NFIII). The adenovirus genome is a linear double-stranded DNA molecule of 35–36 kbp with inverted terminal repeats (ITRs) of about 100 bp, the exact size depending upon serotype. DNA synthesis is initiated at either of the termini by transfer of dCMP, the terminal nucleotide, on to pTP (relative molecular mass (M_r) 80 000) in a template-dependent reaction. The 3′OH of the pTP–dCMP complex serves as a primer for synthesis of the nascent strand by the viral DNA polymerase. Concomitant displacement of the non-template strand generates a single-stranded molecule which then acts as a template for a second round of DNA synthesis.

Adenovirus origins of replication are located at the ends of the genome within the ITRs. A molecule of terminal protein (TP; the mature proteolytically processed form of pTP) is attached covalently to each 5′ end of the DNA; this is likely to be a *cis*-acting protein component of the replication origin. Removal of TP reduces the efficiency of Ad2 and Ad4 DNA replication *in vitro*, but it does not abolish replication, and plasmid templates have 25% of the activity of protein-linked genomes, provided that the origin has been exposed by restriction enzyme cleavage. Using plasmid templates, extensive mutational analysis has revealed that four regions within the terminal 51 bp of the Ad2 genome influence origin activity *in vitro* and *in vivo*. The origin of Ad2 DNA replication consists of a core domain comprising the terminal 18 bp of the genome, which is capable on its own of supporting only a low level of initiation, and an auxiliary region encompassing nucleotides 25–50 which contains recognition sequences for the sequence-specific DNA-binding proteins NFI and NFIII. Occupancy of the recognition site by NFI increases the frequency of initiation of viral DNA replication *in vivo* and *in vitro*. Separating the core and the NFI-binding site is a region of DNA where sequence changes are tolerated, but insertions or deletions are not. Immediately adjacent to the NFI site is the binding site for another cellular DNA-binding protein, NFIII or octamer-binding protein (Oct 1). The mechanisms by which these host factors increase the efficiency of DNA replication have yet to be established, but in each case it has been demonstrated that the DNA-binding domain of the protein is sufficient to stimulate DNA synthesis. Genes for the three viral replicative proteins pTP, pol, and DBP have been identified, and the proteins have been expressed in a variety of heterologous systems thus facilitating large-scale purification and analysis. DBP has an M_r of 68 000, is expressed at high levels in infected cells, and is involved at multiple stages of DNA replication. During elongation of nascent DNA chains, DBP functions by increasing the processivity of pol and by coating displaced single strands. Prior to initiation, DBP binds to template molecules, resulting in a higher affinity of NFI for its recognition site in the replication origin and leading to an increase in the frequency of initiation of viral DNA replication. A direct interaction betweeen NFI and pol can then target the pTP–pol heterodimer to the replication origin. pTP and pol can each specifically recognize DNA sequences within the terminal 18 bp of the viral genome, but the pTP–pol heterodimer binds with enhanced specificity, protecting base pairs 8–17 from DNase I cleavage. Thus the highly conserved DNA sequence present in all human adenovirus genomes within the minimal origin of DNA replication is recognized by the pTP–pol heterodimer (8, 9).

Transfection assays carried out with Ad4 showed that, in contrast to Ad2, only the terminal 18 bp of the genome, which are identical in both serotypes and in Ad2 constitute the core origin, are required for efficient DNA replication *in vivo*. This is also the case *in vitro*, where it was demonstrated that linearized plasmid containing only the terminal 18 bp of the ITR can support initiation of DNA replication as effectively as a template containing a complete Ad4 ITR. The protein requirements for DNA replication differ markedly between Ad2 and Ad4, in that Ad4 appears to have circumvented the need for the host factors NFI and NFIII. It does not possess an NFI site, and whilst it does have a binding site for NFIII, neither factor is required for DNA replication *in vivo* nor are they capable of stimulating DNA replication *in vitro*.

1.2 HSV-1

The HSV-1 genome is a linear double-stranded DNA of approximately 152 kbp which encodes over 70 distinct proteins. To date, no cell-free system for replication of HSV-1 DNA has been described, and most of our present knowledge has been derived from studies *in vivo*. HSV-1 DNA is circularized prior to replication via complementary single-nucleotide 3′-overhangs at the genomic termini. Although there may be an initial amplification of circular templates, the major replication products are long tandem head-to-tail concatemers which are probably generated by a rolling-circle mechanism of synthesis. Packaging of the replicated DNA is tightly coupled to sequence-specific cleavage of concatemers into unit-length genomes. The signals which direct cleavage and packaging reside within the *a* sequence, a region of approximately 400 bp which occurs as a direct repeat at the genomic termini and also in inverted orientation between the L and S segments of the genome.

The use of transient assays in transfected tissue culture cells has enabled the origins of replication to be mapped and viral gene products with a direct role in DNA replication to be identified. The HSV-1 genome contains three origins: two identical copies of ori_S present in the TR_S and IR_S repeat elements, and a single copy of ori_L near the centre of the U_L region. These origins differ in DNA sequence, but the same essential functional elements occur in each, and it is thought (but not yet proven) that they behave similarly during infection. A set of seven HSV-1 genes (*UL5, UL8, UL9, UL29, UL30, UL42*, and *UL52*) encode proteins which have been shown to be both necessary and sufficient for viral origin-dependent DNA synthesis in tissue culture cells.

Biochemical activities have now been ascribed to each of these proteins. UL5, UL8, and UL52 together form a complex which exhibits DNA primase and DNA helicase activities. UL9 binds specifically to the viral origins of replication, and UL29 is a non-sequence-specific single-stranded DNA-binding protein. UL30 and UL42 represent catalytic and accessory subunits of the viral DNA polymerase. HSV-1 thus appears to encode most of the functions required for genome replication, although the unavailability of an *in vitro* system at present prevents the involvement of host-cell proteins from being excluded.

2. Study of viral DNA replication *in vivo*

Several experimental approaches are available which allow the process of viral DNA replication to be studied in tissue culture cells. These are of particular value when suitable *in vitro* assays do not exist, and have contributed to our knowledge of a variety of aspects, including the timing of genome replication and the definition of *cis*-acting signals and *trans*-acting factors essential for DNA synthesis. The protocols presented in this section are based upon experimental procedures which we have used extensively to investigate various aspects of the replication of HSV-1 DNA in baby hamster kidney (BHK) cells in tissue culture. BHK cells are passaged routinely in ETC10 (Eagle's medium containing 10% (v/v) newborn calf serum, 10% (v/v) tryptose phosphate broth, 100 U/ml penicillin, and 100 μg/ml streptomycin; see Chapter 10, *Protocol 1*) at 37 °C in an atmosphere containing 5% (v/v) carbon dioxide. Cell monolayers in plastic Petri dishes are prepared the day before use, and following infection or transfection are maintained in EC5 (Eagle's medium containing 5% (v/v) newborn calf serum, 100 U/ml penicillin, and 100 μg/ml streptomycin). The following protocols have been optimized for this particular cell/virus combination, but may be adapted easily to other systems.

2.1 Assay for viral DNA synthesis

A variety of techniques are available which allow the accumulation of viral DNA in infected cells to be monitored. In the past these have often relied upon radioactive labelling of newly synthesized viral DNA coupled, where possible, to the separation of host and viral DNA by caesium chloride equilibrium density-gradient ultracentrifugation. A frequently raised criticism of this method, when applied to quantitative studies, is that changes in pool size or depletion of the labelled species may occur during infection. The advent of Southern blotting provided an alternative approach which avoided the need to label infected cells or to separate host and viral DNA, it was sufficiently sensitive to enable the accumulation of low levels of viral DNA to be detected in the presence of continuing host chromosomal replication, and avoided complications due to changes in pool sizes. DNA is prepared from infected cells, cleaved with appropriate restriction endonucleases, and the resulting fragments are separated by agarose gel electrophoresis. The fragments are transferred to an appropriate membrane (for example nitrocellulose or nylon) which is then hybridized to a probe which will specifically detect the viral DNA species. Suitable probes are purified viral DNA or cloned fragments of the viral genome, and the approach is in principle suitable for detecting the replication of any viral DNA species. Additionally, both radioactive and non-radioactive detection methods can be used. Most of the techniques employed in this assay are standard and have been described in detail elsewhere (for example Southern blotting and the preparation of probes and hybridization; see refs 10–12). The purification of total cellular DNA from infected cells forms an integral part of several of the other assays described in this section and a detailed description of this procedure is included in *Protocol 1*. *Figure 1* shows a time course for the accumulation of HSV-1 DNA in infected BHK cells performed using this procedure.

Protocol 1. Preparation of total cellular DNA and screening for the presence of viral DNA

Materials

- BHK cell monolayers in 35 mm Petri dishes (approximately 2×10^6 cells/plate), infected with virus as required
- Tris-buffered saline (TBS): 137 mM NaCl, 5 mM KCl, 0.7 mM Na_2HPO_4, 5.5 mM glucose, 25 mM Tris–HCl (pH 7.4)

Protocol 1. *Continued*

- TE: 10 mM Tris–HCl (pH 7.5), 1 mM EDTA
- 20 mg/ml protease (Sigma grade XIV) in TE, predigested for 1 h at 37°C
- cell lysis buffer (CLB): 10 mM Tris–HCl (pH 7.5), 1 mM EDTA, 0.6% (w/v) SDS
- 200 × RNase solution: 1 mg/ml RNase A, 10 000 U/ml RNase T1 in TE; store at 4°C

Method

1. Remove the medium from the cell monolayers and wash the cells with TBS.
2. Add 2 ml of CLB containing 0.5 mg/ml protease to each drained cell monolayer.
3. Incubate at 37°C for 2–6 h.
4. Transfer the cell lysate to polypropylene tubes, extract sequentially with phenol then chloroform, and precipitate nucleic acids with ethanol.
5. Recover the nucleic acids by centrifugation and redissolve the pellet in 100–200 μl of TE containing 1 × RNase solution.
6. Digest a sample of the DNA with an appropriate restriction endonuclease, separate fragments by agarose gel electrophoresis, blot and hybridize to a suitable probe specific for the viral DNA of interest. The recovery of DNA is approximately 15 μg per 2×10^6 cells. In practice, samples of DNA recovered from equivalent numbers of cells (usually 1–2 μg/gel lane) are analysed in parallel. As an alternative to restriction enzyme digestion and Southern blot analysis, viral DNA may be detected by a 'dot blot' method (13, 14).
7. After hybridization, wash the filter, and detect the signal (for example by autoradiography).

2.2 Preparation of nuclei from infected cells

Although *Protocol 1* is simple and the recovery of DNA is highly reproducible, it is often desirable to isolate DNA from only the nuclei of infected cells. This procedure has the advantage for nuclear replicating viruses that backgrounds due to non-replicating genomes can usually be reduced, with a consequent increase in the sensitivity of the assay. Similarly, in transfection experiments a proportion of the input DNA usually remains associated with the cell membrane or cytoplasm and can be removed by first preparing nuclei. *Protocol 2* gives a simple method for preparing nuclei from monolayer cultures by lysis with hypotonic buffer in the presence of the non-ionic detergent NP40.

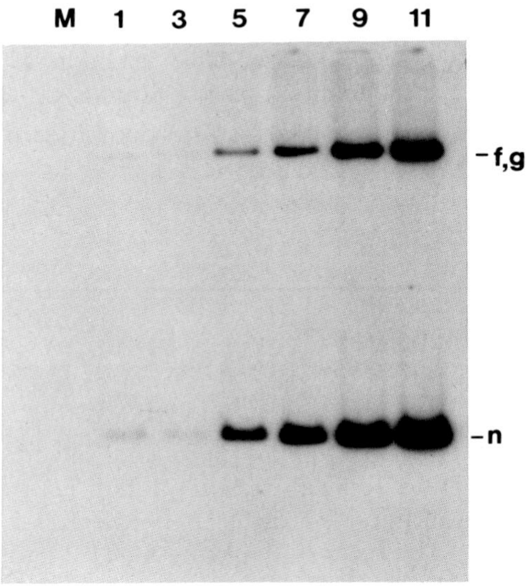

Figure 1. Time course of HSV-1 DNA replication. BHK cells were mock-infected (M) or infected with 10 p.f.u./cell of HSV-1. Total cellular DNA was prepared as described in *Protocol 1* at the indicated times (h) after virus addition, or at 11 h after mock-infection. DNA samples from 2 × 10^5 cells were cleaved with *Eco*RI and the fragments were separated on a 0.8% (w/v) agarose gel. The gel was blotted on to a nylon membrane (Hybond-N, Amersham) and hybridized to plasmid pGX158 (containing the HSV-1 *Bam*HI fragment) which had been ^{32}P-labelled by nick translation. An autoradiograph of the washed filter is shown. The probe detects the *Eco*RI f, g (which co-migrate) and n fragments as indicated. Viral DNA detected in the 1 and 3 h samples represents predominantly unreplicated input DNA. This type of background is reduced substantially if nuclei are isolated prior to the preparation of total cellular DNA (*Protocol 2*). The accumulation of HSV-1 DNA from 5 h onwards can be clearly seen.

Protocol 2. Preparation of DNA from cell nuclei

Materials

- Reticulocyte standard buffer (RSB): 10 mM Tris–HCl (pH 7.5), 10 mM KCl, 1.5 mM $MgCl_2$
- 10% (v/v) NP40
- protease solution: 1 mg/ml protease in 2 × CLB

Method

1. Scrape the cells from the dish into the medium, transfer to tubes, and pellets the cells by centrifuging at 600 *g* for 2 min. Wash the cells once with TBS and pellet them by centrifugation.

Protocol 2. *Continued*

2. Resuspend the cell pellet in 1 ml of RSB and add 10% NP40 to final concentration of 0.5%. Incubate on ice for 10 min.
3. Pellet nuclei by centrifuging at 600 *g* for 2 min and discard the supernatant.
4. Resuspend the pellet of nuclei by vortexing in 1 ml of water and add 1 ml of protease solution.
5. Proceed as described in step 3, *Protocol 1*.

2.3 Assay for *in vivo* HSV-1 origin function

In this section we describe a transient assay which has been used to test for the presence of *cis*-acting sequences required for DNA synthesis in cloned fragments of HSV-1 DNA (15–17). Similar approaches have been employed to define other origins of DNA replication, although direct evidence that such elements contain the initiation sites for DNA synthesis is frequently lacking (as in the case of HSV). The basis of the assay for HSV origin function is as follows. Cells permissive for HSV DNA replication are transfected with circular molecules of the test plasmid and superinfected subsequently with wild-type (WT) HSV-1 helper virus. All the proteins required for viral DNA synthesis will be present within these lytically infected cells, and will be able to recognize and act upon a viral origin carried on a transfected plasmid molecule. As a result autonomous viral origin-dependent replication of the input plasmid occurs, and the amplified plasmid DNA sequences can be detected by hybridization to a plasmid vector-specific probe.

As is the case with replicated HSV-1 genomes, the products of plasmid amplification are tandem head-to-tail copies of the input species. Total DNA prepared from the cells is therefore usually cleaved with a restriction enzyme which reduces these concatemers to monomeric units prior to analysis by agarose gel electrophoresis and Southern blotting. The enzyme *Dpn*I is also included in digests since it allows discrimination between unreplicated and replicated input plasmid DNA. *dam*$^{+}$ strains of *Escherichia coli* routinely used in the laboratory yield plasmids which are completely methylated at A residues within the *Dpn*I recognition sequence (GATC), a prerequisite for *Dpn*I cleavage. Following replication of such DNA in cells of eukaryotic origin, the progeny molecules no longer contain methylated GATC motifs and are consequently resistant to *Dpn*I digestion. Thus plasmid-length molecules detected by gel analysis represent DNA which has replicated after introduction into the eukaryotic host, whereas unreplicated input molecules are cleaved into lower M_r species (*Dpn*I cuts most commonly used cloning vectors at multiple sites).

The assay for HSV-1 origin function is described in detail in *Protocol 3*. *Figure 2* illustrates the type of result obtained in experiments in which differential sensitivity to *Dpn*I is used to distinguish plasmid DNA sequences

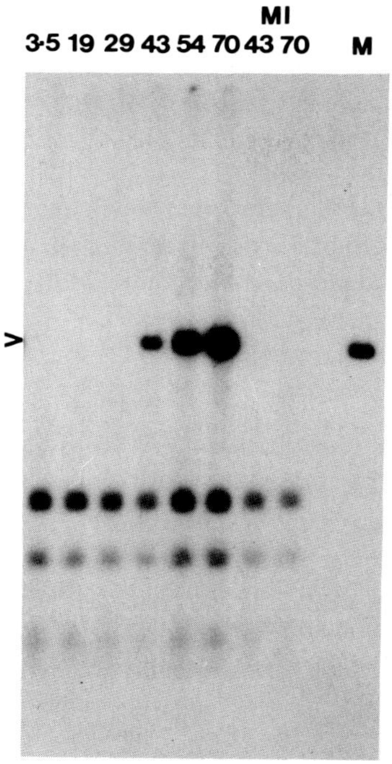

Figure 2. *In vivo* replication of a plasmid containing HSV-1 ori_S. Sf9 insect cells were transfected with ori_S-containing plasmid pST19 and then mock-infected (MI) or infected with a mixture of seven AcNPV recombinants which express the seven viral proteins required for HSV-1 DNA synthesis (see Section 2.5). DNA prepared from cells harvested at the indicated times (h p.i.) was analysed following cleavage with *Eco*RI (which linearizes pST19) and *Dpn*I. The fragments were separated by agarose gel electrophoresis, transferred to a nylon membrane and hybridized to ^{32}P-labelled DNA of the vector pTZ19U. An autoradiograph of the washed filter is shown. The figure illustrates the essential features of *in vivo* replication assays, namely the small DNA fragments produced as a result of *Dpn*I cleavage of unreplicated input plasmid molecules and *Dpn*I-resistant species which co-migrate with the linearized input plasmid DNA (the marker, M, is *Eco*RI cleaved pST19DNA). The latter molecules, indicated by the arrowhead, represent plasmid molecules replicated in the transfected cells. The appearance of the *Dpn*I-resistant molecules requires the presence of a functional HSV-1 origin and the seven viral DNA replication proteins. (Reproduced from ref. 29 with permission.)

replicated following transfection into eukaryotic cells from unreplicated molecules. The experimental approach is applicable to other virus systems in which the substrate for replication is a double-stranded circular DNA (for example SV40 and BPV; refs 18, 19). A similar approach has also been described for adenovirus, in which minichromosomes containing viral origins at each terminus are able to direct amplification of the intervening sequences in cells co-infected with helper virus (20).

Protocol 3. Assay for HSV-1 origin activity

Materials

- a suitable test plasmid
- Hepes-buffered saline (HBS): 137 mM NaCl, 5 mM KCl, 0.7 mM Na_2HPO_4, 5.5 mM D-glucose, 21 mM Hepes–NaOH (pH 7.05)
- 2 mg/ml double-stranded calf thymus DNA
- 25% (v/v) dimethylsulphoxide (DMSO) in HBS

Method

1. Prepare the calcium phosphate–DNA co-precipitate. Add 10 μl of calf thymus DNA and 0.5 μg[a] of plasmid to 1 ml of HBS. Mix gently, add 70 μl of sterile 2 M $CaCl_2$ and mix rapidly. Allow the precipitate to form for approximately 5 min at room temperature.[b]
2. Remove the medium from cell monolayers in 35 mm Petri dishes and add 0.4 ml/plate of precipitated DNA. Incubate at 37°C for 45 min.
3. Add 2 ml/dish of EC5. Continue incubation for a further 3.25 h at 37°C.
4. Decant the medium and wash the cells once with Eagle's medium. Add 1 ml of 25% DMSO, incubate at room temperature for 4 min, decant, wash once more, and add 2 ml of EC5.
5. Incubate for a further 2 h at 37°C. Mock-infect or superinfect the cells with HSV-1 at a multiplicity of infection (m.o.i.) of 5 plaque-forming units (p.f.u.)/cell. Adsorb virus at 37°C for 45 min and then overlay the monolayers with 2 ml of EC5 prior to continuing incubation at 37°C.
6. Prepare total cellular or nuclear DNA 24 h after initial transfection (see *Protocols 1* and *2*).
7. Digest samples of DNA (1–2 μg) with an appropriate enzyme (for example one which linearizes the input plasmid) together with *Dpn*I. To ensure complete digestion, incubate reactions containing 3 U of each enzyme overnight at 37°C. Separate the fragments on an agarose gel, transfer to an appropriate membrane, and hybridize to a probe containing plasmid vector but not viral DNA sequences (this will usually be the plasmid vector itself labelled *in vitro* with ^{32}P). After hybridization, process the filter, and detect the signal as appropriate.

[a] This amount of test plasmid is suitable for a plasmid of approximately 3.5 kbp. Equimolar amounts of other plasmids should be used.

[b] 1 ml of precipitate is sufficient for transfection of two monolayers, one of which should be superinfected with virus and the other serving as a mock-infected control.

The efficiency of the calcium phosphate transfection procedure is known to be influenced greatly by several factors, including the concentration of plasmid and carrier DNAs and the pH of HBS, and these should be optimized individually. In this regard, it is useful to have a quick and convenient method for evaluating and comparing transfection efficiencies which does not require the completion of relatively time-consuming schedules such as *Protocol 3*. We have used for this purpose a test plasmid containing the *E. coli lacZ* gene driven by the human cytomegalovirus (HCMV) major immediate early (IE) promoter, which allows high-level constitutive expression of β-galactosidase in a wide range of mammalian cell types. Expressed protein can be visualized by histochemical staining 24 h after transfection, and the proportion of cells expressing β-galactosidase determined readily by light microscopy (21).

Numerous variations of the calcium phosphate precipitation procedure have been reported, including the use of plasmid DNAs as carrier, modifications in the method of producing the precipitate, and alternatives to DMSO for post-transfection treatment. Many of these are described in detail elsewhere (10, 11). In addition, other transfection procedures, such as lipofection, can be used for the introduction of plasmid DNA (10, 11; see Chapter 10, *Protocol 9*). As with most other transfection-based protocols, the optimal method probably depends upon the cell type and system under investigation.

2.4 Transient assay for origin-dependent DNA replication using cloned HSV-1 DNA replication genes

For many years the use of virus mutants (usually temperature sensitive (ts) mutants) provided the major approach to determining whether specific viral gene products play a role in HSV DNA replication. In 1986, Challberg described a significant modification of the assay described in *Protocol 3* which enabled the minimal set of viral genes necessary for DNA synthesis to be determined (22–24). The rationale behind the assay was that a defined collection of plasmids encoding the necessary *trans*-acting functions should be able to substitute for superinfecting WT helper virus in activating a functional viral origin. Using this approach, Challberg and his colleagues identified a set of seven HSV-1 genes which are both necessary and sufficient for viral DNA synthesis (see Section 1.2), and provided the basis for a powerful assay to test the functionality of mutated versions of any of these genes. Because the seven genes all belong to the early transcriptional class, two further plasmids which encode HSV-1 IE transcriptional activators are also necessary. This assay has been refined by cloning the open reading frames (ORFs) encoding the seven DNA replication proteins downstream of the HCMV major IE promoter (25), thereby allowing high-level constitutive expression in a wide range of mammalian tissue culture cells and circumventing the requirement for additional plasmids encoding transcriptional transactivators.

A procedure for performing a viral origin-dependent DNA synthesis using cloned copies of HSV DNA replication genes is presented in *Protocol 4*. The effect of lesions within any of the DNA replication genes on the ability of the protein product to function in DNA synthesis can be assessed by substituting the WT version with the mutated copy. A similar approach should, in principle, be applicable to other DNA viruses for which the origins of replication and genes encoding products required for DNA synthesis are known. The assay described in *Protocol 4* is suitable for monolayers of cells in 35 mm Petri dishes. The reagents required and procedures are essentially as described in *Protocol 3*. Replication genes cloned under control of the HCMV major IE promoter allow efficient activation of a functional plasmid-borne copy of HSV-1 ori_S. If plasmids in which the replication genes remain under the control of their normal promoters are employed, two additional plasmids encoding the HSV-1 transactivators V_{mw} 110 and V_{mw} 175 must be included in the transfection (23).

Protocol 4. Assay for HSV-1 DNA replication gene function

Method

1. Make up a mixture of the appropriate DNAs in HBS. 1 ml should contain 1 μg of each of seven plasmids encoding the essential HSV-1 DNA replication proteins, 0.5 μg of a plasmid containing a functional origin of replication, and 12 μg of calf thymus DNA. Form a calcium phosphate precipitate of the DNAs (see step 1, *Protocol 3*).
2. Carry out steps 2–4, *Protocol 3*. Incubate the cells for approximately 24 h at 37°C.
3. Carry out steps 6–7, *Protocol 3*.

2.5 Other methods for testing cloned HSV DNA replication genes for *in vivo* functionality

Although the method described in *Protocol 4* provides a convenient assay for the ability of mutated versions of HSV-1 DNA replication genes to participate in viral DNA synthesis, several other approaches are available.

Mutations which may be lethal for virus growth can be introduced into the viral genome if a cell line capable of complementing the mutant virus is available. Suitable cell lines have been obtained by co-transforming permissive cells with a selectable marker and an HSV DNA fragment encoding the gene of interest. To generate virus mutants, these cells are co-transfected with DNA fragments containing the desired mutation and intact HSV-1 genomes. The desired recombinants can be purified and propagated on the complementing cells and their phenotype examined in the parental cell line.

A powerful application of this approach to the study of HSV-1 DNA replication genes, which incorporates the use of a marker to identify virus recombinants, has been made by Weller and her colleagues (26, 27).

The functionality of individual DNA replication genes has also been examined using a complementation assay based upon the method described in *Protocol 3* (28). In this approach, the replication gene to be tested is co-transfected with a plasmid containing a known functional HSV origin, and the helper virus is a mutant with a deletion in the gene of interest. Amplification of the origin-containing plasmid indicates that the product of the transfected gene is able to participate in viral DNA synthesis. It is also possible to employ ts mutants with lesions in DNA replication genes as helper viruses at the non-permissive temperature, although in this instance the possibility of functional activity resulting from recombination or intragenic complementation between plasmid and virus-coded gene copies should be considered.

Another possible approach makes use of recombinant baculoviruses which over-express the HSV-1 DNA replication proteins (29). It was shown that co-infection of *Spodoptera frugiperda* (Sf9) insect cells with a mixture of baculovirus recombinants capable of expressing the seven HSV-1 DNA replication proteins enables viral origin-dependent amplification of a transfected test plasmid. In addition to demonstrating that the heterologously expressed HSV proteins are competent for viral DNA synthesis, these results indicate that any non-HSV proteins which may be required for replication must be present within the insect cells (although they could potentially be encoded by the baculovirus vector). This approach also carries the advantage that, by expressing mutated versions of the replication genes in a baculovirus vector, it should be possible to screen for replicative ability and to produce sufficient amounts of protein for biochemical studies (see below and Chapter 8).

2.6 Other approaches for investigating viral DNA replication *in vivo*

2.6.1 Two-dimensional gel analysis of infected cell DNA

As mentioned in Section 2.3, there is no formal requirement that sequence elements essential in *cis* for DNA synthesis should correspond to the sites of initiation of replication. The possible use of additional or alternative techniques to locate origins of viral DNA replication, therefore, is worth considering. Brewer and Fangman described a technique whereby restriction enzyme digests of cellular DNA are separated by two-dimensional agarose gel electrophoresis in such a way that fragments containing replication bubbles or forks migrate to characteristic positions dependent on both size and structure (30, 31). Specific fragments can be detected by hybridization; replicative intermediates producing arcs with shapes characteristic of whether synthesis was initiated within or outside the fragments. Thus, origin-containing fragments yield a distinctive pattern due to the presence of a replication bubble, whereas fragments which do not include an origin produce a different pattern owing to the presence of Y-shaped molecules only. Combinations of digests allow the direction in which the replication fork passes through such a fragment to be determined.

An alternative two-dimensional gel procedure described by Huberman and colleagues (32, 33) utilizes alkaline electrophoresis in the second dimension to resolve nascent DNA strands. By hybridizing the nascent strands to short specific DNA probes it is possible to deduce on which side of a particular fragment the origin is located, since a probe that lies closest to an origin will detect the shortest nascent strands. Analysis of adjacent fragments allows the position of the origin to be defined more closely.

2.6.2 Analysis of replicative intermediates

Useful information on the mechanisms of viral DNA replication can frequently be obtained by analysing the structure of replicative intermediates (RIs) isolated from infected cells. It is usually possible to obtain a fraction enriched for RIs by passage of cellular DNA over a benzoylated naphthoylated DEAE (BND) cellulose column (Sigma) to which they bind preferentially because of their partially single-stranded character.

SV40 *in vivo* RIs have been studied in detail, and the positions of RNA primers and Okazaki fragments have been determined on the two strands. The results enable the continuously and discontinously synthesized daughter strands to be recognized, and provide information concerning the precise location at which DNA synthesis is initiated. For molecules like SV40 DNA which are replicated bidirectionally, the origins are defined by distinct transition points between continuous and discontinuous synthesis of each of the daughter strands (34).

The use of PCR and strand extrusion techniques to analyse RIs and identify sequences corresponding to origins of replication have also been reported, and these techniques will probably become significant approaches in the future (35, 36).

2.6.3 Viral DNA replication and the nuclear matrix

An increasingly important aspect of cellular DNA replication concerns the association between DNA synthesis and the structural framework of the nucleus. It has been reported that nascent DNA is associated preferentially with the proteinaceous structure (nuclear matrix) which remains following high salt extraction and nuclease digestion of isolated nuclei, suggesting that this might represent a site at which the replicative machinery is located and DNA synthesis occurs (37, 38). Relatively little work has been done to investigate whether viral DNA replication takes place at a similar site, although it should be possible to pulse-label infected cells and determine whether viral DNA restriction fragments remaining attached to the nuclear matrix are labelled to a greater specific activity than those released by digestion, or are enriched for replication forks.

3. Study of the activities of proteins involved in viral DNA synthesis

When certain proteins, whether virus- or host-coded, are identified as playing essential roles in viral DNA synthesis, elucidation of their functions becomes an important task. Although several functions can be assayed in relatively crude preparations, critical study of the enzymatic activity of a particular replication protein invariably demands purification to a level as near homogeneity as reasonably possible. Because many replication proteins are present at very low abundance in infected cells, the traditional approach to their isolation has used several grams of cells as starting material, and has relied upon multistep purification procedures during which a large proportion of the starting activities may be lost.

Modern molecular biology now allows proteins of interest to be expressed at high levels in a plethora of heterologous systems, and, not surprisingly, these are now becoming essential tools in the study of DNA replication proteins. Several factors should be taken into consideration in choosing an expression system for a particular viral DNA replication protein. For example, does the level of expression obtained in that system confer any real advantage over using infected cells? Is the protein processed appropriately? Can it be isolated readily in a soluble and active form? Of the various systems with which we have experience, high-level expression of large DNA replication proteins in a functional form has been most successfully achieved using the insect baculovirus system, although this, of course, is not a guarantee that it will always be the best choice. Further details about the use of baculovirus expression systems are given in Chapter 8.

In our laboratories, recombinants have been constructed in which cDNAs corresponding to host and viral genes required for adenovirus DNA synthesis, or the genomic ORFs necessary for HSV-1 DNA replication are expressed under control of the polyhedrin late promoter (see Chapter 8). All proteins were synthesized in high yield, and on only one occasion was the product of very low solubility. Interestingly, this protein, the product of the HSV-1 *UL*52 gene, is found normally as part of the heterotrimeric helicase–primase complex, and co-expression with the other two components of the complex (the UL5 and UL8 proteins) results in a significant increase in the amount of UL52 protein in the soluble fraction. Another approach which yielded increased amounts of soluble UL52 protein was to harvest singly infected insect cells at an earlier time after infection (about 30 h post infection (p.i.)). Although the total amount of recombinant protein present at this time is much lower than at the later times normally used (for example 48–72 h p.i.), it nevertheless still greatly exceeds the amount present in HSV-1 infected cells (39).

3.1 Preparation of extracts

Many procedures have been described for the preparation of extracts from virus-infected cells, but most essentially yield a soluble cytoplasmic fraction or a high salt eluate of nuclei. A widely-used method is that described by Dignam *et al.* (40). Cells are harvested, swollen by resuspension in a hypotonic buffer, and lysed by Dounce homogenization or a non-ionic detergent (for example 0.5% (v/v) NP40; see ref. 41). (*Non-ionic detergents should be used with caution, since in some instances they may cause an increase in the release of proteases.*) Nuclei are sedimented by low speed centrifugation (for example 600 *g* for 5 min), and cellular organelles can be removed from the cytoplasmic supernatant by higher centrifugal forces (for example 100 000 *g* for 60 min yields a so-called S100 soluble cytoplasmic fraction). The nuclei are washed and extracted with a high salt buffer (for example containing 0.4–0.6 M KCl or NaCl), and a soluble nuclear extract is obtained by centrifugation at 25 000–100 000 *g* for 30 min. Whether proteins of interest are found in the cytoplasmic or nuclear fraction does not necessarily depend on their actual intracellular location, and many known nuclear proteins have been found to leach into the cytoplasmic fraction upon hypotonic swelling of cells.

Table 1 gives recipes for hypotonic and high salt buffers (40). The use of protease inhibitors in extraction buffers and during most subsequent steps is generally considered advisable; phenylmethylsulphonyl fluoride (PMSF) is commonly employed. Subsequent protocols often call for a reduction in the salt concentration of nuclear extracts prior to their use, and dilution in low salt buffer or dialysis is normally appropriate. Proteins which precipitate at lower salt concentrations should be removed by centrifugation.

Cytoplasmic and nuclear extracts can be prepared from insect cells essentially using the above procedures. For most of the baculovirus recombinants described above, the over-expressed DNA replication protein was found to

Table 1. Buffers used for the preparation of nuclear and cytoplasmic extracts

Hypotonic buffer (Buffer A):	10 mM Hepes–NaOH (pH 7.9), 10 mM KCl, 1.5 mM $MgCl_2$, 0.5 mM dithiothreitol (DTT), 0.5 mM PMSF. This buffer may be supplemented with 0.5% (v/v) NP40, which allows cell lysis without homogenization.
High salt elution buffer (Buffer C):	20 mM Hepes–NaOH (pH 7.9), 420 mM NaCl, 25% (v/v) glycerol, 1.5 mM $MgCl_2$, 0.2 mM EDTA, 0.5 mM DTT, 0.5 mM PMSF. Other monovalent cations may replace Na^+, and higher salt concentrations can be employed (for example, 0.6 M KCl).

be the major polypeptide in the soluble extract. This represents a significant degree of 'purification' in comparison with extracts from cells infected with HSV-1 or adenovirus, possibly a relative enrichment of over 1000-fold for replication proteins of low abundance.

3.2 Purification of DNA replication proteins

Numerous procedures have been reported for the purification of viral DNA replication proteins which rely mainly upon permutations of standard ion exchange, sizing, and ligand affinity procedures. These have been described elsewhere (42, 43). However, because crude extracts which contain a high proportion of a protein of interest can often be obtained from cells infected with baculovirus recombinants, simple one-step procedures may suffice to yield an almost homogeneous preparation. In this regard, the affinity of most DNA replication proteins for DNA is a property which can frequently be exploited. The following sections illustrate purification procedures based upon general or sequence-specific interactions of replication proteins with DNA.

3.2.1 DNA-affinity chromatography

As many of the proteins involved in DNA replication interact directly with DNA, affinity chromatography using DNA as the immobilized ligand has been used widely as an efficient step in purification of these proteins. Three types of DNA-affinity matrix can be constructed: non sequence-specific single- and double-stranded DNAs, and recognition site-specific DNA. The synthesis of these matrices has been described fully in a previous volume in this series (43), although the following modification of the procedure described for recognition-site affinity chromatography results in increased DNA loading of the columns. As an alternative to using tandemly ligated oligonucleotides, one of the oligonucleotides is synthesized such that an amino group is attached via a six carbon spacer to the 5′-end of the oligonucleotide (Amino link, Applied Biosystems). The increased accessibility of the amino group ensures efficient coupling of DNA to the Sepharose matrix. The use of specific and non-specific DNA-affinity matrices in the purification of proteins required for replication of adenovirus DNA is described in detail below.

3.2.2 Purification of Ad2 pTP and pol from recombinant baculovirus-infected insect cells

Sf9 insect cells are grown at 28 °C in suspension in TC100 medium supplemented with 5% fetal calf serum and infected with 10 p.f.u./cell of a recombinant *Autographa californica* nuclear polyhedrosis virus (AcMNPV) containing the gene for Ad2 pTP or pol (44, 45; see Chapter 8 for further practical details). The cells are incubated at 28 °C in the same medium for a further 72 h and then collected by centrifugation at 600 *g* for 15 min. The cell pellet is washed in PBS$^+$ (see Chapter 2, *Protocol 1*), and the appropriate protein is purified as described below in *Protocol 5*. Identical procedures may be used for purification of pTP and pol. All manipulations should be carried out at 4°C.

Protocol 5. Purification of Ad2 pTP and pol

Materials

- extraction buffer: 25 mM Hepes–KOH (pH 8.0), 5 mM KCl, 0.5 mM $MgCl_2$, 0.5 mM DTT
- a 10 ml column of denatured calf thymus DNA–Sepharose[a]
- column buffer: 25 mM Hepes–NaOH (pH 8.0), 1 mM EDTA, 2 mM DTT, 10% (v/v) glycerol

Method

1. Resuspend the cells at 0.4 g/ml (wet weight) in ice-cold extraction buffer supplemented with 1 mM benzamidine, 1 mM sodium metabisulphite, 1 mM PMSF, and 10 μg/ml each of antipain, pepstatin, and leupeptin, and incubate at 4°C for 10 min (see also Chapter 1, *Table 1*).
2. Disrupt the cells by ten strokes in a Dounce homogenizer using a type B pestle.
3. Add 5 M NaCl to 0.2 M and incubate the extract at 4°C for 30 min.
4. Remove cell debris by centrifugation at 15 000 *g* for 5 min and clarify the extract by further centrifugation at 100 000 *g* for 15 min.
5. Apply the clarified cell extract in 0.2 M NaCl to the column equilibrated with column buffer containing 0.2 M NaCl. Wash the column extensively with column buffer containing 0.2 M NaCl and elute bound proteins with two column volumes of column buffer containing 0.6 M NaCl.
6. Collect fractions and examine them for the presence of the required protein by SDS–PAGE followed by staining with Coomassie brilliant blue.

[a] Prepared as described in ref. 43, using denatured calf thymus DNA.

An example of the purification of pol and pTP using *Protocol 5* is shown in *Figure 3*. Prominent polypeptide species representing pTP and pol can be seen to be retained quantitatively on the affinity matrix and eluted when the NaCl concentration is raised to 0.6 M. In each case greater than 90% of the total protein eluted from the DNA–Sepharose was either pTP or pol, and both proteins were active in viral DNA replication (9). Removal of minor contaminants was accomplished by chromatography on hydroxylapatite (9). 10 mg of pol and 6 mg of pTP could be purified from 1 litre of infected insect cells.

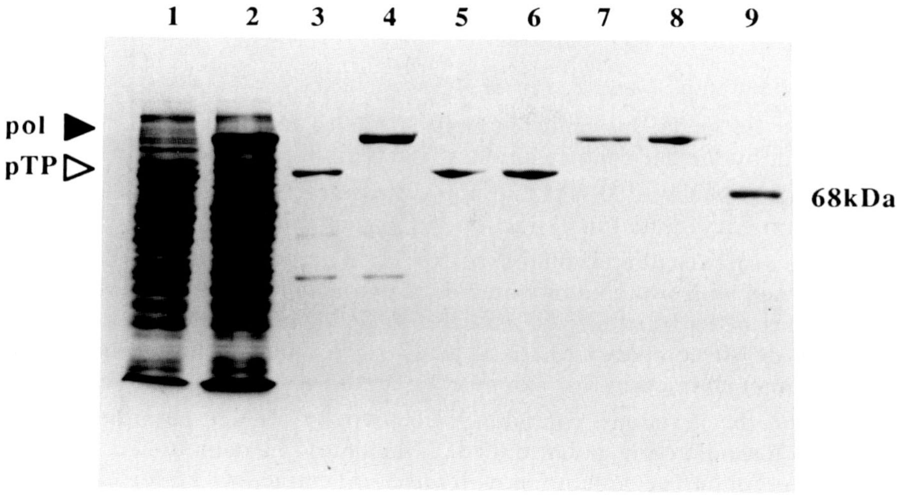

Figure 3. Purification of Ad2 pTP and pol. Sf9 cells were infected with recombinant baculoviruses containing the genes for pTP and pol. At 72 h p.i. cell extracts were prepared and the proteins were purified by column chromatography on denatured DNA–Sepharose and hydroxylapatite (*Protocol 5*). Fractions from each stage of the purification were analysed by SDS–PAGE on 8% (w/v) polyacrylamide gels, and proteins were stained with Coomassie blue. (1) pTP cell extract; (2) pol cell extract; (3) 3 μg of pTP DNA–Sepharose fraction; (4) 5 μg of pol DNA–Sepharose fraction; (5), (6) 3 μg and 5 μg of pTP hydroxylapatite fraction; (7), (8) 3 μg and 5 μg of pol hydroxylapatite fraction; (9) 5 μg of BSA. (Reproduced from ref. 9 with permission.)

3.2.3 Purification of Ad4 DNA replication proteins on DEAE–Sephacel and single-stranded DNA–Sepharose

Protocol 6 describes a method for purifying the Ad4 DNA replication proteins from HeLa cells infected with WT virus (49). The two-column procedure yields a fraction which is very active in viral DNA synthesis and contains predominantly only four polypeptide species. A 1 litre suspension culture of HeLa S3 cells is concentrated by centrifugation, resuspended in a small volume of medium, and infected with 100 p.f.u./cell of Ad4. After incubating for 90 min at 37°C, 1 litre of medium containing 2% (v/v) calf serum and 10 mM hydroxyurea is added, and the cells are incubated for 22 h at 37°C prior to pelleting by low-speed centrifugation.

Protocol 6. Purification of Ad4 DNA replication proteins

Materials

- a 30 ml DEAE–Sephacel column (Pharmacia)

Protocol 6. *Continued*

Method

1. Carry out steps 1 and 2, *Protocol 5*.
2. Pellet the nuclei by centrifugation at 1500 *g* for 5 min and clarify the extract by further centrifugation at 100 000 *g* for 45 min.
3. Add 5 M NaCl to 50 mM and apply the extract to the DEAE–Sephacel column equilibrated in extraction buffer containing 50 mM NaCl.
4. Elute with two column volumes of extraction buffer containing 50 mM NaCl and then with two column volumes of extraction buffer containing 0.2 M NaCl. Collect fractions and assay for *in vitro* DNA replication activity by pTP–dCMP complex formation (46, 47). Determine the protein concentration in each fraction (for example by Bradford's method; see ref. 48).
5. Apply the fractions containing adenovirus DNA replication activity (which should correspond to the 0.2 M eluate) to a denatured calf thymus DNA–Sepharose column equilibrated in extraction buffer containing 0.2 M NaCl.
6. Elute with two column volumes of extraction buffer containing 0.2 M NaCl and then five column volumes of a linear 0.2–2 M NaCl gradient in extraction buffer. Collect fractions and assay them for their ability to initiate adenovirus DNA replication *in vitro* and for their protein concentration.

Figure 4. (a) Purification of a fraction from Ad4-infected cells capable of replicating Ad4 DNA *in vitro*. The DEAE–Sephacel fraction from Ad4-infected HeLa cells (see *Protocol 6*) was applied to denatured DNA–Sepharose, and bound proteins were eluted with a gradient of 0.2–2 M NaCl. Plasmid p4A2 (50 ng), which contains a copy of the Ad4 ITR, was cleaved with *Eco*RI and incubated with 8 μl of each fraction (dialysed against buffer containing 0.1 M NaCl) in a standard assay for initiation of DNA replication. After incubation at 32°C for 90 min and treatment with micrococcal nuclease, polypeptides were fractionated by SDS–PAGE. Labelled species were excised from the gel and the radioactivity in each was determined by liquid scintillation counting (dCMP transfer). Protein concentrations were determined by the method of Bradford (48). The first peak to elute is denoted P1 and the second P2. (b) Elongation of initiated templates by fraction P2. 100 ng of plasmid templates p4A2 (WT origin; lanes 1, 3, 5, 7, 9) and pM18 (origin containing a debilitating point mutation; lanes 2, 4, 6, 8, 10 (were cleaved with *Eco*RI and *Ava*II and incubated with 3 μg of crude Ad4-infected extract (lanes 1, 2), 3 μg of DEAE–Sephacel eluate (lanes 3, 4), 2 μg of DNA–Sepharose eluate P2 (lanes 5, 6), 2 μg of DNA–Sepharose eluate P1 (lanes 7, 8), and a mixture of 2 μg each of P1 and P2 (lanes 9, 10) in a standard elongation reaction containing [α-^{32}P]dCTP (*Protocol 8*). After incubation at 30°C for 90 min, the reaction products were resolved by electrophoresis in a 2% (w/v) agarose gel containing 0.1% (w/v) SDS. Labelled species were detected by autoradiography. Nascent DNA incorporates [^{32}P]dCMP and has pTP attached (black arrow head), giving it a lower electrophoretic mobility than input template (white arrow head), which is labelled at a low level owing to repair synthesis. A second fragment containing only plasmid sequences electrophoresed from the bottom of the gel. (Reproduced from ref. 49 with permission.)

When this procedure was used, two peaks of protein eluting from the single-stranded DNA–Sepharose column at 0.5–0.8 M and 1.0–1.2 M NaCl were obtained, but DNA replication activity was detected only in the fractions eluted with 1.0–1.2 M NaCl (*Figure 4*). Proteins present at each stage of purification were analysed by SDS–PAGE followed by silver staining. The fraction which eluted with 1.0–1.2 M NaCl and which exhibited all the DNA replication activity contained one predominant species with an apparent M_r of 65 000 and three less abundant species with apparent M_rs of 70 000, 85 000, and 95 000. On the basis of its relatively high abundance and elution characteristics, the predominant species was thought to be DBP. Western blotting with polyclonal antisera raised against Ad2 pTP and DBP confirmed the presence of the corresponding Ad4 proteins in the active fraction, while the presence of Ad4 pol was indicated by the detection of an aphidicolin-resistant DNA polymerase activity. In the presence of a linearized plasmid containing

a

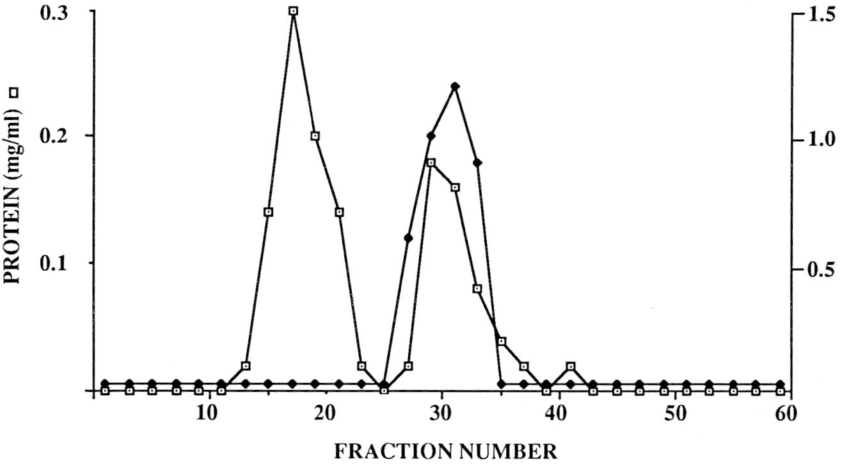

b

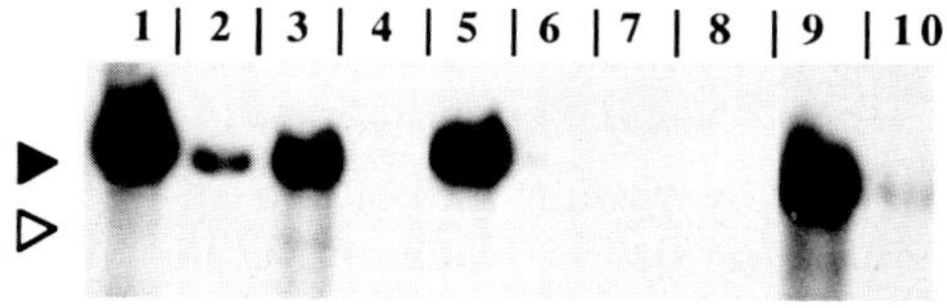

the Ad4 origin of DNA replication, this protein fraction not only initiated DNA replication but efficiently elongated the nascent strands.

3.2.4 Purification of NFI and NFI_{DBD} by DNA recognition-site affinity chromatography

NFI is a cellular transcription factor that binds to the adenovirus origin of DNA replication and increases the frequency of initiation. Recombinant baculoviruses have been constructed which express either full length NFI or the DNA-binding domain of the protein (NFI_{DBD}) after infection of insect cells. Both proteins can be conveniently purified by sequence-specific DNA affinity chromatography. A detailed procedure for purification of NFI_{DBD} is described in *Protocol 7* (50).

Protocol 7. Purification of NFI_{DBD}

Materials

- 1 litre of Sf9 cells (10^9 cells)
- recombinant baculovirus containing NFI_{DBD} cDNA
- resuspension buffer: 25 mM Hepes–NaOH (pH 8.0), 1 mM EDTA, 2 mM DTT, 0.4 M NaCl, 1 mM benzamidine, 1 mM sodium metabisulphite, 1 mM PMSF, 0.5% (v/v) NP40, and 10 μg/ml each of antipain, pepstatin, and leupeptin (see Chapter 1, *Table 1*)
- a 5 ml column of Bio-Rex (Bio-Rad)
- a column containing the immobilized NFI-binding site (prepared as described in references 43 and 50)

Method

1. Infect the cells in suspension with 2 p.f.u./cell of the recombinant baculovirus. Incubate for 72 h at 28 °C.
2. Pellet the cells by centrifugation, wash the pellet with PBS^+ and resuspend it in 5 ml of resuspension buffer. Incubate for 30 min on ice.
3. Pellet nuclei by centrifugation and clarify the supernatant by centrifugation at 45 000 *g* for 30 min at 4°C.
4. Apply the clarified extract directly to the Bio-Rex column equilibrated with column buffer (see *Protocol 5*) containing 0.4 M NaCl.
5. Remove unbound proteins by washing the column with two column volumes of column buffer containing 0.4 M NaCl and elute NFI_{DBD} with column buffer containing 0.7 M NaCl.

6. Dilute the eluate with column buffer to reduce the NaCl concentration to 0.25 M and add 0.4 mg of poly dI:dC.
7. Apply the eluate to the NFI binding site column. Wash the column extensively with column buffer containing 0.25 M NaCl and elute NFI_{DBD} with 1 M NaCl.

Full-length NFI is purified in a similar fashion, except that it is loaded on to the Bio-Rex column at 0.25 M NaCl and eluted with buffer containing 0.5 M NaCl. DNA-binding activities of NFI and NFI_{DBD} can be determined in a gel electrophoresis DNA-binding assay as described in Section 3.3.4.

Figure 5 shows examples of the purification of the two proteins carried out using *Protocol 7*. The polypeptide composition of individual fractions was determined by SDS–PAGE followed by staining with Coomassie blue. The full-length protein was predominantly a single species with an M_r of 62 000, whereas the DNA-binding domain appeared as two species which were possibly generated by limited proteolysis during preparation. One to two milligrams of each protein could be purified from 1 litre of insect cells (50).

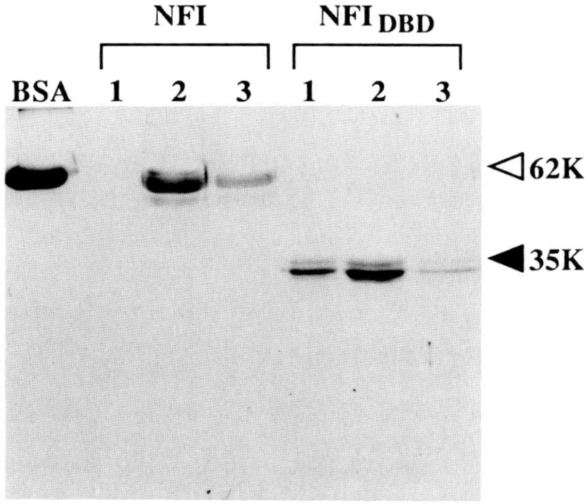

Figure 5. Purification of NFI proteins by recognition-site affinity chromatography. Extracts from Sf9 cells infected with baculoviruses expressing either NFI or NFI_{DBD} were eluted from Bio-Rex as described in *Protocol 7* and applied to DNA-affinity columns which contained a double-stranded synthetic oligonucleotide representing positions 18–41 from the Ad2 origin of DNA replication, linked via a 5′-amino link (Applied Biosystems) to CNBr-activated Sepharose (Pharmacia). Unbound proteins were removed by washing with buffer containing 0.25 M NaCl, and specifically-bound proteins were eluted with buffer containing 1.0 M NaCl. Fractions of 1 ml (1–3) were collected, and 5 μl of each was analysed by SDS–PAGE and Coomassie blue staining. BSA (10 μg) was co-electrophoresed as a standard. (Reproduced from ref. 50 with permission.)

3.3 Assays for the activities of proteins involved in DNA replication

3.3.1 DNA polymerase

The standard assay for DNA polymerase activity (*Protocol 8*) involves determining incorporation of a radioactively labelled deoxyribonucleoside monophosphate (dNMP) into an acid-insoluble DNA product. For most purposes, the standard DNA template is 'activated' DNA, which has a high primer: template ratio. Additional templates such as specifically primed homopolymers or single-stranded circular phage DNAs are often used to investigate particular activities associated with DNA polymerase (see below).

Protocol 8. Standard assay for DNA polymerase activity

Materials

- 10 × DNA polymerase buffer: 500 mM Tris–HCl (pH 8.0), 70 mM $MgCl_2$, 50 mM DTT
- activated DNA: 5 mg/ml DNase I-treated calf thymus DNA in 1 × DNA polymerase buffer (see ref. 51)
- TE(0.1): 10 mM Tris–HCl (pH 8.0), 0.1 mM EDTA
- 10 × dNTPs: 1 mM dTTP, 1 mM dCTP, 1 mM dGTP, 0.2 mM dATP in TE(0.1)
- labelled dNTP: in this case [α-^{32}P]dATP or [^{3}H]dATP, but any dNTP could be used, with an accompanying adjustment in the composition of 10 × dNTPs
- 5% (w/v) trichloroacetic acid (TCA)
- TCA/PPi: 10% (w/v) TCA, 0.1% (w/v) sodium pyrophosphate
- 2.5 cm glass fibre filter (Whatman GF/C)

Method

1. Mix 5 µl of 10 × DNA polymerase buffer, 2 µl of activated DNA, 5 µl of 10 × dNTPs, 1 µCi of labelled dNTP, and 36 µl of water on ice.
2. Add 1 µl of protein extract and incubate at 37 °C for 1 h.
3. Add 4 ml of ice-cold TCA/PPi and incubate on ice for 10 min.
4. Filter the entire contents of the tube through a glass fibre filter mounted on a vacuum filter manifold and wash twice with 4 ml of TCA/PPi, twice with 4 ml of 5% TCA, and once with 4 ml of ethanol.
5. Dry the filter under an infra-red lamp and process it for liquid scintillation counting.

Protocol 8 may not be appropriate for determining the activity of all DNA polymerases. For example, inclusion of 0.2 M KCl in the buffer inhibits the activity of cellular DNA polymerases but stimulates the activity of HSV DNA polymerase. Thus the activity of HSV DNA polymerase can be determined selectively in crude extracts containing cellular DNA polymerases.

The activated DNA used in *Protocol 8* is suitable for determining the catalytic activity of DNA polymerase, but it is not suitable for more specialized assays, such as those for determining the activity of factors involved in processivity or primer recognition. Alternative templates with low primer: template ratios are used in these situations. One type of template is prepared by annealing a short oligonucleotide to a sequence present in the circular single-stranded DNA of a bacteriophage such as M13 or ϕX174. In the absence of primer recognition proteins, many replicative DNA polymerases utilize these templates poorly, since the concentration of primer may be below the K_d for interaction between DNA polymerase and the annealed primer. It is therefore possible to set up an assay containing the primed template plus DNA polymerase and then add factors which allow the template: primer combination to be utilized efficiently (52). Homopolymer templates such as poly dT can be primed with oligo A or oligo dA to create templates that are particularly suitable for assaying factors which increase the processivity of replicative DNA polymerases.

3.3.2 DNA primase

During initiation of DNA synthesis at an origin of DNA replication or during discontinuous DNA replication on the retrograde side of a replication fork, short chains of RNA are made which act as primers for subsequent DNA synthesis. In many instances the enzyme which synthesizes these primers, DNA primase, is intimately associated with the DNA polymerase which utilizes them. Most assays for DNA primase measure DNA synthesis and rely on the requirement of DNA polymerases for a primer that is annealed to the template. DNA polymerase alone is unable to initiate DNA synthesis on an unprimed template, but in the presence of primase activity it can utilize the primers for DNA synthesis. In this way, the coupled activity of DNA polymerase/primase is assayed. Alternatively, primase activity alone can be determined if an exogenous DNA polymerase (such as the Klenow fragment of *E. coli* DNA polymerase I) is provided. Suitable templates for DNA primase activity are the single-stranded DNA form of bacteriophage M13 or the homopolymer poly dT. It is important to ensure that these DNAs are completely free from small contaminating RNA fragments, which might adventitiously prime DNA synthesis. For this reason it is advisable to treat all template preparations with 0.4 M NaOH at 37°C for 1 h. The solution is then neutralized by the addition of acetic acid, and the DNA is ethanol precipitated. *Protocol 9* presents a method which can be employed as a coupled assay for DNA polymerase/primase activity or as an indirect assay for the presence of primase performed in the presence of exogenously added DNA polymerase. It is also possible to determine DNA primase activity directly in the absence of DNA polymerase by the use of radioactive NTPs. However, owing to the small size of the product (average 10 nucleotides), radioactive incorporation is determined after binding the products to DEAE paper under conditions which preclude binding of labelled precursors.

Protocol 9. Assay for DNA primase activity

Materials

- template: 1 mg/ml bacteriophage M13 DNA (or poly dT[a]) in TE(0.1)
- 10 × NTPs: 5 mM ATP, 5 mM CTP, 5 mM GTP, 5 mM UTP in TE(0.1)

Method

1. Mix 5 μl of 10 × DNA polymerase buffer, 1 μl of template, 5 μl of 10 × NTPs, 5 μl of 10 × dNTPs, 1 μCi of labelled dNTP, and 32 μl of water on ice.
2. Add 1 μl of protein extract and incubate at 37 °C for 1 h. Include 0.5 U of the Klenow fragment of *E. coli* DNA polymerase I if primase is being assayed in the absence of endogenous DNA polymerase.
3. Carry out steps 3–5, *Protocol 8*.

[a] If poly dT is used as template then 10 × NTPs is replaced by 5 mM ATP and 10 × dNTPs is replaced by 1 mM dATP.

For each batch of template used, a control containing the Klenow fragment of *E. coli* DNA polymerase I in the absence of primase should be assayed to check for adventitious priming. A problem associated with using *Protocol 9* to screen crude extracts for primase activity is the presence in the protein preparation of nucleic acids which may hybridize to the template and prime DNA synthesis (for example poly A on a poly dT template). However, application of the extracts to DEAE–Sepharose in the presence of 0.3 M NaCl allows the bulk of the protein to pass through the matrix while contaminating nucleic acid is retained.

3.3.3 DNA helicase

DNA helicase activity can be assayed conveniently on model substrates containing regions of single- and double-stranded DNA which resemble (to varying extents) a replication fork. Different helicases vary in their requirements for free DNA ends; in the case of the HSV-1 UL5/UL8/UL52 helicase/primase complex, a free 3′ end is necessary. This enzyme is assayed using a procedure described by Crute *et al.* (53) and which is presented in *Protocol 10*.

To prepare the substrate, a 45 base oligonucleotide is first synthesized, of which only the first 23 nucleotides at its 5′ end are complementary to single-stranded M13mp 18 DNA. The oligonucleotide (30 ng) is labelled either at the 5′ end with T4 polynucleotide kinase in the presence of [γ-^{32}P]ATP, or at the 3′ end with terminal deoxynucleotide transferase in the presence of [α-^{32}P]dATP (the use of a ddNTP to prevent the addition of more than one nucleotide is not necessary at low dATP concentrations). The labelled oligonucleotide is then annealed with an equimolar amount of single-stranded M13mp 18 DNA, and the annealed substrate is purified by exclusion chromatography on a Biogel A1.5m bead column. The action of a DNA helicase on this model substrate causes unwinding of the labelled oligonucleotide, which is able to migrate more rapidly through a polyacrylamide gel than is the hybrid substrate (for example see *Figure 6*).

Other substrates are also employed for assaying DNA helicases. Those lacking a 3′ tail can be prepared more conveniently by using the Klenow fragment of *E. coli* DNA polymerase I and an appropriate [α-^{32}P]dNTP after annealing of oligonucleotide and single-stranded circular DNA. More complex substrates which allow the direction of movement of the helicase to be determined have also been described (54).

Protocol 10. Assay for HSV-1 DNA helicase activity

Materials

- a DNA helicase substrate containing a 3′ tail
- 10 × helicase buffer: 0.1 M Tris–HCl (pH 7.5), 17.5 mM $MgCl_2$, 25 mM DTT, 15 mM ATP, 0.5 mg/ml bovine serum albumin (BSA), 50% (v/v) glycerol
- 10 × TBE: 0.89 M Tris, 0.89 M boric acid, 20 mM EDTA
- gel loading buffer: 50% (w/v) sucrose, 2% (w/v) SDS in 5 × TBE with bromophenol blue to colour

Method

1. Set up reactions in 1 × helicase buffer containing 20 ng of helicase substrate and enzyme in a final volume of 40 μl. Incubate for 2 h at 37 °C.
2. Add 5 μl of gel loading buffer.
3. Separate the products by electrophoresis through an 8% (w/v) polyacrylamide gel electrophoresed in 1 × TBE. Fix and dry the gel and expose it to X-ray film. The gel should include, as markers, the untreated helicase substrate and a sample which has been denatured by boiling for 1 min immediately prior to loading. The latter indicates the position to which the displaced oligonucleotide will migrate.

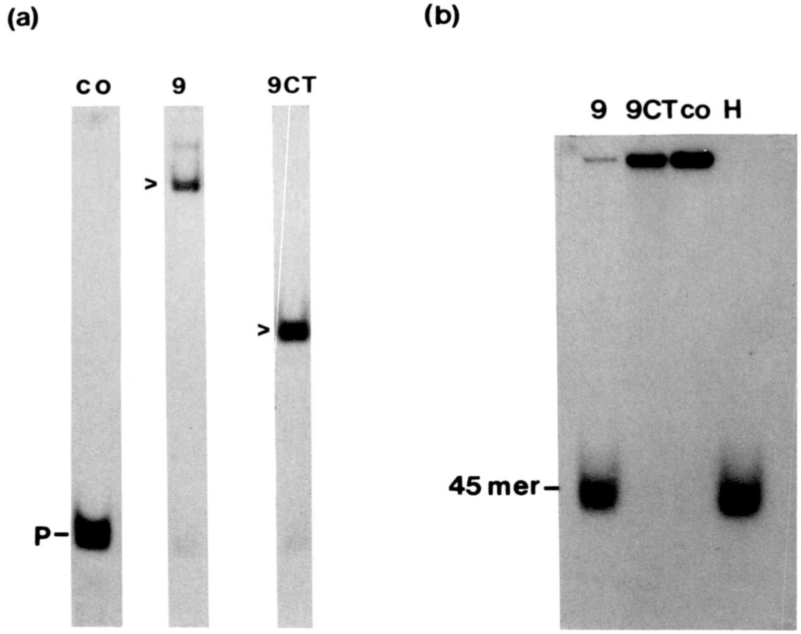

Figure 6. An example of DNA helicase and gel retardation assays. The 851 amino acid residue HSV-1 UL9 protein (9) and a fragment consisting of only the N-terminal 10 and C-terminal 317 amino acids (9CT) were purified by recognition-site affinity chromatography from Sf9 cells infected with recombinant baculoviruses (Section 3.2.4). Aliquots (3 μl) of the purified proteins or of the column buffer (co) were tested for sequence-specific DNA-binding activity and DNA helicase activity as described in the text (Sections 3.3.4 and 3.3.3). (a) Gel retardation assay (see *Protocol 11*). Incubation was performed with a labelled oligonucleotide containing UL9 binding site I (from HSV-1 ori_S), and the products were analysed on a non-denaturing 8% (w/v) polyacrylamide gel. The position of the free probe is labelled P and the retarded complexes are indicated by arrowheads. Both proteins are clearly capable of binding to the probe and retarding its gel mobility. As expected, the larger protein yields a more slowly migrating complex. (b) DNA helicase assay (see *Protocol 10*). The helicase substrate consisted of a labelled 45-base oligonucleotide annealed to single-stranded M13 DNA so as to leave an unannealed 3′ tail. The products were analysed by electrophoresis through a non-denaturing 8% (w/v) polyacrylamide gel. Track H contains substrate which was denatured by boiling immediately prior to electrophoresis to displace the oligonucleotide (45-mer). In contrast, the non-denatured substrate barely enters the gel (co). It can be seen that intact UL9 protein, but not the deleted version, exhibits DNA helicase activity. (Reproduced from ref. 29 with permission.)

3.3.4 Gel retardation

Gel retardation assays (43, 55–58) have become an extensively used approach for studying sequence-specific interactions of proteins with DNA. Extracts containing the protein of interest are incubated with a labelled DNA fragment containing the target binding site. Complexes of the protein and DNA probe may then be separated from unbound probe by electrophoresis through a non-denaturing gel (usually polyacrylamide), during which the mobility of the DNA–protein complex is retarded relative to the unbound probe. Using this assay it is possible to detect sequence-specific DNA-binding activity in relatively crude extracts, as long as molar excesses of unlabelled competitor DNAs are included in the binding reaction to compete for binding of non-sequence-specific DNA-binding proteins. Poly dI:dC and double-stranded calf thymus DNA are frequently used as competitor DNAs. Labelled probe fragments should ideally be as short as possible, so that the likelihood of the presence of binding sites for other proteins is minimized. Synthetic double-stranded oligonucleotides of approximately 20 bp containing the target binding site are usually appropriate, and these may be labelled with ^{32}P at their 5′ ends using polynucleotide kinase, or at their 3′ ends by incorporation of an appropriate dNMP using a DNA polymerase (59). The sequence specificity of any observed binding activity is readily confirmed by including a molar excess (for example 50–100 fold) of an unlablled oligonucleotide containing the binding site, which will compete with labelled molecules in the formation of DNA–protein complexes.

We have used this assay to study the interactions of host and viral proteins with specific sequences within the adenovirus and HSV origins of DNA replication. In addition to providing a convenient screen for the presence of the DNA-binding proteins of interest in extracts and fractions, the gel retardation assay also provides a sensitive method for assessing binding affinity (60) and screening the effects of mutations in the target sequence or protein on DNA-binding. If any antibody against a protein which forms part of the retarded complex is included in the incubation with labelled probe, the mobility of the complex is usually retarded further, as long as the antibody does not inhibit binding. This serves as a convenient method for confirming the presence of a particular protein in a complex. *Protocol 11* gives a procedure which has been employed successfully to detect the HSV-1 origin-binding protein in crude extracts. In practice, very similar methods should be appropriate for other sequence-specific DNA-binding proteins, the major avenues for variation being in the selection of binding buffer, competitor DNA, and gel composition. The results of a typical gel retardation assay are shown in *Figure 6*.

Protocol 11. Gel retardation assay for HSV-1 origin binding activity

Materials

- radioisotopically labelled oligonucleotide containing the binding site
- calf thymus competitor DNA: dissolve native DNA at 2 mg/ml in TE and sonicate it extensively

Protocol 11. *Continued*

- 5 × binding buffer: 100 mM Hepes–NaOH (pH 7.9), 50% (v/v) glycerol, 7.5 mM $MgCl_2$, 1 mM EDTA, 2.5 mM DTT
- loading dyes: 25% (v/v) glycerol, 10 mM DTT, 0.01% (w/v) bromophenol blue

Method

1. Mix 1 ng (approximately 50 fmol) of labelled oligonucleotide, 4 µl of 5 × binding buffer, 1 µl of sonicated calf thymus DNA. Add 1–5 µl (approx. 5–10 µg protein) of the extract to be tested, 1 M NaCl so that the final concentration (including that in the extract) will be 100–150 mM, and water to a final volume of 20 µl. Incubate for 20 min at 25°C.
2. Add 5 µl of loading dyes.
3. Separate protein–DNA complexes from the free probe by electrophoresis through an 8% (w/v) polyacrylamide gel (55:1 acrylamide:*N,N'*-methylenebisacrylamide) containing 1 × TBE.
4. Fix and dry the gel, and analyse by autoradiography. Controls from which extract has been omitted should be included to indicate the position of free probe. Drying the gel without fixation directly on to DEAE paper may be preferable for quantitative work, since this prevents loss of free DNA by leaching.

4. *In vitro* systems for DNA replication

In vitro systems which faithfully mimic events *in vivo* have been invaluable in determining the sequence of events that take place during SV40 and adenovirus DNA replication, from initiation to termination. However, it is only in these two eukaryotic virus systems, and more recently in BPV, that cell extracts have been capable of origin-dependent initiation of DNA synthesis. It is therefore very difficult to describe a method that is applicable to other viruses, although certain principles are common to most methods. In addition to the conditions set out in *Protocol 9* (i.e. the inclusion of $MgCl_2$, NTPs, dNTPs), most *in vitro* systems also incorporate a means of regenerating ATP and maintaining it at a high concentration. The example given in *Protocol 12* uses Ad4 (49) which, as it replicates by a protein-primed strand displacement mechanism, does not require the other three NTPs.

Protocol 12. Assay for elongation of Ad4 DNA initiated *in vitro*

Materials

- 5 × pre-incubation buffer: 125 mM Hepes–KOH (pH 7.5), 20 mM $MgCl_2$, 5 mM DTT, 0.5 mg/ml BSA

- elongation buffer: 80 μM each of dTTP, dATP, and dGTP, 5 μM dCTP, 0.3 μCi [α^{32}P]dCTP (3000 Ci/mmol), 4 mM ATP, 10 mM creatine phosphate, 10 μg/ml creatine phosphokinase, 0.1 mg/ml BSA, 25 mM Hepes–KOH (pH 7.5), 4 mM $MgCl_2$, 1 mM DTT
- termination buffer: 5% (w/v) SDS, 50% (v/v) glycerol, 100 mM EDTA, bromophenol blue to colour

Method

1. Pre-incubate Ad4 template DNA with 1–3 μl of Ad4-infected HeLa cell cytoplasmic extract or purified viral replication proteins and 1.4 μl of 5 × pre-incubation buffer in a total volume of 7 μl at 30°C for 30 min.
2. To allow elongation to proceed, add 7 μl of elongation buffer and incubate at 30°C for 90 min.
3. Add 6 μl of termination buffer and heat at 70°C for 5 min.
4. Resolve the reaction products on a 2% (w/v) agarose gel containing 0.1% (w/v) SDS by electrophoresis at 35 mA for 4 h. Fix the gel in 10% (v/v) acetic acid, dry, and expose it to X-ray film.
5. To determine the total incorporation, carry out steps 3–5, *Protocol 8*.

References

1. Challberg, M. D. and Kelly, T. J. (1989). *Annu. Rev. Biochem.*, **58**, 671.
2. Hay, R. T. and Russell, W. C. (1989). *Biochem. J.*, **258**, 3.
3. Salas, M. (1991). *Annu. Rev. Biochem.*, **60**, 39.
4. Challberg, M. D. (1991). *Seminars in Virology*, **2**, 247.
5. van der Vliet, P. C. (1991). *Seminars in Virology*, **2**, 271.
6. Kornberg, A. and Baker, T. A. (1992). *DNA replication*, (2nd edn). Freeman, New York.
7. Alberts, B., Bray, D., Lewis, J., Raff, M., Roberts, K., and Watson, J. D. (1989). *Molecular biology of the cell*, (2nd edn). Garland, New York.
8. Mul, Y. M. and van der Vliet, P. C. (1992). *EMBO Journal*, **11**, 751.
9. Temperley, S. M. and Hay, R. T. (1992). *EMBO Journal*, **11**, 761.
10. Sambrook, J., Fritsch, E. F., and Maniatis, T. (ed.) (1989). *Molecular cloning, a laboratory manual*, (2nd edn). Cold Spring Harbor Press, Cold Spring Harbor, New York.
11. Ausubel, F. M., Brent, R., Kingston, R. E., Moore, D. D., Seidman, J. G., Smith, J. A., and Struhl, K. (ed.) (1987). *Current protocols in molecular biology*. Wiley, New York.
12. Brown, T. A. (ed.) (1991). *Essential molecular biology: a practical approach*, Vol. II. IRL Press, Oxford.
13. Dyson, N. J. (1991). In *Essential molecular biology: a practical approach*, Vol. II (ed. T. A. Brown), p. 111. IRL Press, Oxford.
14. Kafatos, F. C., Jones, C. W., and Efstratiadis, A. (1979). *Nucl. Acids Res.*, **7**, 1541.

15. Stow, N. D. (1982). *EMBO Journal,* **1,** 863.
16. Weller, S. K., Spadaro, A., Schaffer, J. E., Murray, A. W., Maxam, A. M., and Schaffer, P. A. (1985). *Mol. Cell Biol.,* **5,** 930.
17. Weir, H. M. and Stow, N. D. (1990). *Mol. Cell Biol.,* **71,** 1379.
18. Guo, Z.-S. and DePamphilis, M. L. (1992). *Mol. Cell Biol.,* **12,** 2514.
19. Ustav, M., Ustav, E., Szymanski, P., and Stenlund, A. (1991). *EMBO Journal,* **10,** 4321.
20. Hay, R. T., Stow, N. D., and McDougall, I. M. (1984). *J. Mol. Biol.,* **175,** 493.
21. Ho, D. Y. and Mocarski, E. S. (1988). *Virology,* **167,** 279.
22. Challberg, M. D. (1986). *Proc. Natl Acad. Sci. USA,* **83,** 9094.
23. Wu, C. A., Nelson, N. J., McGeoch, D. J., and Challberg, M. D. (1988). *J. Virol.,* **62,** 435.
24. McGeoch, D. J., Dalrymple, M. A., Dolan, A., McNab, D., Perry, L. J., Taylor, P., and Challberg, M. D. (1988). *J. Virol.,* **62,** 444.
25. Heilbronn, R. and Zur Hausen, H. (1989). *J. Virol.,* **63,** 3683.
26. Goldstein, D. J. and Weller, S. K. (1988). *J. Virol.,* **62,** 2970.
27. Zhu, L. and Weller, S. K. (1992). *J. Virol.,* **66,** 458.
28. Zhu, L. and Weller, S. K. (1992). *J. Virol.,* **66,** 469.
29. Stow, N. D. (1992). *J. Gen. Virol.,* **73,** 313.
30. Brewer, B. J. and Fangman, W. I. (1987). *Cell,* **51,** 463.
31. Gahn, T. A. and Schildkraut, C. L. (1989). *Cell,* **58,** 527.
32. Huberman, J. A., Spotila, L. D., Nawotka, K. A., El-Assouli, S. M., and Davis, L. R. (1987). *Cell,* **51,** 473.
33. Nawotka, K. A. and Huberman, J. A. (1988). *Mol. Cell Biol.,* **8,** 1408.
34. Hay, R. T. and DePamphilis, M. L. (1982). *Cell,* **28,** 767.
35. Anachkova, B. and Hamlin, J. L. (1989). *Mol. Cell Biol.,* **9,** 532.
36. Vassilev, L. and Johnson, E. M. (1989). *Nucl. Acids Res.,* **17,** 7693.
37. Cook, P. R. (1991). *Cell,* **66,** 627.
38. Jackson, D. A. (1991). *BioEssays,* **13,** 1.
39. Calder, J. M. and Stow, N. D. (1990). *Nucl. Acids Res.,* **18,** 3573.
40. Dignam, J. D., Lebovitz, R. M., and Roeder, R. G. (1983). *Nucl. Acids Res.,* **11,** 1475.
41. Preston, C. M., Frame, M. C., and Campbell, M. E. M. (1988). *Cell,* **52,** 425.
42. Deutcher, M. P. (ed.) (1990). *Methods Enzymol.,* **182,** 1.
43. Sorger, P. K., Ammerer, G., and Shore, D. (1989). In *Protein function: a practical approach* (ed. T. E. Creighton), p. 199. IRL Press, Oxford.
44. Watson, C. J. and Hay, R. T. (1990). *Nucl. Acids Res.,* **18,** 1167.
45. Bosher, J., Robinson, E. C., and Hay, R. T. (1990). *New Biologist,* **2,** 1083.
46. Temperley, S. M., Burrow, C. R., Kelly, T. J., and Hay, R. T. (1991). *J. Virol.,* **65,** 5037.
47. Harris, M. P. G. and Hay, R. T. (1988). *J. Mol. Biol.,* **201,** 57.
48. Bradford, M. M. (1976). *Anal. Biochem.,* **72,** 248.
49. Temperley, S. M. and Hay, R. T. (1991). *Nucl. Acids Res.,* **19,** 3243.
50. Bosher, J., Leith, I. R., Temperley, S. M., Wells, M., and Hay, R. T. (1991). *J. Gen. Virol.,* **72,** 2975.
51. Fanster, B. S. and Loeb, L. A. (1974). *Methods Enzymol.,* **29,** 52.
52. Gottlieb, J., Marcy, A. I., Coen, D. M., and Challberg, M. D. (1990). *J. Virol.,* **64,** 5976.

53. Crute, J. J., Mocarski, E. S., and Lehman, I. R. (1988). *Nucl. Acids Res.,* **16,** 6585.
54. Fierer, D. S. and Challberg, M. D. (1992). *J. Virol.,* **66,** 3986.
55. Fried, M. and Crothers, D. M. (1981). *Nucl. Acids Res.,* **9,** 6505.
56. Scheider, R., Gander, I., Muller, U., Mertz, R., and Winnacker, E. L. (1986). *Nucl. Acids Res.,* **14,** 1303.
57. Weir, H. M., Calder, J. M., and Stow, N. D. (1989). *Nucl. Acids Res.,* **17,** 1409.
58. Rhodes, D. (1989). In *Protein function: a practical approach* (ed. T. E. Creighton), p. 177. IRL Press, Oxford.
59. Munday, C. R., Cunningham, M. W., and Read, C. A. (1991). In *Essential molecular biology: a practical approach,* Vol. II (ed. T. A. Brown), p. 57. IRL Press, Oxford.
60. Cleat, P. H. and Hay, R.T. (1989). *FEBS Lett.,* **258,** 51.

4

Analysis of viral sequence variation by PCR

PETER SIMMONDS and SHIU-WAN CHAN

1. Introduction

Sequence analysis has played a fundamental part in virological research since methods for the determination of nucleotide sequences were developed in the late 1970s (1, 2). However, these techniques require relatively large amounts of template DNA which are often obtainable only by virus culture *in vitro* and DNA cloning. The dependence on virus isolation and the time required to construct and screen libraries have always been major impediments to such studies.

The polymerase chain reaction (PCR) provides the means to amplify DNA specifically from any viral nucleotide sequence (3–6). The method is sufficiently sensitive and specific to allow direct amplification and sequence analysis of viral genetic material present *in vivo*, eliminating the need for prior virus isolation and cloning. The limitation of PCR for viral analysis is its dependence on detailed sequence information to specify the primers necessary for amplification. Therefore, the majority of PCR studies have been aimed at comparing large numbers of sequences over a relatively short region of the viral genome. PCR is intrinsically unsuitable for sequencing a complete viral genome; construction of libraries and screening with cloned viral DNA is likely to remain the method of choice for large sequencing projects for the foreseeable future.

1.1 Analysis of viral sequence variation

A large number of PCR-dependent methods for sequence analysis have been developed in recent years, many of which have been applied to viral studies. Some are designed to detect the presence or absence of nucleotide substitutions at a specific position in the amplified sequence (Sections 1.1.2 and 1.1.3); others exploit variability in the target sequence to cause mismatching with the PCR primer to prevent amplification (Section 1.1.1) or use a probe and thus prevent detection of the amplified product (Section 1.1.4). Finally, the amplified DNA can be analysed by RNase A digestion (Section 1.1.5), restriction fragment length polymorphisms (Section 1.1.6), or can be sequenced (Section 1.1.7).

1.1.1 Type-specific primers

PCR amplification is critically dependent on the specificity with which the oligonucleotide primers hybridize to their target (viral) sequence and thus initiate DNA synthesis (7). This specificity can be exploited by designing primers that hybridize to variable regions (between strains or types of a virus) in the viral genome to allow selective amplification of those sequences that match the primer and prevent amplification of those that differ. The discrimination of viral sequence variants by this method has been frequently used for the identification of different types of human papillomavirus (HPV) in cervical biopsies (8–10). Other examples include typing of herpes simplex virus (11), human T-lymphotropic virus (HTLV) types 1 and 2 (12, 13), and human immunodeficiency virus (HIV) types 1 and 2 (14). The reader is referred to these references for further practical details.

1.1.2 Oligonucleotide ligation

In this method of sequence analysis, two oligonucleotides are hybridized to template DNA such that the 3′ end of one primer is immediately adjacent to the 5′ base of the other. In the presence of DNA ligase, the two oligonucleotides are covalently joined. However, if there is a mismatch between the template and the 3′ base of the upstream or the 5′ base of the downstream oligonucleotide, ligation will not occur. Thus, detection of joined oligonucleotides indicates the presence of a particular sequence in the template DNA; it is usual to carry out these reactions in parallel with oligonucleotides that match or mismatch either of the two possible alternative nucleotides at a particular site. Specific protocols (applicable for viral analysis) for detecting point mutations in PCR-amplified DNA have been previously described (15, 16).

More recently, a ligation assay (ligase chain reaction; LCR) has been developed that dispenses with PCR (17, 18). This method uses *two* pairs of abutting oligonucleotides that hybridize to opposite strands of template DNA (i.e. are complementary in sequence to each other). After denaturation of the template DNA, oligonucleotides hybridize to the target, permitting ligation if there is a perfect match of their 3′ and 5′ ends. Following ligation, the sample is denatured again, and cooled to allow a second round of reactions to occur. The difference between this method and that described previously is the use of complementary oligonucleotides; not only will the original target DNA serve as a template for ligation, but the joined oligonucleotides from the first reaction will serve as a template for ligation of the other (complementary) oligonucleotide pair. Thus, the presence of a single perfectly matched target sequence at the start of thermal cycling leads to exponential accumulation of ligated primers, and under ideal conditions produces detectable amounts of product after 25–35 cycles. The first paper describing this method used T4 DNA ligase, which needed to be replenished after each cycle (17). In the same way that *Taq* DNA polymerase simplified and improved the specificity of PCR, LCR can be carried out using a thermostable DNA ligase, also from *Thermus aquaticus* (18). A considerable amount of work is now being carried out to optimize the sensitivity of LCR and to make amplification by this method reliably sensitive to single base mismatches.

1.1.3 Differential hybridization assays

These are carried out under conditions of high stringency such that a probe will anneal to an amplified sequence of one virus type but not to another. Using oligonucleotide probes under carefully controlled conditions, it is possible to distinguish sequences that differ in sequence by only one base (19). However, more reliable and technically simpler methods for differentiating virus variants use longer probes that contain a number of heterogeneous sites. For example, amplified sequences in the NS-5 region of hepatitis C virus (HCV) type I can be differentiated from those of type II by hybridization of labelled type I and type II probes in a slot blot assay (20). Other examples include human rotavirus electropherotypes (21), HTLV-1 and -2 (22), Dengue fever virus serotypes (23), and hepatitis B virus mutants (24).

1.1.4 Labelled nucleotide incorporation

Amplified DNA can be analysed for sequence variation immediately downstream from a primer by the addition of specific labelled dNTPs and a DNA polymerase (25). PCR products are purified from residual dNTPs and primers. A primer whose 3′ base after annealing is one base upstream from the variable site is annealed, and a single biotinylated or radiolabelled dNTP complementary to one or other of the sequence variants is added in the presence of DNA polymerase. Addition of a labelled nucleotide to the primer takes place if the target sequence matches, whereas little or no incorporation of label occurs if a different base is present. The primer can then be separated from unreacted dNTPs by polyacrylamide gel electrophoresis (25). Using biotinylated or digoxygenin- or radio-labelled dNTPs, it should be possible to use a separation method not based on gels to detect and quantify incorporated nucleotides. This method can be regarded as complementary to the use of type-specific primers and LCR, in that it can measure relative proportions of two or more sequence variants in a mixed population. In contrast, type-specific primers and the LCR can detect scarce sequences present in a mixture which are undetectable by non-selective amplification methods.

1.1.5 RNase A digestion

Sequence mismatches between amplified DNA and an RNA probe derived from a recombinant viral sequence can be detected by RNase A digestion. Amplified DNA is denatured and annealed to a single-stranded RNA probe.

After incubation with RNase A, cleavage of the RNA probe resulting from mismatches in the DNA/RNA hybrid can be detected by polyacrylamide gel electrophoresis and autoradiography. This method has been used to detect nucleotide substitutions in HIV *pol* which confer resistance to the antiviral agent, zidovudine (26).

1.1.6 Analysis of amplified DNA by restriction enzymes

Restriction fragment length polymorphisms (RFLPs) can be detected by digestion of the PCR product with suitable restriction endonucleases. A simple and rapid method is given in *Protocol 6*. RFLP analysis has been used extensively for virus typing and for investigation of sequence heterogeneity between epidemiologically or geographically separated virus variants. Examples of such methods include those used for differentiating human cytomegalovirus variants (27, 28), Epstein–Barr virus (29), hepatitis B (30), hepatitis C (31) and hepatitis delta (32) viruses, respiratory syncytial virus (33), poliovirus (34), rabies virus (35), and flaviviruses (36). Identification of restriction sites may also be made by hybridization, in solution, of a labelled oligonucleotide to the PCR product, followed by the addition of a restriction enzyme that recognizes a site within the duplex formed. Electrophoresis of the product of the restriction digest on a high percentage polyacrylamide gel indicates whether the labelled probe is cut (therefore, the amplified DNA sequence contained the restriction site) or uncut (site not present). Such probe-shift assays have been used more extensively for sensitive and specific identification of amplified DNA (5, 6, 37) than for the analysis of viral sequence variation.

1.1.7 Nucleotide sequencing

Determination of the nucleotide sequence gives a complete description of the amplified DNA, and for many purposes, such as evolutionary and epidemiological studies, is more informative than any of the methods described above. It is most commonly carried out by strand extension in the presence of dideoxynucleoside triphosphates (ddNTPs) that terminate synthesis at specific bases (1). The alternative method (Maxam–Gilbert sequencing; ref. 2) is technically more complex, and is less suited to the analysis of PCR-amplified DNA. Sequencing reactions can be carried out directly using amplified DNA as the template (see Section 3.2); alternatively, the PCR product can be ligated into a suitable vector and cloned into bacteria (see Section 2.2).

1.2 Samples for PCR

The reader is referred elsewhere (for example *Medical virology: a practical approach*, edited by U. Desselberger; in press) for specific protocols for the preparation of viral nucleic acids from lymphocytes, plasma, biopsy material, etc. PCR can be readily adapted for use with RNA sequences, i.e. by synthesis of cDNA from genomic viral RNA using reverse transcriptase (RT) from Moloney murine leukaemia virus (Mo-MLV) or avian myeloblastosis virus (AMV). Thermostable DNA polymerases are now available that can also use an RNA template for DNA synthesis (for example *Tth* polymerase, refs 38, 39), and combined protocols for the reverse transcription and amplification reactions have been devised by the manufacturers.

1.3 Single versus consensus sequencing

The analysis of amplified DNA described in Sections 1.1.6 and 1.1.7 can be carried out in two ways, depending on the degree of sequence heterogeneity of the original viral DNA or RNA sequences. In general, viruses with RNA genomes show considerable sequence variability, not only between strains and over time, but, more importantly for analysis, within a particular sample. This is especially true of viral sequences *in vivo*; for example, HIV *env* sequences in lymphocytes of infected individuals may show up to 10% variability within the same sample (40, 41). In contrast, some DNA viruses show very little heterogeneity; to take another example of a human virus, parvovirus B19 shows greater than 99.9% sequence conservation between geographically and epidemiologically separated variants (42, 43), and none within an infected individual.

Several of the methods described above may produce uninterpretable results if the amplified sample contains a complex mixture of sequence variants. In particular, the pattern of bands generated by a restriction enzyme cleavage of the PCR product may resemble a partial digest, while the sequencing reaction may terminate in two or more of the tracks at some of the nucleotide positions (see Section 3.2).

2. Separation of viral sequences

2.1 Limiting dilution

There are two methods for isolating single sequences to enable the analysis of heterogeneous samples. In the first method, template DNA is sufficiently diluted prior to amplification such that only single target sequences are amplified in each reaction. To separate the sequences, it is necessary to amplify DNA in multiple replicates at a dilution where only a small proportion of the reactions yield amplified DNA. As the distribution of very dilute DNA between samples by macroscopic pipetting is a completely random process, the Poisson formula can be used to calculate the likelihood of positive samples having originally contained one or more molecules of target DNA. From *Table 1*, it can be seen that when the frequency of PCR-positive reactions is 0.2, approximately 90% of reactions originally contained a single target sequence, whereas only half of the reactions are derived from single sequences when the frequency of positives is 0.7.

Table 1. Quantitation and separation of sequences by limiting dilution

Observed frequency of positives	Calculated number of DNA sequences[a]	Proportion (%) of single copies[b]
0.001	0.001	99.9
0.01	0.010	99.5
0.05	0.051	97.5
0.10	0.105	94.8
0.15	0.163	92.1
0.20	0.223	89.3
0.25	0.288	86.3
0.30	0.357	83.2
0.40	0.511	76.6
0.50	0.693	69.3
0.60	0.916	61.1
0.70	1.204	51.2
0.80	1.609	40.2
0.90	2.302	25.6
0.95	2.996	15.8

[a] Actual frequency, f (in target molecules/replicate tube) is calculated according to the formulae $f = -\ln(f_0)$, where f_0 is the observed frequency of negative reactions.
[b] Proportion of positive replicates, f_1, derived from a single copy of target DNA.

This method of separation is critically dependent on the reliable detection of single molecules of the appropriate viral sequence. Under ideal circumstances, it may be possible to detect positive reactions by hybridization of the PCR products to a probe of high specific activity (44), although the maximum sensitivity of such methods is generally around 3–100 copies of template DNA (45–51).

In our laboratory, we use nested PCR in preference to hybridization (52). In this method, the PCR is repeated using a pair of primers lying within the sequence originally amplified (*Protocol 1*). A standard PCR is carried out on the DNA sample over 25 cycles using 'outer' primers (steps 1–5). A small aliquot of the product is then transferred to a second tube and subjected to PCR using a second pair of primers (the 'inner' primers) that correspond to part of the amplified sequence (step 6). Thermal cycling provides a further 10^5- to 10^6-fold amplification of the viral DNA. Calculation shows that this should produce sufficient amounts of DNA from a single template molecule for sequence analysis. For example, a 300 bp target sequence has a relative molecular mass (M_r) of 200 000 and a mass of 0.32 ag (1 ag = 10^{-18} g); amplification by a factor of 10^7 with outer primers and then by a factor of 10^6 with inner primers would produce 3.2 μg of PCR product. In contrast, single amplification of one copy of target DNA by as much as 10^8-fold would produce barely enough product (32 pg) to be detectable by the most sensitive hybridization techniques.

Protocol 1. Amplification of DNA by nested PCR

Materials

- 10 × PCR buffer: 200 mM Tris–HCl (pH 8.8), 500 mM KCl, 15 mM $MgCl_2$, 0.5% (v/v) Triton X-100
- 100 × dNTPs: 3 mM each of dATP, dCTP, dGTP, dTTP
- 15–60 μM sense primers[a] (for a 20mer, 100–400 μg/ml)
- 15–60 μM anti-sense primers
- mineral oil (BDH)
- *Taq* polymerase (Cetus, Promega, Northumbria Biologicals or Boehringer)
- thermal cycler (for example Techne, Perkin-Elmer, or Hybaid)
- 0.01 μg/ml orange G dye in water (optional)
- 10 × TBE: 0.89 M Tris–HCl, 0.89 M boric acid, 0.02 M EDTA (pH 8.0)
- 1.5% (w/v) agarose[b] gel in 1 × TBE containing 0.66 μg/ml ethidium bromide[c]
- electrophoresis buffer: 1 × TBE containing 0.66 μg/ml ethidium bromide
- size markers spanning the expected size of the second PCR product (for example pBR322 digested with *Hae*III; Boehringer)
- protective goggles or (preferably) full face mask

Method

1. For each sample, mix the following. (For amplifying multiple samples, it is easier to make up a stock containing all but the DNA and add 45 μl to each tube.) Cover with a drop of mineral oil.

• 10 × PCR buffer	5 μl
• water	39 μl
• 100 × dNTPs	0.5 μl
• outer sense primer approximately 4–16 pmol)	0.25 μl
• outer anti-sense primer	0.25 μl
• *Taq* polymerase	1 U (usually 0.1–0.25 μl)
• DNA or cDNA sample	5 μl

2. Transfer the reactions immediately to a thermal cycler programmed to the following temperatures and times: 94°C for 35 sec, 50°C for 40 sec, 68°C for 150 sec.
3. Amplify for 25–35 cycles (25 cycles is sufficient with outer or inner primers in nested PCR).
4. Heat the samples at 68–72°C for 5 min at the end of the last cycle to terminate uncompleted strands.

Protocol 1. *Continued*

5. Amplified DNA is stable: store at 4°C (or at −20°C long-term) before carrying out the second amplification reaction or cloning (steps 6–9).
6. For each reaction, prepare 20 μl reactions containing the following. Cover with a drop of mineral oil.

• 10 × PCR buffer	2 μl
• water	up to 19 μl
• orange G dye[d] (optional)	0.1 μl
• 100 × dNTPs	0.2 μl
• inner sense primer (1.5–6 pmol)	0.1 μl
• inner anti-sense primer (1.5–6 pmol)	0.1 μl
• *Taq* polymerase	0.4 U (0.05–0.2 μl)
• amplified DNA from step 5[e]	1 μl

7. Repeat steps 2 and 3.
8. Analyse the entire sample by agarose gel electrophoresis; include size markers. View the gel on a UV transilluminator. Bands ranging in size from 100–500 bp are resolved adequately by this procedure.
9. Clone (*Protocol 2*) or sequence (*Protocol 8*) the amplified DNA product.

[a] Oligonucleotides are available from commercial and academic institutions in many countries. We obtain ours from Oswel DNA Service at a cost of £3.50/base (0.2 μmol scale). HPLC purification costs an additional £20.

[b] High concentrations of agarose impair the transparency of the gel. Low melting point agarose produces clearer gels, but is more expensive and is not necessary unless gels are to be photographed.

[c] Ethidium bromide is highly mutagenic—always wear gloves when handling gels or solutions.

[d] Orange G helps to monitor electrophoresis on an agarose gel. It does not interfere with PCR.

[e] Most laboratory workers experience the urge to add more than 1 μl to the second PCR reaction. In fact, this amount is more than adequate; addition of larger amounts will lead to the appearance of multiple non-specific bands in the product, which can reduce the sensitivity of the PCR.

To demonstrate the detection of single molecules of target DNA, serial tenfold dilutions of a cloned HIV sequence (pBH10.R3) were amplified with conserved *gag*-specific primers as described in *Protocol 1* (*Figure 1A* and *B*). Whereas 65 fg of target DNA yielded a visible band after the first round of amplification (*Figure 1A*, lane 3), as little as 6.5 ag produced a positive signal after the second round (*Figure 1B*, lane 7). A control which contained 1 μg of herring sperm DNA was completely negative after the second round of amplification (*Figure 1B*, lane 9). The mass of one molecule of the target DNA used in this titration was 13 ag, and thus the amount detected at the final dilution may have been as little as one molecule.

To demonstrate how dilution can separate single molecules of target

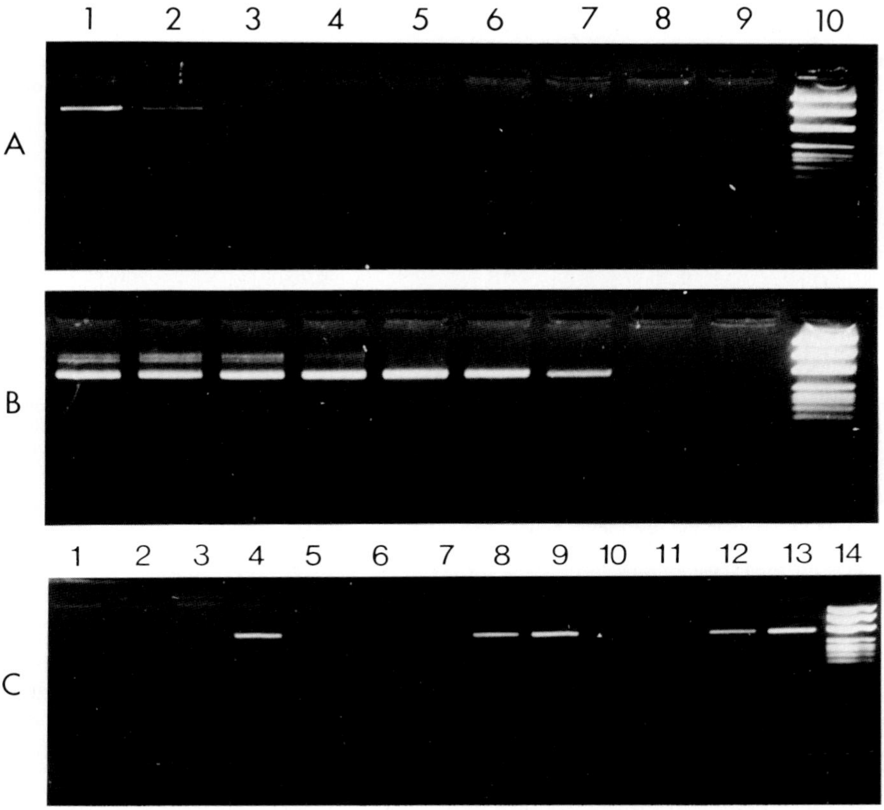

Figure 1. Titration of viral DNA sequences by nested PCR (reproduced with permission from the American Society for Microbiology). (A and B) Dilution series of recombinant HIV-1 DNA amplified by outer *gag* primers (A), followed by amplification of 1 μl of the product with inner *gag* primers (B). Lanes 1–8, 10-fold dilutions of pBH10.R3 in 200 μg/ml of herring sperm DNA, from 6.5 pg in lane 1 (5×10^5 copies of target sequence) to 0.65 ag in lane 8 (nominally 0.05 copies); lane 9, negative control (1 μg of herring sperm DNA); lane 10, size markers (pTZ18R digested with *Hae*III). Note that 1 ag = 10^{-18} g. Addition of 65 fg (5000 copies) of DNA produced a faint band after the first PCR (lane 3), while 6.5 ag (nominally 0.5 copies) produced a band after the second PCR (lane 7). (C) Limiting dilution: amplification of replicate samples, each nominally containing 0.5 molecules of target HIV-1 sequence, with nested *gag* primers (6.5 ag of pBH10.R3; lanes 2–12). Lane 1, negative control (1 μg of herring sperm DNA); lane 13, positive control (65 ag of pBH10.R3); lane 14, size markers. See *Table 2* for frequencies of positives at other input concentrations of HIV DNA.

Table 2. Frequency of positives with different amounts of target DNA

Input amount of pBH10.R3 [a] (ag [b])	Average input copies [c]	Observed frequency of positives [d]	Estimated copies per reaction [e]	Efficiency of amplification (%)
104	8	6/6	n.a. [f]	n.a.
52	4	11/12	2.49	62
26	2	6/12	0.69	35
13	1	6/12	0.69	69
6.5	0.5	7/19	0.46	92
3.25	0.25	4/18	0.25	100
1.63	0.13	2/18	0.12	92
0.81	0.06	0/18	0	n.a.
0 (control)	0	0/40	0	n.a.

[a] pBH10.R3 is a recombinant plasmid containing an almost full-length HIV-1 genome. The DNA concentration was determined from OD_{260}. The plasmid was serially diluted in TE, supplemented with 200 μg/ml sonicated herring sperm to give the indicated amounts of target DNA in the 5 μl sample.
[b] 1 ag = 10^{-18} g.
[c] Predicted number of molecules in sample from the first column.
[d] Number of positive reactions from the number of replicates tested, as obtained by amplification with nested primers in the *gag* region (52).
[e] Calculated number of molecules in the sample, using the Poisson formula.
[f] n.a. = not applicable.

sequence, replicates containing different nominal amounts of target sequence were amplified by nested PCR, and the products were analysed by agarose gel electrophoresis (*Figure 1C*; ref. 52). Multiple replicates of 6.5 ag amounts of target DNA yielded some positive and some negative signals. Their frequency varied with the amount of target DNA amplified (*Table 2*).

These experiments were carried out using recombinant DNA as target. However, the method can be readily used for separating viral sequences present *in vivo*. In order to ensure that the target viral sequences are distributed randomly on dilution, it may be necessary in practice to disaggregate DNA samples that are viscous after extraction by ultrasonication or with a restriction enzyme that cleaves at rare sites (for example *Sfi*I or *Not*I). There are also several situations where it may be essential to cleave genomic DNA to separate viral sequences (for example where cellular DNA may contain more than one copy of an integrated retroviral provirus on each chromosome). Separation of viral RNA sequences can be achieved by limiting dilution of the RNA itself or the cDNA synthesized from it. In practice, it is normally more convenient to titrate cDNA, as it is less likely to degrade on storage (53, 54).

2.2 PCR product cloning

The second method for separating sequences requires cloning of the PCR product; RFLP or sequence analysis can then be carried out on the progeny

DNA from one or more of the inserted viral sequences. The rationale for this method is that the individual colonies detected on hybridization have been transfected with a single plasmid containing one of the amplified viral sequences from the PCR, which is ultimately a copy of a single viral sequence present in the original sample.

DNA produced in a PCR differs from ordinary DNA in that the terminal 5′ bases of the sense and anti-sense strands are not phosphorylated. As the phosphate group is required for ligation, it is necessary either to phosphorylate amplified DNA or to cleave it with a suitable restriction enzyme to expose internal 5′-phosphate groups. Normally, an artificial restriction site that may not be present in the amplified sequence is incorporated into the 5′ end of the primer sequence, since amplified DNA with overhanging ends generated by restriction endonuclease cleavage can be ligated more efficiently than blunt-ended DNA fragments to the vector. To ensure that the PCR product inserts into the vector in only one orientation, different restriction sites that generate different overhanging ends can be incorporated into the primer sequences. Techniques for pre-treating the two types of finished PCR products before cloning are described in *Protocol 2*, step 4.

Protocol 2. Cloning of PCR products into a plasmid vector

Materials

- Glassmilk and NaI solution (for example Geneclean kit II; Bio101 Inc.)
- 10 × pol I buffer: 0.5 M Tris–HCl (pH 7.5), 0.1 M $MgCl_2$, 10 mM dithiothreitol (DTT), 0.5 mg/ml bovine serum albumin, 200 μM each of dATP, dCTP, dGTP, dTTP
- T4 polynucleotide kinase, DNA polymerase I, T4 DNA ligase (Boehringer)
- 10 × TE: 100 mM Tris–HCl (pH 8.0), 10 mM EDTA
- 10 × ligation buffer: 500 mM Tris–HCl (pH 7.5), 100 mM $MgCl_2$, 100 mM DTT, 100 mM ATP
- L-broth: 10 g/l Difco Bacto tryptone, 5 g/l Difco Bacto yeast extract, 5 g/l NaCl (pH 7.2); autoclave
- L-amp plates: poured using 15 g/l Difco agar autoclaved in L-broth with ampicillin added to 50–100 μg/ml when cooled to 55°C
- L-amp plates supplemented with 5-bromo-4-chloro-3-indolyl β-D-galactopyranoside (X-gal) and isopropyl β-D-thiogalactopyranoside (IPTG): prepared by streaking 30 μl of 2% (w/v in dimethylformamide) X-gal and 20 μl of 100 mM IPTG on to each plate
- competent bacterial cells (for example *Escherichia coli* JM83 made competent according to *Protocol 3*)

Protocol 2. *Continued*

Method

1. Amplify the DNA to be cloned by following steps 1–5 of *Protocol 1*. In most circumstances, it is sufficient to amplify for 25 to 35 cycles with a single pair of primers.
2. Add 100 μl of chloroform to the PCR product. Vortex briefly and centrifuge in a microcentrifuge at 12 000 *g* for 2 min.
3. Add the extracted PCR product to 2.5 volumes of NaI solution, add 10 μl of Glassmilk and incubate on ice for 15 min. Wash and elute the DNA from the Glassmilk according to the manufacturer's instructions. Resuspend the DNA in a small volume of water (for example 20 μl). Clone the DNA by following either step 4a or 4b.

4a. To obtain a fragment from a PCR product with restriction sites in the primer sequences, add 4 μl of the appropriate 10 × restriction buffer and 1 U of restriction enzyme(s) to the DNA. Add water to a final volume of 40 μl. To digest with two enzymes simultaneously, use the restriction buffer which gives 100%, or near to 100%, activity for both enzymes. Boehringer's 5-buffer system serves as a good guide for the choice of suitable restriction buffers. Incubate the reactions at the optimal temperature for 1–4 h. Heat-inactivate the enzymes at 65°C for 10 min. Proceed to step 5.

4b. To phosphorylate the PCR product, add 10 μl of 10 × pol I buffer, 1 μl of 100 mM ATP, and 10 U each of T4 polynucleotide kinase and DNA polymerase I. Adjust the final volume to 100 μl and incubate at 37°C for 1 h. Stop the reaction by adding 1 μl of 0.5 M EDTA (pH 8.0). Proceed to step 5.

5. Purify the DNA using Glassmilk as described in step 3. Resuspend the DNA in 20 μl of 1 × TE.
6. Ligate 50 ng of insert DNA to 100 ng of appropriately cleaved vector DNA (for example pUC18 or pUC19) in a total volume of 10 μl of 1 × ligation buffer containing 1 U of T4 DNA ligase by incubating at 15°C overnight. Heat-inactivate the ligase at 70°C for 10 min.
7. Mix 1 μl of the ligation mixture with 100 μl of competent *E. coli* JM83 cells. Incubate on ice for 30 min and then heat-shock at 42°C for 5 min.
8. Add 0.5 ml of L-broth and continue incubation by shaking at 37°C for 1–2 h.
9. Plate 100 μl, 50 μl, and 20 μl of the bacteria on to L-amp plates supplemented with X-gal and IPTG. Invert the plates and incubate at 37°C overnight to allow the colonies to grow.
10. Screen the colonies for positive clones (*Protocol 4*).

Protocol 3. Preparation of competent bacteria cells

Method

1. Inoculate a single colony of *E. coli* JM83 into 5 ml of L-broth and incubate at 37°C overnight with shaking.
2. Add 1 ml of the overnight culture to 50 ml of warm (37°C) L-broth. Continue incubation at 37°C with shaking until the culture has reached the logarithmic phase of growth. This normally takes 2 h.
3. Harvest the bacterial cells by centrifuging at 10 000 *g* for 15 min at 4°C.
4. Remove the supernatant and resuspend the pellet in 25 ml of ice-cold 0.1 M $CaCl_2$. Pellet the bacterial cells by centrifuging at 10 000 *g* for 15 min at 4°C.
5. Resuspend the pellet in 2.5 ml of ice-cold 0.1 M $CaCl_2$. Store the treated cells on ice for 30 min or longer (up to 24 h) before use.

Commonly used vectors include the single-stranded bacteriophage M13 series and the plasmid pUC series, but the latter are often preferable as they allow both strands of DNA to be sequenced from single clones, and, in addition, the cloned inserts are maintained more stably (see below). The pUC vectors contain an ampicillin-resistance gene and a multiple cloning site at the 5′ end of the *lacZ* gene. Insertional inactivation of *lacZ* allows colour detection of recombinants using X-gal prior to screening colonies by hybridization (*Protocol 4*). Identical PCR fragments are usually labelled as probes to screen for positives, but synthetic oligonucleotides can also be used. Methods for labelling probes and using them for hybridization have been described elsewhere (55, 56).

Protocol 4. Detection of positive clones

Materials

- nylon membrane (for example Hybond-N; Amersham)
- 20 × SSC: 3 M NaCl, 0.3 M trisodium citrate
- hybridization solution: see refs 55 and 56 for details

Method

1. Using sterile toothpicks, streak colourless colonies in duplicate on to a master L-amp plate supplemented with X-gal and IPTG, and on to a nylon membrane overlaid on a similar L-amp plate. Incubate the plates at 37°C for at least 6 h to overnight.

Protocol 4. *Continued*

2. Seal and store the master plate at 4°C until the results of the screening are known.

3. Remove the filter and place the membrane, colony side-up, on to Whatman 3MM filter papers saturated with 0.5 M NaOH, 1.5 M NaCl for 5 min.

4. Transfer the membrane on to filter papers saturated with 0.5 M Tris–HCl, pH 8.0, 1.5 M NaCl, pH 7.5 for a further 5 min.

5. Transfer the membrane on to filter papers saturated with 2 × SSC for 5 min.

6. Dry the membrane at 37°C for 5 min. Place it on to a UV transilluminator with the DNA side facing the UV light and fix the DNA on to the membrane by irradiating for 5 min.

7. The membrane is now ready for hybridization. Prepare a probe as described in either step 8a or 8b.

8a. For probing with double-stranded DNA, make the probe by labelling recombinant or PCR-amplified DNA by nick translation or (preferably) by random priming using [α-^{32}P]dATP or non-isotopic labels such as dUTP-digoxigenin (but not biotin). For initial experiments and small scale work, it may be convenient to buy a labelling kit (for example Boehringer).

8b. For probing with a labelled oligonucleotide, label the 5′ ends of synthetic oligonucleotides (20–40 bases) complementary to the amplified DNA enzymatically using [γ-^{32}P]ATP and polynucleotide kinase, or chemically using an appropriate non-isotopic label.

9. Hybridize the probe to the filter. Hybridization, washing, and detection steps depend on the nature of the probe. The reader should consult methods published in this series and elsewhere (40, 55, 56, 65) for further details of solutions and conditions.

Possible reasons for failing to obtain the cloned PCR product include the following:

(a) Ligation of blunt-ended PCR products to cleaved plasmid DNA is very inefficient. It is preferable to include restriction sites in the primers. Where this is not possible, linkers can be ligated to the phosphorylated PCR products and then treated with restriction endonucleases. Alternatively, phosphorylated PCR products can be ligated to a synthetic adaptor whose 5′ ends can then be phosphorylated.

(b) Restriction sites in the primers may not be cleaved efficiently. The site must be at least 5 bp from the end of the fragment. DNA for cloning should be purified from the reaction mixture using Glassmilk or phenol: chloroform extraction to remove possible enzyme inhibitors.

(c) The vector may self-ligate without an insert. This is a particular problem with blunt-ended ligation, but can also occur when the vector DNA has been cleaved with a single restriction enzyme. To prevent this, and thus increase the efficiency of insertion of amplified DNA, the vector can be dephosphorylated prior to ligation. Self-ligation can also be reduced by using two restriction sites with incompatible termini for cleavage of the vector DNA and PCR product.

(d) DNA ligase loses its activity at room temperature. Ensure that it is kept at −20°C or on ice at all times.

(e) Some viral sequences are prone to rearrangement and partial deletion in transfected cells. This is a particular problem with M13 cloning, and thus we recommend the use of pUC plasmids in *Protocol 2*. Suitable alternative plasmids include pBluescript (Stratagene) or the pGEM series (Promega).

3. Sequence analysis

3.1 Restriction fragment length polymorphism (RFLP)

Comparison of fragment sizes after restriction enzyme digestion is a rapid and convenient method of sequence analysis that can be readily used to compare large numbers of DNA samples amplified by PCR. As DNA sequences produced by this latter method are generally short (< 1 kbp), restriction enzymes that have 4 base recognition sites are usually more informative than those with 5 or 6. The G+C content of the sequence being amplified also has a dramatic effect on the frequency of restriction sites. As restriction fragments generated by 4 base enzymes can often be extremely small, it is not normally possible to obtain adequate resolution or sensitivity using an agarose gel stained with ethidium bromide. In *Protocols 5* and *6* we describe a convenient and rapid method for carrying out an RFLP analysis by polyacrylamide gel electrophoresis of viral DNA amplified by nested PCR.

The method uses ^{35}S-labelled nucleotides for the visualization of amplified bands. ^{32}P-Labelled nucleotides are also suitable (add 5 μCi/reaction), but are more hazardous, have a shorter half-life, and give poorer resolution. The advantage of using a radiolabel is that only DNA actually copied during PCR becomes labelled (thus non-target DNA, unreacted primers, and primer dimers will not appear), and only a very small volume needs to be added to the restriction digest. Thus, no form of DNA purification is required. Finally, as such a small amount of DNA is present, one can reliably achieve complete digestion with all common 4 and 6 base restriction enzymes.

The M_rs of the fragments can be estimated by including standard DNA size markers in the two outermost tracks of the gel. After electrophoresis, the marker tracks can be excised, stained with ethidium bromide, and the migration distances measured. A plot of M_r versus relative migration (Rf) can be used to estimate the size of the bands on the autoradiograph after exposure and development. Alternatively, 5′ end-labelled DNA markers, prepared using [γ-^{32}P]ATP and polynucleotide kinase, can be used for sizing on the autoradiograph. While this is convenient, it is an expensive and wasteful option if the radiolabel needs to be ordered specially for the purpose. Unfortunately 5′ end-labelling DNA with [γ-^{35}S]ATP under standard conditions is very inefficient.

Some laboratories may lack the facilities to store and handle radioactive chemicals, and may not be equipped with a gel dryer. In these circumstances, silver staining is a suitable alternative method, although this requires far more amplified DNA (from 5–20 μl of PCR product) to produce clear bands. Addition of such a large volume of PCR buffer to the restriction enzyme buffer may interfere with the cleavage reaction, necessitating prior extraction of the DNA by ethanol precipitation or Glassmilk treatment. Reliable and relatively inexpensive silver staining kits for this purpose are available commercially (for example from Amersham). Sizing of the fragments can be carried out with unlabelled M_r standards.

Protocol 5. Preparation of a polyacrylamide gel for RFLP analysis

Materials

- apparatus suitable for electrophoresis of proteins
- stock acrylamide solution: 25% (w/v) acrylamide, 1.25% (w/v) *N,N′*-methylenebisacrylamide in water (stable for 1 month at room temperature)
- 10% (w/v) ammonium persulphate (stable for 1 week at 4°C)
- *N,N,N′,N′*-tetramethylethylenediamine (TEMED)

Method

1. Clean and assemble the glass plates for the gel. Spacers and comb of thickness 0.5–0.75 mm are recommended.
2. Mix together the following for a 5% (w/v) polyacrylamide gel:
   - stock acrylamide solution 20 ml
   - 10 × TBE 10 ml
   - water 70 ml
   - 10% ammonium persulphate 250 μl
   - TEMED 25 μl
3. Without delay, pour the solution into the glass plate assembly, taking care to avoid air bubbles. Insert a comb to a depth of 1–2 cm. Leave for 1 h to allow polymerization to occur.

Protocol 6. Analysis of amplified viral sequences by RFLP

Materials

- [α-^{35}S]dNTP[a]
- polyacrylamide gel (*Protocol 5*)
- 10 × gel loading solution: 25% (w/v) Ficoll, 5 mM EDTA, 0.25% (w/v) bromophenol blue
- methanol/acetic acid: 5% (v/v) of each in water

Method

1. Amplify viral DNA (either a single molecule isolated by limiting dilution, or undiluted DNA; see Section 1.3) with nested primers as described in *Protocol 1*. Reduce the concentration of dNTPs in the second amplification reaction (step 6 of *Protocol 1*) from 30 μM to 7.5 μM (i.e. 0.125 μl of 100 × dNTPs/20 μl reaction) and supplement it with 2 μCi of [α-^{35}S]dNTP.
2. After amplification, transfer 1 μl of the PCR product to 50 μl of the appropriate restriction enzyme buffer. Add 1–10 U of restriction enzyme and incubate at the appropriate temperature for 1 h.
3. Add 5 μl of 10 × gel loading solution and heat the sample at 65°C for 5 min.
4. Electrophorese the entire sample on a polyacrylamide gel (60 V overnight or 150 V for 4 h), using 1 × TBE as electrophoresis buffer.
5. When the bromophenol blue has reached the bottom of the gel, fix the gel for 1 h in two changes of methanol/acetic acid.
6. Dry the gel and expose it to X-ray film for 2–7 days.

[a] [α-^{35}S]dATP (SJ1304; Amersham) is used, it can also be employed for sequencing reactions (*Protocol 8*).

The optimal concentration of polyacrylamide used for electrophoresis depends on the sizes of the DNA fragments to be separated: 8% is normally suitable for digests generated by 4 base restriction enzymes (30–500 bp fragments), 3% is suitable for larger fragments (100–1000 bp), and 10% for smaller fragments (20–200 bp). These alternative concentrations can be produced by altering the amount of stock acrylamide solution and water in step 2 of *Protocol 5*. A particular problem with this type of RFLP analysis is in trying to distinguish incomplete cleavage of the DNA (from sub-optimal conditions or inactivation of the restriction enzyme) from superimposition of two or more cleavage patterns generated from heterogeneous sequences in the amplified DNA. However, there should be equimolar amounts of each band when homogeneous DNA has been digested completely, and thus a proportionate reduction in the intensity of staining as the fragments get smaller. The sum of the sizes of the bands should be the same as that of uncut DNA (i.e. equalling the spacing between the primers used in the PCR). If all sequence variants present in the PCR product have at least one restriction site, then the two possibilities can be distinguished by including a track containing uncut DNA on each gel. The presence of co-migrating DNA in samples that have been cut indicates partial digestion; if all bands are smaller than this, then it is likely that sequence heterogeneity is responsible.

An example of RFLP analysis that demonstrates the high resolution and sensitivity of the method described in *Protocol 6* is shown in *Figure 2*. Sequence variation of HCV between different infected individuals is shown by digestion with *Scr*FI.

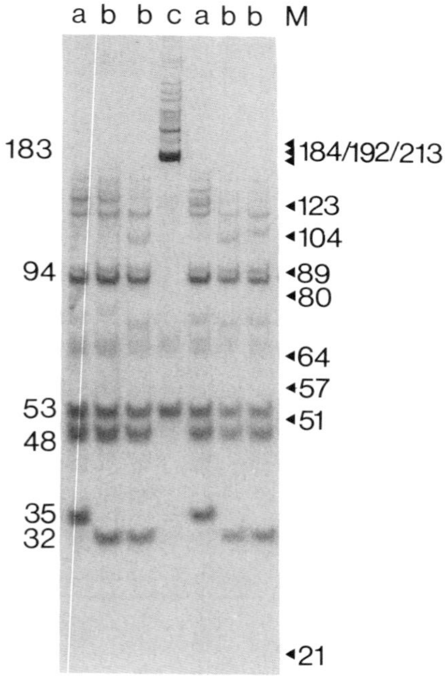

Figure 2. Example of RFLP analysis of the 5′-non coding region of HCV. Different HCV variants were identified by digesting a 251 bp amplified fragment with *Scr*FI, and resolving the radiolabelled fragments on a polyacrylamide gel (see *Protocol 6*). Migration distances and sizes (in bp) of DNA M_r markers (*Hae*III-digested pBR322 DNA) stained separately by ethidium bromide are shown in M. Three different HCV electropherotypes (designated a, b, and c) can be identified among the six samples analysed. Inferred sizes of bands are shown on the left (in bp). The high resolution of polyacrylamide gel electrophoresis allows differentiation of variants a and b on the basis of a 3 bp difference in size of the smallest band.

3.2 Nucleotide sequence determination

Nucleotide sequencing can be carried out on DNA amplified from single viral sequences obtained by limiting dilution or from cloning. Alternatively, if sequence heterogeneity within the sample is absent or minimal and unlikely to affect the analysis, then sequencing may be carried out on amplified undiluted samples (see Section 1.3). Sequencing using the Sanger method (1) requires strand extension from a primer; this can be one of the primers previously used in the PCR, or it can correspond to sequences within the amplified fragment. The sequencing reaction is highy sensitive to dNTPs, primers, and buffer components present in the PCR product; therefore, it is necessary to purify the DNA (*Protocol 7*). For viral DNA sequences inserted into pUC-derived vectors, sense and anti-sense M13 sequencing primers that match sequences in the plasmid linker are normally used.

Protocol 7. Preparation of DNA for sequencing

Materials

- 50 × TAE buffer: 2 M Tris, 1 M acetic acid, 0.5 M EDTA (pH 8.0)
- 10 × loading buffer: 25% (w/v) Ficoll, 0.25% (w/v) bromophenol blue

Method

A. *PCR products*

1. Amplify viral sequence(s) by nested PCR (*Protocol 1*). Carry out the second PCR, (step 6 of *Protocol 1*), but in a volume of 50 μl. Follow either step 2a or 2b.

2a. Electrophorese the PCR product on a low melting point agarose gel using 1 × TAE as the gel and electrophoresis buffers. (Clean the gel apparatus carefully prior to use if it has previously contained TBE buffer.) Excise the band corresponding to the amplified DNA. Weigh the agarose to estimate the volume. Proceed to step 3.

2b. Add 100–300 μl of chloroform directly to the PCR product. Vortex and centrifuge briefly to disperse the paraffin overlay. Proceed to step 3.

3. Transfer the excised gel fragment or extracted PCR product to a tube containing 2.5 volumes of NaI solution and 5–10 μl of Glassmilk. Follow the manufacturer's instructions for washing and eluting DNA.

4. Resuspend the eluted DNA in 20 μl of water.

B. *Cloned PCR products*

1. Remove, from the master plate, the agar plug corresponding to a positive colony identified in *Protocol 4* and transfer it to 5 ml of L-broth supplemented with 50–100 μg/ml of ampicillin. Incubate overnight at 37°C with shaking.

Protocol 7. *Continued*

2. Harvest the cells by centrifuging in a microcentrifuge at 12 000 *g* for 30 sec. Prepare DNA from the bacterial pellet according to the method previously described, involving boiling (40).
3. Sequence the DNA as described in *Protocol 8*, using M13 sense and anti-sense primers.

Protocol 8 describes a method of sequencing that has been used successfully on a wide variety of combinations of viral DNAs and primers. Double-stranded DNA (either the PCR product or its cloned derivative) is heat-denatured in the presence of a single primer, and strand synthesis is allowed to take place initially in the presence of unlabelled dGTP, dCTP, and dTTP and a limiting concentration of radiolabelled dATP, followed by incubation in the presence of each ddNTP to allow termination. The partial transcripts formed in each of the four termination reactions (i.e. with ddGTP, ddATP, ddTTP, and ddCTP) can be sized by electrophoresis on a high-resolution denaturing polyacrylamide gel to allow the nucleotide sequence to be read. Double-stranded sequencing reactions are readable from 10–20 bp downstream from the 3′ base of the sequencing primers to about 250 bp (see *Figure 3a*).

Protocol 8. Direct sequencing of amplified DNA

Materials[a]

- 5 × reaction buffer: 200 mM Tris–HCl (pH 7.5), 100 mM $MgCl_2$, 250 mM NaCl
- 5 × labelling mixture: 7.5 μM dGTP; 7.5 μM dTTP; 7.5 μM dCTP
- [α-^{35}S]dATP (SJ1304; Amersham)
- four termination solutions: 80 μM dATP, dTTP, dGTP, dCTP in 50 mM NaCl supplemented with 8 μM ddATP, ddTTP, ddGTP, or ddCTP
- stop solution: 95% (v/v) formamide, 20 mM EDTA, 0.05% (w/v) xylene cyanol, 0.05% (w/v) bromophenol blue

Method[b]

1. For each sequencing reaction, mix together the following and place in a boiling water bath or heating block for 3 min at 95°C to denature the DNA:

• dimethylsulphoxide (DMSO)	1 μl
• 5 × reaction buffer	2 μl
• primer (10 ng)	2 μl
• DNA prepared as in *Protocol 7*	5 μl

2. Immediately transfer the tube to an ice–water bath. When tube is cold, centrifuge it briefly to remove condensation and replace it on ice.
3. Add DMSO to each of the termination solutions to 10% (w/v). For each sequencing reaction, pipette 2.5 μl of each termination solution into separate wells of a 96-well round-bottom microtitre plate.
4. Warm the microtitre plate in a water bath at 37–40°C for at least 5 min before adding the DNA.
5. Add the following sequentially to the DNA sample:
   - 0.1 M DTT — 1 μl
   - labelling mixture, diluted 1:10–20 in water — 1 μl
   - [α-^{35}S]dATP (approximately 20 μCi) — 1 μl
   - T7 DNA polymerase (2 U in TE) — 1 μl
6. Mix the tube contents and add 3 μl to each of the four termination solution aliquots.
7. After approximately 5 min, terminate the reactions by adding 4 μl of stop solution.

[a] The reagents can be bought in ready-made form as part of the USB Sequenase 2.0 kit. This is especially recommended for initial experiments and for small-scale sequencing projects.
[b] This is a modification of the USB protocol for sequencing double-stranded DNA; the use of DMSO to reduce secondary structure follows the method of Winship (66).

Protocol 9. Denaturing PAGE

Materials

- 30% stock acrylamide solution: 30% (w/v) acrylamide, 1.5% (w/v) *N,N'*-methylenebisacrylamide
- sequencing gel apparatus (BRL Model S2 is recommended)
- ultrafine tips or drawn glass capillary tubes for loading the gel

Method

1. Mix the following, dissolve the urea by heating while stirring:
   - 10 × TBE — 10 ml
   - urea — 50 g
   - 30% stock acrylamide solution — 26 ml
   - 10% (w/v) ammonium persulphate — 1 ml
   - water to — 100 ml

Protocol 9. *Continued*

2. Carefully clean glass plates for the gel by swabbing them with 70% (v/v) ethanol then acetone, and assemble them using 0.4 mm flat or wedge spacers.
3. When all the urea has dissolved, add 20–25 μl of TEMED[a]. Immediately pour the gel solution into the assembly, taking care to avoid air bubbles. Leave the gel for at least 1 h to allow complete polymerization to occur.
4. Pre-electrophorese the gel at the manufacturer's recommended voltage for 10–15 min before loading the samples. Use 1 × TBE as the electrophoresis buffer.
5. Rinse out the space above the gel with 1 × TBE when the comb is removed, and again immediately before loading the samples.
6. Heat the samples from *Protocol 8* at 70–90 °C for 2–5 min prior to loading.
7. Load the entire sample into the well—start loading from the centre of the gel, as these wells are less likely to leak. Normally, all samples can be loaded at the same time without risk of renaturation. If leakage is a problem, however, loaded samples can be electrophoresed for 5 min before adding subsequent samples. Electrophorese the samples until the bromophenol blue reaches the bottom of gel. The outer gel plate should be warm to the touch.
8. Fix the gel in two changes of 1 litre of methanol/acetic acid (see *Protocol 6*).
9. Dry the gel and expose it to X-ray film for 24–72 h.

[a] Gel solution must be at room temperature before adding TEMED.

The following section illustrates a number of problems specifically associated with the use of PCR-amplified DNA as the template for sequencing reactions, and outlines some preventative steps.

3.2.1 Sequence heterogeneity

As described for RFLP analysis (*Protocol 6*), heterogeneity within the amplified product hinders correct reading of the sequence and may lead to considerable inaccuracies. Minor sequence variability can cause the appearance of two or more bands at the same position on the gel (*Figure 3b*). In theory, the marked position (a) could be read as T or C, but as heterogeneity can occur elsewhere in the gel (at b, G/A; c, A/C; d, T/C; e, A/C; f, G/A), one cannot determine which of the variants T or C is associated with the alternatives at b and elsewhere (c–f). A second problem is that ambiguous sites are often difficult to distinguish from strong stops; these are positions where strand extension is inhibited by secondary structure within the single-stranded template and leads to bands appearing in two, three, or four lanes of the sequence.

Where insertions or deletions occur between sequences, the gel can become unreadable downstream from the site (*Figure 3c*). The four tracks on the left show the termination product from a single sequence. Those on the right contained two sequences that are identical as far as the arrow, but one contained a deletion beyond it, leading immediately to a double reading and thus to an unreadable gel image downstream.

If sequence heterogeneity is found on direct sequencing of uncloned and undiluted DNA, it is necessary to follow the steps described in *Protocols 1* and *2* to ensure that the DNA to be sequenced is homogeneous. If sequence ambiguities are encountered on sequencing after limiting dilution, it is probable that the target DNA was insufficiently diluted to separate single molecules of target DNA (for example, if 50% of replicates at a given dilution are positive, only 30% of these correspond to single sequences; see *Table 1*).

3.2.2 Inappropriate priming

The majority of DNA sequences amplified by PCR can be successfully sequenced using one of the PCR primers for the sequencing reaction. However, as primers are present in a vast molar excess over product DNA at the end of the PCR, even minor carry-over of primer can cause a significant level of unwanted priming during the sequencing reaction. The gel in *Figure 3d* shows the typical appearance of double reading due to priming of DNA strand termination reactions from both ends of the sequence by oligonucleotides carried over from the PCR reaction mixture. If this is a problem, electrophorese the PCR product on a low melting point agarose gel prior to DNA extraction to remove unreacted primers (steps 2a and 3 of *Protocol 7*). Alternatively, an asymmetric PCR can be carried out to generate single-stranded DNA (57).

4. Choice of method

Almost all published sequence data from PCR-amplified viral sequences have been obtained either by sequencing the amplified product directly (where heterogeneity of sequences within the sample is ignored) or from cloned DNA. In deciding which method to use, it is necessary to consider the nature of the sample being analysed, and the possibility of obtaining inaccurate or non-representative sequences or artefacts introduced by the amplification process. Finally, the choice of method may often be decided by the laboratory facilities available. An increasing number of laboratories using PCR for viral analysis do not have suitable designated areas in order to carry out the genetic manipulation necessary for cloning of PCR products.

The ability of PCR to amplify viral sequences directly from clinical specimens eliminates the need for virus isolation and cloning which were previously required for sequence analysis. Sequences obtained by PCR are thus relatively free from the errors and inaccuracies associated with artificial selection, recombination, and sequence changes occurring under conditions of passage

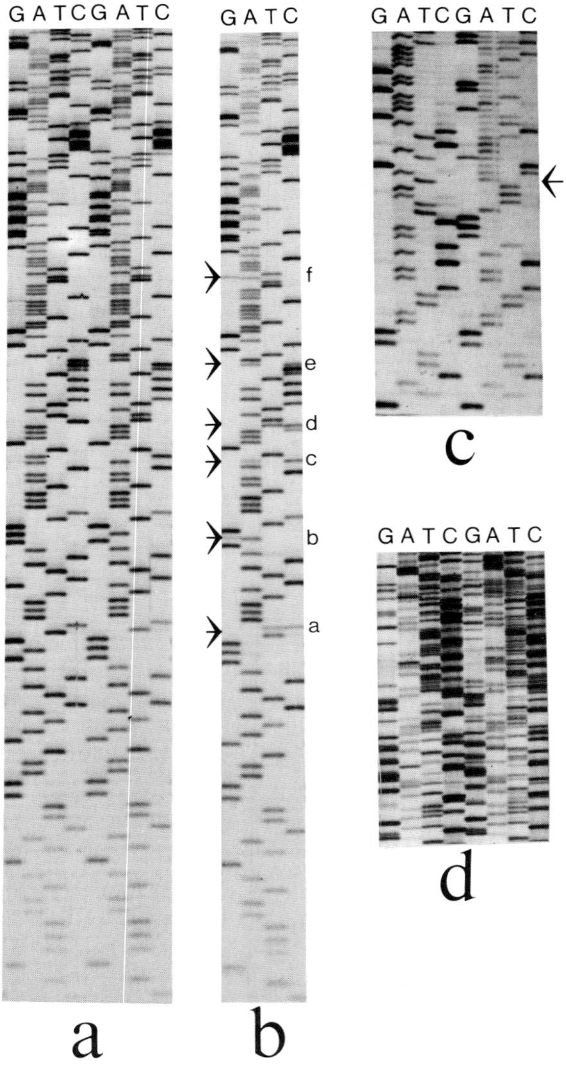

in vitro. However, the process of amplification by PCR can lead to similar problems, as described in the following sections.

4.1 Selection of sequences

The sequences of oligonucleotides used as primers for PCR are normally designed to match a highly conserved region of the viral genome. Mismatches impair hybridization of the primer to its target sequence, and can hinder or

Figure 3. Double-stranded sequencing reactions: common problems. (a) Two successful sequencing reactions. Two molecules of provirus separated by limiting dilution of DNA extracted from an HIV-infected haemophiliac were amplified using primers spanning the V4/V5 hypervariable regions of *env* and sequenced using the sense primer. The two sequences are readable from 9 bp from the primer (at the bottom of the gel) for at least 250 bp (beyond the top of the gel shown). They differ at several sites, illustrating the effectiveness of limiting dilution at separating target sequences. (b) Sequencing reaction with multiple ambiguities resulting from amplification of two molecules of heterogenous target DNA in the same reaction (insufficient dilution of sample). Six polymorphic sites (a–f) are indicated by arrows. (c) Sequencing reaction of DNA derived from a single molecule (left) and from two molecules (right), one of which contains an insertion. In the latter sequencing reaction, the sequences are the same up to the arrow and are readable; beyond the insertion point, the sequences are uninterpretable. (d) Two unreadable sequencing reactions in HIV *gag*, the result of double reading from both ends of the sequence due to carry-over of primer from the PCR into the sequencing reaction.

---

prevent amplification (7). Therefore, if a mixture of sequences is amplified, those that best match the primers will be over-represented in the final PCR products. This is a problem when amplifying highly variable sequences, or where insufficient sequence data are available to establish whether a particular primer is conserved among all of the sequences to be amplified. Amplification of DNA by PCR is an exponential process, and even very small differences in the rate of amplification can dramatically affect the yield of a partially mismatched sequence in the product. Selection is not a problem where the target DNA or cDNA is diluted prior to amplification to separate individual sequences, as mismatches will, at worst, reduce the yield of amplified DNA rather than lead to under-representation in the sequences obtained.

4.2 Introduction of sequence errors during copying

The DNA polymerase commonly used for PCR (from the thermophile *T. aquaticus*) does not 'proof-read' the newly synthesized DNA. As a result, it has a relatively high misincorporation rate of nucleotides (58–60). A number of *in vitro* studies carried out under PCR conditions have shown that an average of one in 10 000 nucleotides incorporated into copied DNA is mismatched with the template. The nature and actual frequency of such errors are dependent on the nucleotide sequence being copied (60), damage to the DNA (61), pH, and dNTP concentration (62). If the amplified product is cloned, the inserted DNA is derived from sequences that may already have been copied 20–30 times. Several studies have estimated this cumulative effect by sequence analysis of cloned DNA amplified from a homogeneous target sequence. Misincorporation rates ranging from 1 in 350 (ref. 59), 1 in 400 (ref. 58) and 1 in 1000–1800 (ref. 60) have been reported over 25–30 cycles of PCR using *Taq* polymerase. Under many circumstances this can significantly interfere with phylogenetic, evolutionary, or functional analysis of the viral sequences. Indeed, if polypeptides were expressed from the cloned DNA, errors may lead to aberrant or non-functional viral proteins.

In contrast to the cloning method, nested PCR can produce sufficient product to allow direct sequencing (ref. 52; *Figure 1*). As a result, the frequency of *Taq* polymerase-induced errors in the final sequence is considerably reduced. Starting with a single target molecule and using the same primer spacing as above, the probability of incorporating an incorrect base in the first round of copying is $10^{-4} \times 250$ bp (i.e. in 1 in 40 PCR amplifications). Should this relatively rare event occur, and double-stranded DNA was used as the initial template, then only one in four of the DNA strands would actually be incorrect at the end of the first cycle. Assuming that amplification rates are equal, then a quarter of the product DNA at the end of the PCR would then be incorrect; this may cause a minor ambiguity on the sequencing gel, but would not normally be sufficient to cause an incorrect base to be scored. It can be readily seen that errors that occur in the second, third, and subsequent cycles would not affect the final sequence.

In most published protocols reverse transcription of viral RNA produces a single-stranded DNA template. If a copying error occurred during reverse transcription (AMV RT has a misincorporation rate of 1 in 24 000; ref. 63) or in the first cycle of PCR, then a half, rather than a quarter of the amplified product would be incorrect in sequence. Such errors, although rare (for example one in 30 sequences of length 250 bp), would lead to problems with reading the gel. As these errors cannot be distinguished from those obtained by amplifying a heterogeneous viral sequence, it is advisable to discard sequences containing all such ambiguities.

Recently, thermostable DNA polymerases that possess a 3′,5′ exonuclease activity have been isolated from *Thermus litoralis* and *Pyrococcus furiosus*. Such enzymes 'proof-read' DNA and show a reduced rate of nucleotide misincorporation during PCR (62). The use of such enzymes will improve the accuracy of sequence analysis of cloned PCR products, and permit more reliable expression of viral sequences.

4.3 *In vitro* recombination

The most significant problem with amplification of heterogeneous viral sequences by PCR is artefactual recombination. This occurs during the later cycles, where there is insufficient DNA polymerase to complete the synthesis of all primed DNA strands in the allotted 2–3 min elongation step. Partially completed DNA detaches from its template during the following denaturation step, and then on cooling anneals to any available complementary strand; synthesis of the partial sequence is then completed by copying the 'wrong' sequence (60, 64). It has been estimated that 1 in 5 sequences is an artificial recombinant after amplification of a 300 bp fragment of DNA over 25 cycles (64). Amplification of DNA that is damaged (for example by ageing or exposure to high temperatures) is even more prone to template jumping during PCR, as the polymerase is more likely to pause at a chemically modified base (61). The frequency of recombinants increases dramatically with longer primer spacings; when using primers over 1500 bp apart, almost all sequences at the end of PCR will be hybrid copies of two, three, or more of the original viral sequences.

Again, separation of DNA, RNA, or cDNA sequences prior to amplification completely obviates this problem. Although recombination between strands will occur during nested PCR, as all the sequences are the same (ultimately derived from the same viral sequence), this will have no effect on the final sequence obtained.

5. Summary

In this chapter, we have attempted to review existing methods for the analysis of viral sequence variation based on PCR and related methods. The two main issues that influence the choice of method, namely sequence heterogeneity within the viral sample to be amplified and accuracy of the sequences obtained, have been discussed and should be borne in mind when choosing methods. In summary, the heterogeneity of viral genomes present *in vivo*, particularly those of RNA viruses, require that cloning or limiting dilution should be carried out to achieve a satisfactory RFLP analysis or nucleotide sequence determination. Cloning of amplified DNA has historically been the method of choice. However, direct sequencing of single molecules of target DNA amplified by nested primers provides a quick and accurate method for sequence determination that is remarkably free from problems of selection, copying errors, and *in vitro* recombination, and which requires the minimum of laboratory facilities.

Acknowledgements

The authors are grateful to Fiona McOmish and Lin-Qi Zhang for making available the gels shown in *Figures 2* and *3*. Many of the methods described were developed in collaboration with the aforementioned colleagues, and with Sandy Cleland and Peter Balfe.

References

1. Sanger, F., Nicklen, S., and Coulson, A. R. (1977). *Proc. Natl Acad. Sci. USA*, **74,** 5463.
2. Maxam, A. M. and Gilbert, W. (1977). *Proc. Natl Acad. Sci. USA*, **74,** 560.
3. Saiki, R. K., Scharf, S. J., Faloona, F. A., Mullis, K. B., Horn, G. T., Erlich, H. A., and Arnheim, N. (1985). *Science,* **230,** 1350.
4. Mullis, K. B. and Faloona, F. A. (1987). *Methods Enzymol.,* **155,** 335.

5. Kwok, S. Y., Mack, D. H., Mullis, K. B., Poiesz, B. J., Ehrlich, G. D., Blair, D., Friedman-Kien, A. S., and Sninsky, J. J. (1987). *J. Virol.,* **61,** 1690.
6. Ou, C. Y., Kwok, S., Mitchell, S. W., Mack, D. H., Sninsky, J. J., Krebs, J. W., Feorino, P., Warfield, D., and Schochetman, G. (1988). *Science,* **239,** 295.
7. Kwok, S., Kellogg, D. E., McKinney, N., Spasic, D., Goda, L., Levenson, C., and Sninsky, J. J. (1990). *Nucl. Acids Res.,* **18,** 999.
8. McNicol, P. J. and Dodd, J. G. (1990). *J. Clin. Microbiol.,* **28,** 409.
9. Nuovo, G. J., Darfler, M. M., Impraim, C. C., and Bromley, S. E. (1991). *Am. J. Pathol.,* **138,** 53.
10. Chow, V. T., Tham, K. M., and Bernard, H. U. (1990). *J. Virol. Methods,* **27,** 101.
11. Kimura, H., Shibata, M., Kuzushima, K., Nishikawa, K., Nishiyama, Y., and Morishima, T. (1990). *Med. Microbiol. Immunol. (Berl.),* **179,** 177.
12. Chen, Y. M., Lee, T. H., Wiktor, S. Z., Shaw, G. M., Murphy, E. L., Blattner, W. A., and Essex, M. (1990). *Lancet,* **336,** 1153.
13. Ehrlich, G. D., Glaser, J. B., LaVigne, K., Quan, D., Mildvan, D., Sninsky, J. J., Kwok, S., Papsidero, L., and Poiesz, B. J. (1989). *Blood,* **74,** 1658.
14. Rayfield, M., De Cock, K., Heyward, W., Goldstein, L., Krebs, J., Kwok, S., Lee, S., McCormick, J., Moreau, J. M., Odehouri, K., *et al.* (1988). *J. Infect. Dis.,* **158,** 1170.
15. Landegren, U., Kaiser, R., Sanders, J., and Hood, L. (1988). *Science,* **241,** 1077.
16. Nickerson, D. A., Kaiser, R., Lappin, S., Stewart, J., Hood, L., and Landegren, U. (1990). *Proc. Natl Acad. Sci. USA,* **87,** 8923.
17. Wu, D. Y. and Wallace, R. B. (1989). *Genomics,* **4,** 560.
18. Barany, F. (1991). *Proc. Natl Acad. Sci. USA,* **88,** 189.
19. Richman, D. D., Guatelli, J. C., Grimes, J., Tsiatis, A., and Gingeras, T. (1991). *J. Infect. Dis.,* **164,** 1075.
20. Enomoto, N., Takada, A., Nakao, T., and Date, T. (1990). *Biochem. Biophys. Res. Commun.,* **170,** 1021.
21. Flores, J., Sears, J., Schael, I. P., White, L., Garcia, D., Lanata, C., and Kapikian, A. Z. (1990). *J. Virol.,* **64,** 4021.
22. Hjelle, B., Scalf, R., and Swenson, S. (1990). *Blood,* **76,** 450.
23. Laille, M., Deubel, V., and Sainte Marie, F. F. (1991). *J. Med. Virol.,* **34,** 51.
24. Li, J., Tong, S., Vitvitski, L., Zoulim, F., and Trepo, C. (1990). *J. Gen. Virol.,* **71,** 1993.
25. Kuppuswamy, M. N., Hoffman, J. W., Kasper, C. K., Spitzer, S. G., Groce, S. L., and Bajaj, S. P. (1991). *Proc. Natl Acad. Sci. USA,* **88,** 1143.
26. Lopez Galindez, C., Rojas, J. M., Najera, R., Richman, D. D., and Perucho, M. (1991). *Proc. Natl Acad. Sci. USA,* **88,** 4280.
27. Zaia, J. A., Gallez Hawkins, G., Churchill, M. A., Morton Blackshere, A., Pande, H., Adler, S. P., Schmidt, G. M., and Forman, S. J. (1990). *J. Clin. Microbiol.,* **28,** 2602.
28. Tendler, C. L., Greenberg, S. J., Blattner, W. A., Manns, A., Murphy, E., Fleisher, T., Hanchard, B., Morgan, O., Burton, J. D., Nelson, D. L., *et al.* (1990). *Proc. Natl Acad. Sci. USA,* **87,** 5218.
29. Cen, H., Breinig, M. C., Atchison, R. W., Ho, M., and McKnight, J. L. (1991). *J. Virol.,* **65,** 976.

30. Cantin, E. M., Chen, J., McNeill, J., Willey, D. E., and Openshaw, H. (1991). *Curr. Eye Res.,* **10,** 15.
31. Nakao, T., Enomoto, N., Takada, N., Takada, A., and Date, T. (1991). *J. Gen. Virol.,* **72,** 2105.
32. Xia, Y. P., Chang, M. F., Wei, D., Govindarajan, S., and Lai, M. M. (1990). *Virology,* **178,** 331.
33. Cane, P. A. and Pringle, C. R. (1991). *J. Gen. Virol.,* **72,** 349.
34. Balanant, J., Guillot, S., Candrea, A., Delpeyroux, F., and Crainic, R. (1991). *Virology,* **184,** 645.
35. Sacramento, D., Bourhy, H., and Tordo, N. (1991). *Mol. Cell. Probes,* **5,** 229.
36. Eldadah, Z. A., Asher, D. M., Godec, M. S., Pomeroy, K. L., Goldfarb, L. G., Feinstone, S. M., Levitan, H., Gibbs, C. J., Jr., and Gajdusek, D. C. (1991). *J. Med. Virol.,* **33,** 260.
37. Kumar, R., Goedert, J. J., and Hughes, S. H. (1989). *AIDS Res. Hum. Retroviruses,* **5,** 345.
38. Elie, C., Salhi, S., Rossignol, J. M., Forterre, P., and de Recondo, A. M. (1988). *Biochim. Biophys. Acta,* **951,** 261.
39. Salhi, S., Elie, C., Jean-Jean, O., Meunier Rotival, M., Forterre, P., Rossignol, J. M., and de Recondo, A. M. (1990). *Biochem. Biophys. Res. Commun.,* **167,** 1341.
40. Balfe, P., Simmonds, P., Ludlam, C. A., Bishop, J. O., and Brown, A. J. (1990). *J. Virol.,* **64,** 6221.
41. Simmonds, P., Balfe, P., Ludlam, C. A., Bishop, J. O., and Brown, A. J. (1990). *J. Virol.,* **64,** 5840.
42. Shade, R. O., Blundell, M. C., Cotmore, S. F., Tattersall, P., and Astell, C. R. (1986). *J. Virol.,* **58,** 921.
43. Blundell, M. C., Beard, C., and Astell, C. R. (1987). *Virology,* **157,** 534.
44. Li, H., Gyllensten, U. B., Cui, X. F., Saiki, R. K., Erlich, H. A., and Arnheim, N. (1988). *Nature,* **335,** 414.
45. Kaneko, S., Feinstone, S. M., and Miller, R. H. (1989). *J. Clin. Microbiol.,* **27,** 1930.
46. Coutlee, F., Yang, B. Z., Bobo, L., Mayur, K., Yolken, R., and Viscidi, R. (1990). *AIDS Res. Hum. Retroviruses,* **6,** 775.
47. Keller, G. H., Huang, D. P., Shih, J. W., and Manak, M. M. (1990). *J. Clin. Microbiol.,* **28,** 1411.
48. Loche, M. and Mach, B. (1988). *Lancet,* **ii,** 418.
49. Arthur, R. R., Dagostin, S., and Shah, K. V. (1989). *J. Clin. Microbiol.,* **27,** 1174.
50. Olive, D. M., Simsek, M., and al Mufti, S. (1989). *J. Clin. Microbiol.,* **27,** 1238.
51. Puchhammer Stockl, E., Popow Kraupp, T., Heinz, F. X., Mandl, C. W., and Kunz, C. (1990). *J. Med. Virol.,* **32,** 77.
52. Simmonds, P., Balfe, P., Peutherer, J. F., Ludlam, C. A., Bishop, J. O., and Brown, A. J. (1990). *J. Virol.,* **64,** 864.
53. Simmonds, P., Zhang, L. Q., Watson, H. G., Rebus, S., Ferguson, E. D., Balfe, P., Leadbetter, G. H., Yap, P. L., Peutherer, J. F., and Ludlam, C. A. (1990). *Lancet,* **336,** 1469.
54. Zhang, L. Q., Simmonds. P., Ludlam, C. A., and Leigh Brown, A. J. (1991). *AIDS,* **5,** 675.

55. Sambrook, J., Fritsch, E. F., and Maniatis, T. (ed.) (1989). *Molecular cloning: a laboratory manual.* Cold Spring Harbor Laboratory Press, Cold Spring Harbor.
56. Ausubel, F. M., Brent, R., Kingston, R. E., Moore, D. D., Seidman, J. G., Smith, J. A., and Struhl, K. (ed.) (1987). *Current protocols in molecular biology.* Wiley Interscience, New York.
57. Gyllsenten, U. B. and Erlich, H. A. (1988). *Proc. Natl Acad. Sci. USA,* **85,** 7652.
58. Saiki, R. K., Gelfand, D. H., Stoffel, S., Scharf, S. J., Higuchi, R. G., Horn, G. T., Mullis, K. B., and Erlich, H. A. (1988). *Science,* **239,** 487.
59. Dunning, A. M., Talmud, P., and Humphries, S. E. (1988). *Nucl. Acids Res.,* **16,** 10393.
60. Ennis, P. D., Zemmour, J., Salter, R. D., and Parham, P. (1990). *Proc. Natl Acad. Sci. USA,* **87,** 2833.
61. Paabo, S., Higuchi, R. G., and Wilson, A. C. (1989). *J. Biol. Chem.,* **264,** 9709.
62. Ling, L. L., Keohavong, P., Dias, C., and Thilly, W. G. (1991). *PCR Methods and Applications,* **1,** 63.
63. Roberts, J. D., Bebenek, K., and Kunkel, T. A. (1991). *Science,* **242,** 1171.
64. Meyerhans, A., Vartanian, J. P., and Wain Hobson, S. (1990). *Nucl. Acids Res.,* **18,** 1687.
65. D'Addario, M., Roulston, A., Wainberg, M. A., and Hiscott, J. (1990). *J. Virol.,* **64,** 6080.
66. Winship, P. R. (1989). *Nucl. Acids Res.,* **17,** 1266.

5

Molecular analysis of immunodeficiency viruses

THOMAS F. SCHULZ and BRUNO SPIRE

1. Introduction

The rapid expansion of our knowledge of human and animal immunodeficiency viruses over the last decade would not have been possible without the availability of methods developed by molecular biologists. In this chapter we summarize some of the more commonly used molecular biological techniques employed with these viruses. As it would go beyond the remit of this chapter to list every technique that could fall under this definition, we refer the reader to some of the excellent molecular biology manuals (1, 2) for the more basic methods.

The term immunodeficiency virus usually refers to animal or human lentiviruses, although a number of other viruses can also cause transient states of immunodeficiency. *Table 1* contains a list of these lentiviruses. From a technical point of view, the techniques discussed in this chapter can be applied to all the viruses in this group, although some details (for example tissue culture conditions) may differ with individual viruses or virus strains. Most of the protocols in this chapter are based on our experience with the human immunodeficiency viruses HIV-1 and HIV-2.

2. Culturing primate immunodeficiency viruses and monitoring virus production

In this chapter we will only touch on techniques used to grow immunodeficiency viruses. Lentiviruses exhibit a wide range of properties *in vivo* and *in vitro*. So-called 'virus isolates' or 'virus strains' are always a mixture of closely related viruses and the representation of individual viruses in such a mixture can change after prolonged culture *in vitro* (3). In addition, growth of some of these viruses in human T-cell lines can select for variants which are not present in the original isolate made in primary peripheral blood mononuclear cells (PBMC). For example, HIV-2 and SIV_{mac} (simian immunodeficiency virus from macaques) isolates adapted to grow in human T-cell

Table 1. Human and animal lentiviruses

Human lentiviruses	
HIV-1	
HIV-2	
Primate lentiviruses	
SIV_{mac}	Cause of AIDS-like disease in captive macaques, not found in free-living animals. Related to HIV-2 and SIV_{sm}.
SIV_{sm}	Isolated from sootey mangabeys. Related to HIV-2 and SIV_{mac}. This virus, or a close relative, is thought to represent the ancestors of HIV-2 and SIV_{mac}.
SIV_{agm}	Isolated from free living African green monkeys (vervets), no disease association known.
SIV_{mn}	Isolated from mandrills, no disease association known.
SIV_{cpz}	Isolated from a few chimpanzees. Related to HIV-1 isolates. Possible ancestor of HIV-1?
Other lentiviruses	
Visna virus	Cause of visna/maedi in sheep
EIAV	Equine infectious anaemia virus
CAEV	Caprine arthritis and encephalitis virus
FIV	Feline immunodeficiency virus, cause of an immunodeficiency disease in cats
BIV	Bovine immunodeficiency virus

lines such as H9, c8166, Sup T1, HUT78, and CEM often contain premature stop codons in their transmembrane (TM) envelope protein gp36 (4, 5). In the case of SIV_{mac}, this truncated TM protein results from the adaptation of SIV_{mac} isolates to human cell lines, and it reverts upon passage of this virus *in vivo* (6). Therefore, the possibility of culture-induced changes should be kept in mind when growing lentiviruses *in vitro* to obtain nucleic acid preparations for cloning experiments.

HIV isolates which are adapted to growth in human T-cell lines usually produce a cytopathic effect (c.p.e.) and kill the cells after a few days in culture. Such cultures have to be re-fed continuously with uninfected cells to maintain virus production. We usually add four parts of uninfected cells (at about 5×10^5 cells/ml) to one part of infected cells plus four parts of tissue culture medium. Virus production is usually maximal about 48–72 h later, and this is also the optimal time point for preparing DNA (see Section 3). As an alternative, 'chronic producer cell lines' can be established. These are cell lines derived from a subpopulation of acutely infected cells which have become resistant to the c.p.e. of the virus and are therefore able to outgrow the lytically infected cells. The H9 line, a subclone derived from the HUT78 T-cell line (7), as well as CEM and MOLT4 are often used successfully in attempts to establish a chronic producer line. Chronically infected cell lines continue to produce virus even in the absence of c.p.e. Virus production can also be boosted by co-cultivation with uninfected cells of the same line.

The extent of syncytia formation is a poor guide to the amount of virus produced in culture. The latter can be assayed by determining the infectious titre of the tissue culture supernatant or by measuring the level of viral core antigen or reverse transcriptase (RT) activity in the supernatant. The infectious titre and RT or p24 levels need not necessarily correlate. Assay kits for HIV-1 p24 antigen, the major core protein, are available commercially (Abbot, Dupont). These kits are very expensive, however, and a much cheaper, but just as sensitive, method (8) is outlined in *Protocol 1*. This p24 assay is easier to perform than measuring RT and does not require radioactive isotopes. For HIV-2 and the SIVs, however, the available p24 assays are not sufficiently sensitive, and we prefer to use an RT method which is described in *Protocol 2*. This protocol is particularly useful if many samples have to be handled at the same time. The background is much lower than with other methods (usually below 100 c.p.m.). As some cellular DNA polymerases can also catalyse this reaction if present in high concentrations, it may be advisable to include a control template (i.e. poly dA/oligo dT) which is recognized preferentially by these polymerases.

Many primary isolates of HIV-1 or HIV-2 will grow only in human PBMC or primary human monocyte cultures, and in these cases virus production is best monitored by p24 or RT assay.

Protocol 1. ELISA for HIV-1 p24

Materials

- ELISA plates (Dynatech)
- antibody D7320[a] (Aalto Bioreagents)
- Tris-buffered saline (TBS): 25 mM Tris–HCl (pH 7.5), 144 mM NaCl
- dried skimmed milk (DSM)[b]
- Empigen (Surfachem)
- alkaline phosphatase-conjugated EH12 monoclonal antibody against p24[c]
- Ampak ELISA detection kit (Dako Diagnostics)
- alkaline phosphatase-conjugated affinity-purified antibody against human IgG (Seralab)

Method

1. Coat the ELISA plates with 100 μl/well of antibody D7325 diluted to 5 μg/ml in 100 mM $NaHCO_3$ (pH 8.5). Adsorb the antibody overnight.
2. Block the plates twice for 10 min in TBS containing 2% (w/v) DSM, using 250 μl/well.

Protocol 1. *Continued*

3. Dilute the samples[d] in TBS containing 0.1% (v/v) Empigen and add 100 μl of diluted sample to each coated well. If a standard curve is desired, use a dilution series of recombinant p24 protein and apply 0.01, 0.03, 0.1, 0.3, 1, or 10 ng/well in a total of 100 μl/well. Leave the plates at room temperature for 2–3 h.
4. Wash the plates four times with 250 μl/well of TBS.
5. Add 100 μl/well of conjugated EH12 antibody diluted 1:2000 in TBS containing 2% DSM, 0.1% (v/v) Tween 20, 10% (v/v) sheep serum. Leave for 1 h.
6. Wash the plates four times in the wash buffer provided in the ELISA detection kit.
7. Add 50 μl/well of substrate buffer from the detection kit. Leave for 1 h.
8. Add 50 μl/well of amplifier buffer from the detection kit. Stop the reaction with 50 μl/well of 0.5 M HCl after 10 min or as soon as the desired colour intensity has developed.
9. Read ELISA plates at OD_{490}.

[a] This polyclonal sheep antibody was raised against and affinity-purified on a synthetic peptide derived from HIV-1 p24.

[b] Not all commercial brands of dried skimmed milk give equally good results. We prefer Marvel (Premier Brands Ltd).

[c] This conjugated monoclonal antibody, originally described in ref. 9, is available in the UK through the AIDS Directed Programme Reagent Repository. If this antibody is not available, a patient's serum with a good antibody titre against p24 can be used instead at a dilution of 1:1000–2000 in the buffer used in step 5. In this case, at step 5 it is necessary to incubate the plates with the patient's serum for 1 h at room temperature and then, after washing four times with 250 μl/well of TBS, to add 100 μl/well of conjugated antibody against human IgG diluted 1:2000 in the buffer used in step 5. Then continue with step 6.

[d] To reduce background to a minimum, tissue culture supernatants should be diluted at least 1:5, equivalent to 20 μl/well of neat supernatant. For accurate quantitation of viral antigen it is often advisable to set up several dilutions in replicate so that readings will be obtained in the linear range of the ELISA.

Protocol 2. RT assay

Materials

- NTE: 100 mM NaCl, 10 mM Tris–HCl (pH 7.5), 1 mM EDTA
- 40% (w/v) polyethylene glycol 8000 (PEG) in NTE
- 5 OD_{260} U/ml poly rA:oligo dT
- 1 mCi/ml (40 Ci/mmol) [^{3}H]dTTP
- 2 × SSC: 300 mM NaCl, 30 mM trisodium citrate
- DE81 paper (Whatman)

Method

1. Pellet cellular debris from 1.5 ml of the supernatant from each infected culture at 12 000 *g* for 5 min in a benchtop microcentrifuge. Add 1 ml of the cleared supernatant to 380 μl of 40% PEG. Mix and incubate overnight in an ice bath.

2. Centrifuge the mixture in a benchtop microfuge at 12 000 *g* for 10 min and remove the supernatant. Wash the pellet twice with 1 ml of ice-cold NTE and resuspend it in 20 μl of NTE by vortexing or pipetting vigorously.

3. For each sample to be assayed, mix the following on ice:

• 1 M Tris–HCl (pH 7.8)	2.5 μl
• 0.1 M $MgCl_2$	2.5 μl
• 0.5 M KCl	2.0 μl
• 0.1 M DTT	1.0 μl
• 2% (v/v) Triton X-100	1.0 μl
• water	21 μl
• poly rA:oligo dT	5 μl
• [^{3}H]dTTP	5 μl

4. Aliquot 40 μl/well of the reaction solution into a 96-well plate and add 10 μl of PEG-precipitated virus[a] samples to the wells. Incubate the plate for 2 h at 37°C.

5. Harvest radioactively labelled DNA on DE81 paper in a 96-well plate harvester and wash with 2 × SSC.

6. Determine the level of RT in each sample by counting bound radioactivity in a scintillation counter.

[a] The amount of PEG-precipitated virus can be varied from 1 to 10 μl. Adjust volume of water accordingly.

3. Molecular cloning of immunodeficiency viruses

The lentiviruses discussed in this chapter have an RNA genome of about 10 kb. A single virus particle carries two copies of a positive-stranded RNA molecule which is polyadenylated at its 3′ end and carries a repeated region, termed R, at both ends and unique regions at the 5′ (U5) and 3′ ends (U3). After entry into the cell, the RNA genome serves as a template for the virus-encoded RT. For the details of reverse transcription, the reader should see ref. 10. The resulting double-stranded DNA copy, termed the provirus, has the U3 and U5 regions duplicated at both ends in the order U3–R–U5; this complex is referred to as the long terminal repeat (LTR; see *Figure 1*). The

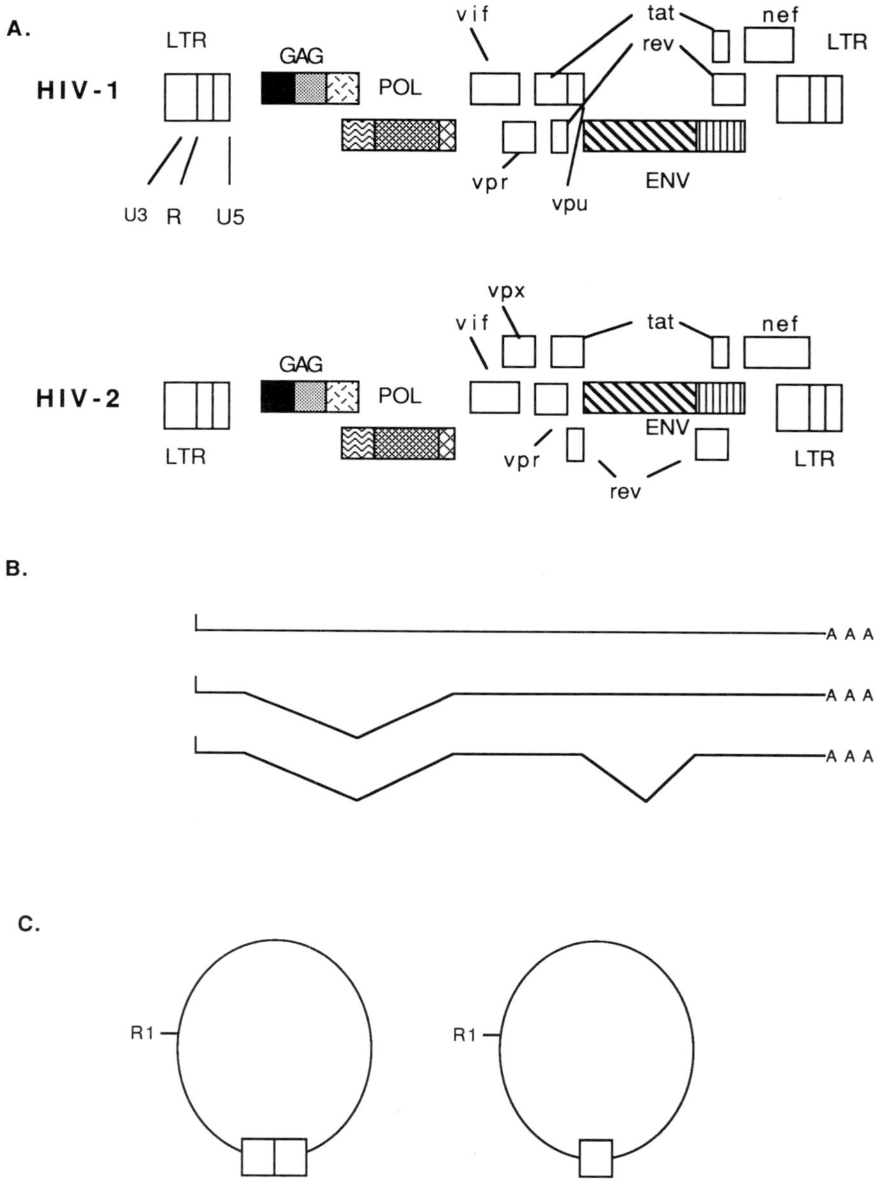

Figure 1. Genome organization of human immunodeficiency viruses. (A) Genome organization of HIV-1 and HIV-2. Boxes denote the individual viral genes. (B) Splicing pattern of viral RNAs. The full-length genomic RNA which also codes for gag and pol proteins, the singly-spliced RNA encoding env and one of several multiply-spliced RNAs encoding control proteins are shown. (C) Structure of circular proviruses containing one or two LTRs.

proviral DNA is then circularized transiently before integrating into the host cell genome in a linear form. The virus thus persists in the infected cell as an integrated copy of proviral DNA. One of the characteristic features of primate lentiviruses is the presence of large amounts of unintegrated proviral DNA in the infected cell a few days after infection. Unintegrated proviral DNA consists of complete circularized proviruses containing two LTRs, in addition to incomplete circularized molecules containing only one LTR and linear molecules containing one or two LTRs.

3.1 Preparing infected cell DNA

In order to obtain full-length clones of primate immunodeficiency viruses it is best to start from DNA and establish libraries from infected cells. Although it is possible to clone from viral RNA by generating cDNA libraries, it is difficult to obtain full-length or nearly full-length clones in this way because of the size of the viral genome. The most commonly used starting material is high molecular weight chromosomal DNA or extrachromosomal unintegrated DNA (so-called Hirt DNA). Methods for preparing these DNAs are given in *Protocols 3* and *4*.

Protocol 3. Extraction of high molecular weight DNA from infected cells

Materials

- 200–300 ml of infected cell suspension (about 1×10^8 cells)
- phosphate-buffered saline (PBS): 170 mM NaCl, 3.4 mM KCl, 10 mM Na_2HPO_4, 1.8 mM KH_2PO_4 (pH 7.2)
- TNE: 10 mM Tris–HCl (pH 8.0), 100 mM NaCl, 10 mM EDTA
- VS membrane (Millipore)

Method

1. Pellet the cells by low speed centrifugation, wash them once in PBS, and discard the supernatant.
2. Resuspend the pellet in 18 ml of TNE and add 2 ml of 10% (w/v) sarcosyl. Add RNase A to 100 μg/ml and incubate at 25°C for 2 h.
3. Add proteinase K to 250 μg/ml and incubate at 55°C in a tightly capped tube for 5–16 h in a waterbath. The solution should be clear at the end of this treatment.
4. Gently extract the solution twice with an equal volume of 1:1 (v/v) phenol: chloroform.
5. Dialyse the DNA extensively, using two or three changes of sterile water (1 litre) for 1 h each and a final change (2 litre) overnight. For measuring the DNA concentration it is advisable to dialyse a small aliquot even more

Protocol 3. *Continued*

extensively, using a VS membrane, as the presence of EDTA interferes with spectrophotometric determination of DNA concentrations.

6. Check an aliquot (5 μg) of the DNA by electrophoresis on a 0.5% (w/v) agarose gel. DNA should migrate as a single band of high relative molecular mass (M_r) without smearing. Store the DNA at 4°C.

Protocol 4. Extraction of unintegrated viral DNA (Hirt DNA) from infected cells

Materials

- 50 ml of infected cells (at about 5×10^5 cells/ml)
- TE: 10 mM Tris–HCl (pH 8 and pH 7.4), 1 mM EDTA
- 20% (w/v) sodium dodecyl sulphate (SDS)

Method

1. Co-cultivate the cells for 3 days with four parts of uninfected cells. Harvest the culture by low speed centrifugation, wash the cell pellet once in PBS and discard the supernatant.
2. Loosen the cell pellet by tapping the tube on the bench and resuspend the cells in 8 ml of TE (pH 8.0). Transfer the resuspended cells into a Beckman SW41 or equivalent ultracentrifuge tube.
3. Add 300 μl of 20% SDS, cover the tube and mix by inversion. Add 2 ml of 5 M NaCl, mix by inversion, and place the tube on ice for 2 h to allow precipitation of chromosomal DNA.
4. Centrifuge the tube in a Beckman SW 41 rotor or equivalent at 35 000 r.p.m. (210 000 *g*) for 2 h at 4°C.
5. Remove the supernatant without disturbing the pellet and extract it gently three times with phenol:chloroform and once with chloroform.
6. Add two volumes of ethanol and leave at −20°C overnight to precipitate DNA. Centrifuge in Corex tubes at 10 000 r.p.m. (14 000 *g*) and wash the pellet three times in 70% (v/v) ethanol. Dry the pellet briefly and resuspend it in 200–300 μl of TE (pH 7.4).
7. Add RNAse A to 100 μg/ml and incubate at 25°C for 2 h. Extract the solution twice with phenol:chloroform and once with chloroform. Ethanol precipitate the DNA. Pellet the DNA by centrifugation at 12 000 *g* for 15 min at 4°C. Wash the pellet twice with 70% ethanol, dry it briefly, and resuspend the DNA in 100 μl of TE (pH 7.4).

3.2 Cloning infected cell DNA

When starting with high M_r chromosomal DNA, the aim is to clone an integrated provirus. Ideally, therefore, the DNA is digested with a restriction enzyme which does not cut within the viral genome but whose site is present in commercially available bacteriophage λ vectors. A fragment of genomic DNA containing a full provirus would be expected to have a size of about 12–15 kbp, but may be larger. For inserts of this size, we prefer vectors such as EMBL 3 and EMBL 4. Restriction enzymes which can be used with these vectors are *Bam*HI, *Eco*RI, and *Sal*I. Unfortunately, many HIV-1 and HIV-2 isolates contain these sites internally. In these cases, it is advisable to establish the presence of internal restriction sites in the provirus by probing a Southern blot of the genomic DNA used as starting material. This also indicates the amount of viral DNA present. *Protocol 5* describes the strategy to be followed if none of the restriction sites to be used for cloning is present in the isolate.

Protocol 5. Separation of restricted DNA on sucrose gradients

Materials

- 150 μg of infected cell DNA obtained in *Protocol 3*
- 10% and 40% (w/v) sucrose in 20 mM Tris–HCl (pH 8.0), 1 M NaCl, 5 mM EDTA

Method

1. Digest the DNA to completion, or partially (*Protocol 7*), with a suitable restriction enzyme. Use 5 U of enzyme/μg DNA in a total volume of 200–300 μl for 16 h at 37°C.
2. Check for complete digestion by electrophoresing an aliquot (1 μg) of digested DNA on a 0.5% (w/v) agarose gel.
3. Extract the digested DNA twice with phenol:chloroform and once with chloroform. Ethanol precipitate the DNA and wash the pellet twice in 70% (v/v) ethanol. Allow the DNA to dry, then dissolve it in 500 μl of TE (pH 7.4).
4. Prepare a 38 ml linear 10–40% sucrose gradient in a Beckman SW27 ultracentrifuge tube. Heat the DNA sample for 10 min at 68°C and leave it to cool to room temperature for 10 min. Load the DNA on to the gradient and centrifuge at 26 000 r.p.m. (118 000 *g*) for 20 h at 16°C.
5. Collect 0.5 ml fractions from the bottom of the tube and electrophorese 20 μl aliquots fractions from the bottom half of the gradient on a 0.5% agarose gel. Use λ DNA digested with *Hin*dIII as markers, mixed with 30% (w/v) sucrose and made 1 M in NaCl.

Protocol 5. *Continued*

6. Pool the fractions containing DNA in the 12–15 kbp range and dilute with two volumes of TE (pH 7.8). Ethanol precipitate the DNA, wash in 70% ethanol, dry and resuspend it in 100 μl of TE (pH 7.4). Adjust the DNA concentration to 500 μg/ml.

DNA of the desired size is ligated into phage λ arms digested with the same restriction enzyme as was used to digest the cellular DNA (*Protocol 6*). The ligated DNA is then packaged into λ phage using commercially available packaging extracts according to the manufacturer's instructions.

Protocol 6. Preparation of phage arms and ligation to restricted DNA from infected cells

Materials

- an appropriate λ DNA vector (for example EMBL 3 or 4; Stratagene)
- digested infected cellular DNA from *Protocol 5*
- 5 × ligase buffer: 250 mM Tris–HCl (pH 7.5), 50 mM $MgCl_2$, 5 mM ATP, 50 mM DTT

Method

1. Anneal the molecular ends of 50 μg of phage DNA by incubating it in 200 μl of 10 mM $MgCl_2$ for 1 h at 42°C. Compare an aliquot on a 0.5% agarose gel with a sample which has been heated for 5 min at 68°C. Digest the annealed DNA with the desired restriction enzyme, using 5 U/μg of phage DNA for 1 h. Heat an aliquot of 0.5 μg to 68°C for 5 min to dissociate the annealed ends and check for complete digestion on a 0.5% agarose gel. Extract the digested DNA once with phenol:chloroform and once with chloroform. Ethanol precipitate the DNA and wash the pellet in 70% ethanol. Resuspend it in 0.5 ml of TE (pH 7.4). Reanneal the phage DNA by incubating for 1 h at 42°C in the presence of 10 mM $MgCl_2$ and check it by electrophoresing an aliquot on a 0.5% agarose gel.
2. Load the digested λ DNA on to a 10–40% sucrose gradient as described in step 4 of *Protocol 5* and centrifuge the gradient at 26 000 r.p.m. (118 000 *g*) for 20 h at 15°C. Collect 0.5 ml fractions from the bottom of the tube and analyse them as described in step 5 of *Protocol 5*. If the annealing was successful, the small and large arms should be present in the same fractions, well separated from the smaller internal fragments. Pool the fractions containing the phage arms and adjust the concentration to 1 μg/μl.

3. Mix 1 μg of purified phage arms with 2 μg of cellular DNA in a total volume of 10 μl of 1 × ligase buffer containing 2.5 U of T4 DNA ligase. Incubate overnight at 16°C in a screw-capped tube. Store the ligated DNA at −20°C.

If all the restriction enzymes that can be used with a given phage vector cut within the viral genome, it is necessary to prepare a genomic library from partially digested DNA as described in *Protocol 7*. In order to obtain an optimal representation of the total genome, three different sets of conditions are determined in which incubation time and enzyme concentration are varied.

Protocol 7. Partial digestion of chromosomal DNA from infected cells

Materials

- infected cell DNA prepared according to *Protocol 3*

Method

1. Add 10 μg of DNA to 15 μl of the appropriate 10 × restriction enzyme buffer and adjust the volume to 150 μl with water. Dispense 30 μl into the first of eight 1.5 ml tubes and 15 μl into the remaining seven. Place the tubes on ice.
2. Add 4 U of restriction enzyme to the first tube, mix and transfer 15 μl of the reaction mixture to the second tube. Mix and transfer 15 μl into the third tube. Repeat this until the seventh tube is reached. Do not add anything to the eighth tube. This gives 4 U of enzyme/μg of DNA in the first tube, 2 U/μg in the second tube, and so on. Incubate the tubes at 37°C for 1 h.
3. Stop the reactions by adding EDTA to a final concentration of 20 mM and analyse the digested DNAs on a 0.5% agarose gel electrophoresed at 1–2 V/cm. Visualize the ethidium bromide-stained gel and identify the samples showing the most intense fluorescence around 15–20 kbp.
4. Repeat the same experiment twice, incubating the samples for 30 min and 2 h, respectively.
5. Digest three aliquots of 50 μg of chromosomal DNA, using the same incubation times as above but only half the amount of enzyme/μg of DNA that produced the most intense fluorescence in steps 3 and 4. Check the digestion products on a 0.5% agarose gel.
6. Pool the partially digested DNA samples obtained using these conditions and proceed to *Protocols 5* and *6*.

Another alternative is to clone from non-integrated circular viral DNA. In this case the aim is to linearize a circular proviral DNA with a unique restriction enzyme which can be used for cloning (i.e. *Eco*RI, *Bam*HI or *Sal*I for EMBL 3 or EMBL 4) and, as a consequence, to clone it as a circularly permuted genome. Extrachromosomal DNA, prepared from infected cells as described in *Protocol 4*, is digested to completion and fragments of 10–15 kbp are separated on a sucrose density gradient and ligated into the phage vector as described in *Protocols 5* and *6* for genomic DNA. *Protocol 8* explains how to plate and screen such a library.

3.3 Screening and growing recombinant λ clones

Probes for screening a library can be obtained by restriction enzyme digestion of plasmid or phage clones of a similar or closely related lentivirus. Generally, all HIV-1 isolates are sufficiently closely related to allow detection with a hybridization probe generated from one of the commonly used strains (for example IIIB, LAV, SF-2). HIV-2 isolates also cross-hybridize with each other as well as with SIV_{mac} and SIV_{sm} isolates. SIV_{agm} isolates can be quite divergent and may only cross-hybridize under conditions of reduced stringency. If no suitably related cloned virus is available, hybridization probes can be generated from purified virus particles using the 'endogenous RT reaction' to prime the synthesis of the first strand of a cDNA, and finishing and cloning the cDNA by conventional methods. This approach is described in ref. 11. It is technically more difficult and is used only for cloning a novel retrovirus.

If the library has been established from partially digested genomic DNA, at least 10^6 plaques must be screened. If the starting material was completely digested DNA, screening 3×10^5 plaques should be sufficient to obtain the desired clone. Positive clones obtained in the first round of screening are plaque-purified by repeated plating of appropriately diluted phage eluted from positive plaques, as described in *Protocol 8*. For manipulating the cloned provirus it is often necessary to subclone the insert from the phage clone into a plasmid vector. *Protocol 9* describes methods which usually generate sufficient amounts of phage DNA for this purpose.

Protocol 8. Plating and screening a library

Materials

- *E. coli* strain C600ΔHFL (Stratagene)
- LB broth: 10 g/l bactotryptone, 5 g/l yeast extract, 5 g/l NaCl (autoclaved)
- SM buffer: 50 mM Tris–HCl (pH 7.5), 10 mM $MgSO_4$, 10 mM NaCl
- ligated DNA from *Protocol 6*
- λ packaging extract (for example Gigapack Gold from Stratagene)

- top agarose: 8 g/l NaCl, 10 g/l bactotryptone, 10 mM $MgSO_4$, 8 g/l agarose (autoclaved)
- LB-agar: 15 g/l agar in LB (autoclaved)
- Colony/Plaque Screen membranes (DuPont)
- an appropriate [^{32}P]DNA probe

Method

1. Inoculate a single bacterial colony into 50 ml of LB broth supplemented with 0.2% (w/v) maltose and 10 mM $MgSO_4$ and grow overnight at 37°C. In our experience, bacteria grown to high density produce the best results. Determine the OD_{600} of the culture. Pellet the bacteria for 2000 *g* for 10 min, resuspend at 4 OD_{600} U/ml in SM buffer and keep them on ice.
2. Package the ligated DNA and titrate the library according to the manufacturer's instructions. We usually pool six packaged ligations, using three ligation reactions, since each ligation reaction is sufficient for two packaging reactions. Do not amplify the library, as this may cause loss of slowly replicating phages.
3. Mix 7.5 ml of bacteria (about 30 OD_{600} U) and 3×10^5 plaque-forming units (p.f.u.) of phage and incubate for 20 min at room temperature. Aliquot the absorbed bacteria into 30 15 ml polypropylene tubes. Melt the top agarose and cool it to 50°C, add 12 ml to each tube, mix by swirling and pour the suspension on to a 140 mm LB-agar plate. Allow the agarose to set at room temperature, incubate the plates for 10–12 h at 37°C and store them at 4°C overnight. Each plate should contain about 10 000 plaques.
4. Screen the library by DNA hybridization. We prefer Colony/Plaque Screen membranes, used according to the manufacturer's instructions.
5. Remove an agar plug containing a positive plaque from the plate using a Pasteur pipette and place it into 500 µl of SM buffer. Incubate for 2 h at 4°C to elute the phage. Plate the eluted phage on 90 mm LB plates as described in step 3, using 0.4 OD_{600} U/plate of bacteria. In order to obtain the desired density of plaques, set up several dilutions for each positive clone. Several rounds of rescreening are usually required to obtain a pure population of positive phage. It is possible to hybridize the membranes in the same solution as was used for the primary screen.

Protocol 9. Growth and purification of phage

Materials

- a single plaque from a pure positive λ clone
- λ DNA purification kit (Quiagen)

Protocol 9. *Continued*

- GTG agarose (FMC)
- STM buffer: 50 mM Tris–HCl (pH 7.5), 10 mM $MgSO_4$, 100 mM NaCl

Method

A. *Growing a phage stock*

1. Elute phage from the plaque into 0.5 ml of SM buffer and adsorb it to 0.4 OD_{600} U of bacteria (step 5 of *Protocol 8*). Apply the bacteria to a single 90 mm plate and incubate at 37°C overnight. The plate should show confluent lysis.
2. Add 10 ml of SM buffer and shake the plate gently for 2 h at 4°C. Retain the surface liquid and repeat this step.
3. Add chloroform to a final concentration of 0.3% (v/v) and pellet bacterial debris by centrifuging at 5000 *g* for 10 min. Titrate the supernatant (step 5 of *Protocol 8*).

B. *Growth of phage in large scale culture for DNA preparation*

1. Prepare bacteria as in step 1 of *Protocol 8* and resuspend them at 80 OD_{600} in SM buffer. Adsorb phage at a ratio of 6×10^6 p.f.u./OD_{600} U of bacteria
   in a small volume for 20 min at room temperature. Add LB broth supplemented with 0.2% maltose and 10 mM $MgSO_4$ to give a final bacterial concentration of 0.5 OD_{600} U/ml. Incubate the culture at 37°C.
2. After 4 h, the culture should become clear. Add 1 μl/ml of chloroform and purify phage DNA using a λ DNA purification kit according to the manufacturer's instructions.

Some phage grow very slowly and do not lyse the culture efficiently under the conditions described above. In this case, increase the amount of phage added in step 1. Alternatively, use the following plate lysis method.

C. *Isolation of DNA from phage plate lysates*

1. Pour two 140 mm plates containing 1.5% (w/v) GTG agarose, 10 mM $MgSO_4$ and 0.2% maltose in LB. Adsorb phage for 20 min to 250 μl of bacteria (4 OD_{260} U/ml) adjusted to a total volume of 500 μl with SM buffer, and apply to the plates in top agarose containing GTG agarose instead of agarose as described in step 3 of *Protocol 8*. Incubate the plates at 37°C overnight.
2. Elute the phage from each plate in 10 ml of SM buffer by incubating overnight at 4°C. Recover the surface liquid and add a few drops of chloroform. Alternatively, scrape the top agarose into 10 ml of SM buffer containing 3% (v/v) chloroform and elute the phage for at least 2 h at 4°C.

3. Clarify the lysate by centrifugation at 10 000 *g* for 10 min and then elute from a 20 ml column of DEAE–Sephadex A-50 in STM buffer. Recentrifuge the eluate at 10 000 *g* to remove debris.
4. Pellet the phage by centrifuging at 100 000 *g* for 2 h at room temperature (for example 26 000 r.p.m. in a Beckman SW28 rotor). The pellet should have a glassy appearance.
5. Resuspend the pellet in 500 μl of STM buffer and incubate with 100 μg/ml RNAse for 10 min at 37°C. Lyse the phage by adding 167 μl of 4% SDS and heating at 70°C for 10 min. Add 667 μl of SM buffer containing 200 μg/ml proteinase K and incubate for at for at least 1 h at 56°C.
6. Cool the lysed phage to room temperature, extract extensively with phenol:chloroform, and add ammonium acetate to a final concentration of 2 M, and ethanol precipitate. Wash the pelleted DNA twice with 70% ethanol and dissolve it in TE (pH 8.0) overnight.

3.4 Isolating plasmid clones

Unfortunately, it is not rare to observe deletions and recombinations in the cloned provirus after attempting to clone it into a plasmid vector. One possible cause of this is the presence of two identical LTRs in the insert. There are several ways to get round this problem. One way is to subclone the phage insert as two partially overlapping restriction fragments which can then be manipulated separately. In order to obtain virus from proviral DNA subcloned in this way, it is necessary to linearize the plasmids containing the two fragments with the same restriction enzyme (which must cut only once within the provirus), treat one of the digested plasmids with phosphatase to prevent self-ligation, ligate the two linearized plasmids together and transfect. The choice of restriction enzymes depends on the individual provirus. Ref. 12 gives an example of this approach which can be modified to suit individual circumstances.

Another way is to exploit the fact that many LTRs contain unique restriction sites absent from the rest of the viral genome. It is, therefore, possible to isolate an insert from a cloned full-length provirus which contains only a partial LTR on either end. Circularizing such an insert by ligation reconstitutes one complete LTR. Linearizing this circular genome at a unique restriction site outside the LTR results in a permuted molecular clone carrying one part of the viral genome in the opposite orientation from the other part (*Figure 2*). In this form, the proviral DNA cannot generate infectious virus after transfection into cells. This is a great advantage from the point of view of safety considerations which are necessary in dealing with infectious molecular clones (see Section 7). However, infectious DNA can easily be regenerated from permuted proviruses by ligating them to form long concatamers. Within these concatamers some proviral molecules will be positioned such that a complete

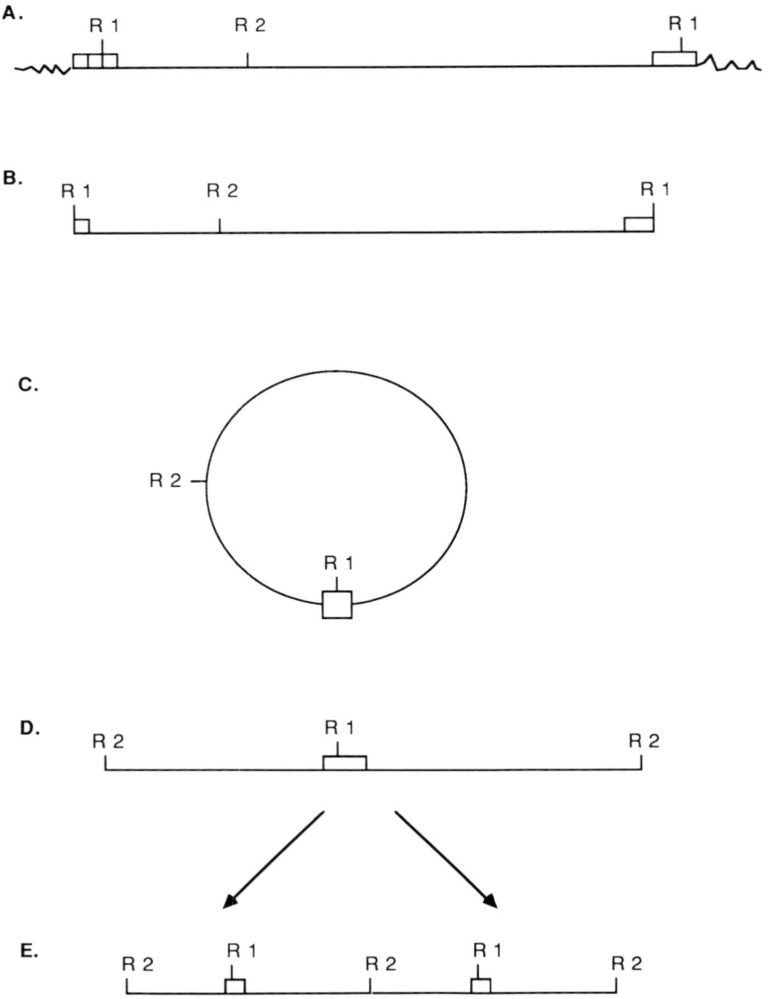

Figure 2. Strategy to clone a permuted proviral genome. (a) Cloned integrated provirus with two LTRs and flanking cellular DNA. (B) Provirus lacking parts of its LTRs, obtained from (A) by digestion with the hypothetical restriction enzyme R1. (C) Circularization by ligation of the provirus shown in (B). (D) Digestion of the circularized provirus shown in (C) with hypothetical restriction enzyme R2 leads to a circularly permuted linear provirus with one reconstituted LTR. (E) Ligation of several of the permuted forms reconstitutes the original genome organization in at least some concatamers.

proviral coding region is re-established, thus allowing transcription of functional viral RNA molecules. As only a proportion of the ligated proviral DNA molecules will be positioned correctly, the virus yield after transfection is lower with this method than after transfection of an infectious plasmid or phage. This is not a problem when transfecting into permissive T-cell lines which can replicate and amplify the virus emerging after transfection. It may, however, be a handicap under certain circumstances, for example if one is analysing virus genomes which have been rendered non-infectious by the introduction of mutations in critical regions and are therefore incapable of reinfecting certain tissue culture cells.

4. Polymerase chain reaction

As in many other areas of biomedical research, the polymerase chain reaction (PCR), originally described by Saiki *et al.* (13), has been employed for many purposes in molecular virology, including diagnostic applications, attempts to clone parts of the viral genome, mapping of RNA splice sites, and site-directed mutagenesis. For the purpose of this chapter we assume that the reader is familiar with the basic principles of PCR. PCR can be performed using DNA or RNA, and methods for these two applications are given in *Protocols 10* and *11*.

Protocol 10. PCR using DNA

Materials

- 10 × *Taq* buffer: 100 mM Tris–HCl (pH 8.3), 500 mM KCl, 20 mM $MgCl_2$, 200 μg/ml bovine serum albumin (BSA)
- 10 mM dNTP solution: 10 mM each of dATP, dTTP, dCTP, and dGTP
- 1 mg/ml oligonucleotide primers (see Section 4.1 and *Table 2*)
- DNA template (plasmid or chromosomal)

Method

1. For a 50 μl PCR reaction, mix the following:

• 10 × *Taq* buffer	5 μl
• 10 mM dNTP solution	1 μl
• each primer	120–250 ng
• *Taq* polymerase	2.5 U
• water to a final volume of	30 μl[a]

2. Add the DNA template in a volume of 20 μl. Use 0.2–1 μg of genomic DNA or up to 1 ng of plasmid DNA. Mix thoroughly and overlay with 70–100 μl of paraffin oil in order to prevent evaporation. Handle the positive control template last, after the other tubes have been capped.
3. Subject the reaction mixtures to 30 cycles in a thermal cycler. Times and temperatures will vary depending on the primers and the size of the

Protocol 10. *Continued*

amplified product, but the following gives a reasonable guideline for a reaction yielding a product approximately 200–1000 bp in size: denaturation temperature, 96°C for 1 min; annealing temperature around 5°C below the T_m[b] for 1 min; extension temperature 72°C for 2 min. Prolong the extension step of the last cycle for 7 min.

4. Check 25 μl of the reaction mixture on a 1–1.5% agarose gel.

[a] As a precaution against contamination, this mixture can be irradiated for 5 min on a short wavelength UV transilluminator.

[b] $T_m = 4 \times$ (number of G+C residues) + $2 \times$ (number of A+T residues), with respect to the primer sequence.

Protocol 11. PCR using RNA

Materials

- 0.75–2.5 mg/ml total cell RNA
- 1 mg/ml antisense oligonucleotide primer[a]
- 10 × annealing buffer: 100 mM Hepes–NaOH (pH 6.9), 2 mM EDTA
- 5 × RT buffer: 250 mM Tris–HCl (pH 7.5), 375 mM KCl, 15 mM $MgCl_2$, 50 mM DTT
- 200 U/μl Moloney murine leukaemia virus RT

Method

1. Mix the following, heat at 90°C for 2 min, and then chill on ice for 5 min.
   - RNA 2 μl
   - antisense primer 0.1 μl
   - 10 × annealing buffer 1 μl
   - water 7 μl
   - overlay with paraffin oil 2 drops
2. For each sample, mix the following. Add 10 μl of the mixture to 10 μl of annealed RNA and incubate at 37°C for 90 min.
   - 5 × RT buffer 4 μl
   - 10 mM dNTP solution 1 μl
   - RT 0.1 μl
   - water 5 μl
3. Amplify 5 μl of the cDNA from step 2 by adding the following and performing 30 PCR cycles (see *Protocol 10*).

- 10 × *Taq* buffer 5 μl
- 10 mM dNTP solution 1 μl
- each primer 0.2 μl
- water 38 μl
- *Taq* polymerase 2.5 U

4. Carry out step 4 of *Protocol 10*.

[a] Alternatively, the cDNA synthesis can be primed with oligo dT or a random primer.

The sensitivity of PCR can be increased even further by re-amplifying the product of a standard PCR in a second PCR using internally nested primers. This method, usually referred to as nested PCR, allows the detection of a single target molecule per reaction tube, provided that optimal pairs of primers are used.

4.1 Diagnostic applications and analysis of virus variability

PCR is enjoying increasing popularity for detecting viral nucleic acids in patients' or laboratory samples. The choice of PCR primers obviously depends on the individual experiment. *Table 2* lists just a few primers from conserved regions of the viral genome (for example *gag* or *pol*) which have been used successfully to demonstrate the presence of viral DNA or RNA in a sample. For the analysis of the variability of virus populations, variable regions should be amplified using primers anchored in neighbouring conserved regions. Some examples for primers which are useful for this application are also listed in *Table 2*, and refs 14 and 15 serve as a further guide to regions which are informative in this respect.

In order to verify that the amplification product corresponds to the target sequence and is not spurious, we usually analyse it by Southern blotting and probing with an end-labelled internal oligonucleotide (*Protocol 12*).

Protocol 12. Abbreviated Southern transfer for PCR products

Materials

- nylon membrane: for example Genescreen (Dupont), Hybond N (Amersham)
- 10 × T4 kinase buffer: 0.5 M Tris–HCl (pH 7.5), 0.1 M $MgCl_2$, 50 mM DTT, 0.5 mg/ml BSA or gelatin
- 10 U/μl T4 polynucleotide kinase
- 20 × SSPE: 3.6 M NaCl, 0.2 M NaH_2PO_4, 20 mM EDTA (pH 7.4)

Protocol 12. *Continued*

- 50 × Denhardt's solution: 1% (w/v) BSA, 1% (w/v) Ficoll, 1% (w/v) polyvinylpyrrolidone
- prehybridization buffer: 5 × SSPE, 5 × Denhardt's solution, 0.5% (w/v) SDS, 50 μg/ml salmon sperm DNA

Method

1. Electrophorese the PCR samples on a 1–1.5% agarose gel, depending on the expected size of the products.
2. Wash the gel twice in 0.4 M NaOH for 20 min and blot for 2 h or overnight in 0.4 M NaOH on to a nylon membrane.
3. Rinse the nylon membrane in 2 × SSC and then expose it to short wavelength UV radiation for 3 min to crosslink the DNA.
4. Incubate the membrane in prehybridization buffer in a plastic bag (or hybridization bottle) for 1–3 h at 65 °C.
5. To obtain an oligonucleotide probe, end label the oligonucleotide with T4 polynucleotide kinase by mixing the following. Incubate at 37 °C for 45 min, then separate the labelled oligonucleotide from unincorporated label using a spun column containing Sepharose G-50 (1).

• the appropriate oligonucleotide	100 ng
• 10 × T4 kinase buffer	2 μl
• [γ-^{32}P]ATP	5 μl
• T4 polynucleotide kinase	1 μl
• water to a total volume of	20 μl

6. Add the probe in 5 ml of prehybridization buffer to the prehybridized membrane and incubate with shaking (or rotating) overnight at a temperature 5 °C below the T_m (see *Protocol 10*).
7. Wash the filter twice in 2 × SSPE containing 0.1% SDS for 10 min at room temperature, then once in 1 × SSPE for 20 min at room temperature. If the filter contains an excessive background level of radioactivity, wash it in 0.5 × SSPE for 1 h at 5 °C below the T_m and, if necessary, in 0.25 × SSPE at the same temperature.
8. Expose the membrane to X-ray film.

When analysing several patients' or tissue culture samples at the same time, the conventional DNA isolation technique (*Protocol 3*) is too laborious, although it produces DNA of higher quality. The simplified DNA extraction method given in *Protocol 13* produces DNA of sufficient quality to be used in PCR. It is especially useful when isolating DNA from small numbers of cells. For isolating RNA from small numbers of cells (fewer than 10^6), we use a

Table 2. Some successfully used PCR primers for HIV-1 and HIV-2

Virus	Primer name (position)	Primer sequence	Product length	Region	Ref.
HIV-1	P1 (764–786)	5′-TACATCAGGCCATATCACCTAC-3′			16
	P2 (1041–1066)	5′-TGAAGGGTACTAGTACTTCCTGC-3′	(with P1) 302 bp	Gag	
	P3 (2356–2381)	5′-TGGGAAGTTCAATTAGGAATACCAC-3′			
	P4 (2637–2663)	5′-CCTACATACAAATCATCCATGTATTC-3′	(with P3) 307 bp	Pol	
	P5 (4085–4103)	5′-ATTAGCAGGAAGATGGCC-3′			
	P6 (4207–4335)	5′-TACTCCTTGACTTTGGGG-3′	(with P5) 140 bp	Pol	
HIV-1	I (325–344)	5′-GAAGGAGAGAGATGGGTGCG-3′			17
	II (518–537)	5′-CCCTGTCTATGTTGGTAGGG-3′	(with I) 212 bp	Gag	
HIV-1	881 (1214–1234)	5′-GGTACATCAGGCCATATCACC-3′			18
	882 (1652–1669)	5′-ACCGGTCTACATAGTCTC-3′	(with 816) 455 bp	Gag	
	883 (1407–1425)	5′-GAGGAAGCTGCAGAATGG-3′			
	990 (1620–1646)	5′-GGTCCTTGTCTTATGTCCAGAATGCTG-3′	(with 817) 239 bp	Gag	
HIV-1	SK29 (501–518)	5′-ACTAGGGAACCCACTGCT-3′			19
	SK30 (589–605)	5′-GGTCTGAGGGATCTCTA-3′	(with SK29) 104 bp	LTR	
	SK38 (1551–1578)	5′-ATAATCCACCTATCCCAGTAGGAGAAAT-3′			
	SK39 (1638–1665)	5′-TTTGGTCCTTGTCTTATGTCCAGAATGC-3′	(with SK38) 114 bp	Gag	
	SK68 (7801–7820)	5′-AGCAGCAGGAAGCACTATGG-3′			
	SK69 (7922–7942)	5′-CCAGACTGTGAGTTGCAACAG-3′	(with SK68) 141 bp	Env	
	CO1 (7855–7864)	5′-ACAATTATTGTCTGGTATAG-3′			
	CO2 (7970–7989)	5′-AGGTATCTTTCCACAGCCAG-3′	(with CO1) 79 bp	Env	
HIV-1	CO11 (8462–8442)	5′-CGGGCCTGTCGGGTTCCCTC-3′			20
	CO12 (5990–6009)	5′-CTTAGGCATCTCCTATGGCA-3′	(with CO11) 127 bp	Tat	
HIV-2	SK100 (1132–1150)	5′-ATCAAGCAGCCATGCAAAT-3′			21
	SK104 (1401–1422)	5′-CCTTTGGTCCTTGTCTTATGTC-3′	(with SK100) 290 bp	Gag	
	SK89 (9432–9449)	5′-AGGAGCTGGTGGGGAACG-3′			
	SK90 (9577–9596)	5′-GTGCTGGTGAGAGTCTAGCA-3′	(with SK89) 146 bp	LTR	
HIV-2	Tat/1(5834–5854)	5′-CTATACTAGACATGGAGACA-3′			22
	Tat/2 (6082–6062)	3′-CTATACAATACTTGCTTTCC-5′	(with Tat/1) 249 bp	Tat	
	TM/1 (8070–8090)	5′-GAACAGGCACAAATTCAGCAA-3′			
	TM/2 (8330–8310)	5′-GTATAGGTGTTCCTGGCCCCT-3′	(with TM/1) 221 bp	Env	

modified phenol/guanidinium isothiocyanate method, marketed as the RNAzol kit by Cinna/Biotecx.

Protocol 13. Rapid DNA extraction for PCR

Materials

- lysis buffer: 10 mM Tris–HCl (pH 8.3), 50 mM KCl, 2.5 mM $MgCl_2$, 0.5% (v/v) NP40, 0.5% (v/v) Tween 20, 50 μg/ml proteinase K

Method

1. Add 100 μl of lysis buffer to approximately 10^5 cells. Resuspend the cells by pipetting up and down.
2. Incubate for 1 h at 56°C and inactivate the proteinase K by heating the samples to 95°C for 10 min.
3. Use 5–25 μl for PCR (*Protocol 10*).

4.2 Cloning PCR products

Cloning PCR amplification products into plasmid vectors follows the standard protocols of molecular cloning. However, PCR fragments tend to clone much more inefficiently than conventional restriction fragments. This is largely due to the uneven ends produced by *Taq* polymerase which render blunt-end ligation more difficult. *Protocol 14* describes how to repair these uneven ends.

Protocol 14. Cloning blunt-ended PCR products

Materials

- TM buffer: 100 mM Tris–HCl (pH 8.0), 50 mM $MgCl_2$
- 2 mM dNTP solution: 2 mM each of dATP, dTTP, dCTP, and dGTP
- 5 U/μl Klenow fragment of *E. coli* DNA polymerase I
- an appropriate plasmid vector linearized with a restriction enzyme which produces blunt ends and resuspended at 50 μl/ml

Method

1. Separate PCR products on an agarose gel, cut out the product of interest and elute it from the gel using Geneclean (Bio101). Resuspend the DNA in 7 μl of water.
2. Treat the DNA with T4 polynucleotide kinase and the Klenow fragment of *E. coli* DNA polymerase I. For the kinase reaction, mix the following and incubate at 37°C for 30 min:

- eluted PCR product 7 μl
- 10 × T4 kinase buffer 1 μl
- 100 mM ATP 2 μl
- T4 polynucleotide kinase 0.5 μl

Then add the following and incubate for a further 20 min at 37°C. Inactivate the Klenow fragment by heating at 85°C for 5 min.

- TM buffer 1.5 μl
- 2 mM dNTP solution 0.5 μl
- water 2.5 μl
- Klenow fragment 0.5 μl

3. Add 7.5 μl of blunt-ended PCR product to 0.5 μl of digested plasmid, 1 μl of 10 × ligation buffer, 1 μl of 10 mM ATP, and 0.5 μl (5 U) of T4 DNA ligase and incubate for 16 h at 16°C.
4. Transform 5 μl of the ligated mixture into competent bacteria by standard methods (1). As blunt-ended PCR fragments clone rather inefficiently, use highly competent bacteria.

Another alternative is to include restriction sites in the primers which will be incorporated into the amplification product. For this purpose we usually design primers such that they include 15–18 bases at the 3′ end matching the target sequence, a convenient restriction site and about 6 additional bases at the 5′ end. The latter will absorb many of the imperfect ends produced by *Taq* polymerase and will thus permit digestion of the included restriction site. As some of the uneven ends in PCR products are due to dA residues added by *Taq* polymerase, another alternative is to clone PCR fragments into dT-tailed vectors. Ref. 23 describes how to prepare these vectors.

4.3 Direct sequencing of PCR products

PCR products can be sequenced directly (i.e. without cloning), using about 200 ng of DNA and one of the PCR primers, or other internal primers, to prime the sequencing reactions. A method for this, which is essentially identical to standard protocols of double-stranded sequencing, but includes 10% DMSO in all reaction steps, is given in ref. 24. Another method, relying on the generation of single-stranded sequencing template molecules, using *Taq* polymerase and an excess of one primer over the other, can be found in ref. 25. The advantage of directly sequencing a PCR product is that occasional errors introduced by *Taq* polymerase do not affect the resulting sequence, since they occur on a minority of molecules and therefore are not apparent. As this approach also eliminates the need for cloning PCR products, it has become very popular.

5. Analysis of viral gene function

The functional analysis of individual viral genes is usually first approached by constructing deletion mutants and testing their biological phenotype. Since the proviral genome can be handled as a DNA clone, such deletions are usually introduced by standard recombinant DNA techniques, using suitable restriction sites. Another approach is to introduce more specific mutations by site-directed mutagenesis. Ref. 2 contains protocols which are relevant to this approach.

On the other hand, the generation, by recombinant DNA technology, of viral proteins which are functionally active *in vitro* has played a major role in analysing the function of different viral gene products. Without going into too much technical detail, this section is intended to give a rough outline of what can be achieved with different expression systems. The reader should refer to specific publications for further information.

5.1 Core proteins

Expression of the complete HIV-1 *gag* gene in *E. coli*, insect cells, or mammalian cells (using stably or transiently transfected cell lines as well as vaccinia vectors) leads to a functionally active gag precursor. Co-expression of the viral proteinase gene, located at the 5′ end of the *pol* gene, results in processing, by proteolytic cleavage, of the gag precursor protein, p55, to the three core proteins, p17, p24, and p15. The recombinant gag precursor proteins are capable of assembling into viral capsid particles which can bud from producer cells as non-enveloped particles. Budding requires the presence of an amino-terminal glycine residue, which has to be myristylated to allow insertion into the cell membrane. Studies of this kind have shown that proteolytic processing of p55 is not necessary for budding but is required for morphological maturation of nascent capsid particles, that is, condensation of the viral core to give the classical appearance found in mature virions. Refs 26–28 serve as examples for these studies.

5.2 Pol proteins

The three pol-derived enzymes of HIV, proteinase, RT, and integrase, can be expressed in a functionally active form in *E. coli*. The commercially available expression vector pKK233-2 can be used to express unfused RT (29). Plasmids of the pGEX series (30) can be used to generate a fusion protein between glutathione S-transferase and integrase, from which functionally active integrase can be excised using thrombin or factor Xa. As these enzymes are normally produced from the gag–pol precursor proteins by proteolytic cleavage, their individual expression requires the introduction of start and stop codons into the expression vector. This can be achieved easily using PCR, employing a 5′ oligonucleotide incorporating an ATG codon and a 3′

oligonucleotide incorporating a stop codon. Introduction of these artificial sequences does not interfere with the functional activity of the recombinant protein.

5.3 Env proteins

Unlike gag- and pol-derived proteins, it is not possible to express functionally active envelope proteins in *E. coli*, as glycosylation is an important prerequisite for correct folding of envelope proteins in the endoplasmic reticulum. The envelope proteins are synthesized as precursor glycoproteins of about 160 kDa which assemble to form tetrameric structures in the endoplasmic reticulum. The outer envelope glycoprotein, gp120, is then cleaved from the precursor complex by a cellular proteinase and remains non-covalently associated with the transmembrane protein complex, gp41. The rev protein controls the extent of splicing and its presence is required to produce sufficient amounts of full-length unspliced genomic RNA and of the 4.5 kb env-encoding RNA (*Figure 1*). Expression of gp160 in stable mammalian cell lines therefore requires the concomitant expression of rev. In the case of HIV-1 this has been achieved by inserting a genomic fragment encompassing the first and second *rev* exon, as well as the *env* coding region downstream from the human cytomegalovirus immediately early gene promoter. Constructs of this kind can be used to transiently transfect COS cells, provided that they contain the SV40 origin of DNA replication. To increase production levels, an amplifiable selection marker, such as the dihydrofolate reductase or glutamine synthetase gene, can be inserted into such a plasmid, and stably transfected chinese hamster ovary cell lines can then be established in the presence of methotrexate or methylsulphoximide, respectively (see refs 31 and 32 for examples). The outer envelope glycoproteins of HIV-1 and HIV-2, gp120 and gp105, respectively, can be expressed on their own, presumably because they are not covalently associated with the TM proteins. Expression of gp120 does not depend on the presence of rev as does the expression of gp160, presumably because the target site for rev is located in the region of the RNA coding for gp41.

Functionally active HIV-1 gp120 and HIV-2 gp105 have also been expressed in insect cells (34, 35) and vaccinia-based expression systems (36). In these systems, co-expression of rev is not required.

5.4 Expression of diagnostic antigens

Short immunogenic fragments of all viral structural proteins can be expressed easily in a variety of expression systems, including *E. coli*, and can provide antigens which are useful for serodiagnostic purposes. In the case of the HIV-1 envelope, residues 55–65, 135–148, 303–339, 504–518, 583–609, 654–666, and 735–752 have been identified as B-cell epitopes (reviewed in ref. 37). Not every region, however, is recognized by all patients' sera. Similar regions in the HIV-2 envelope can also be used, in particular residues 230–388 and 530–769 (38, 39).

5.5 Analysis of viral proteins

A large number of techniques are used to analyse the function and some of the biochemical properties of viral proteins, but it would be impossible to list them all in this chapter. The two most commonly used techniques are immunoprecipitation of radioactively labelled viral proteins and the Western blot. While core proteins are readily detected by immunoprecipitation or immunoblotting, envelope proteins are more easily visualized by immunoprecipitation. A method for the labelling and immunoprecipitation of virus-producing cells is given in *Protocol 15*.

Protocol 15. Isotopic labelling and immunoprecipitation of virus-producing cells

Materials

- virus-producing suspension cells (approx. 5×10^5 cells/ml)
- lysis buffer: 10 mM Tris–HCl (pH 7.6), 140 mM NaCl, 2 mM $MgCl_2$, 1 mM DTT, 1 mM phenylmethylsulphonyl fluoride (PMSF), 1% (v/v) NP40
- buffer A: 20 mM Tris–HCl (pH 7.6), 500 mM NaCl, 1 mM EDTA, 1% (w/v) sodium deoxycholate, 0.5% (v/v) NP40
- buffer B: 10 mM Tris–HCl (pH 7.6), 10 mM NaCl
- SDS loading buffer: 50 mM Tris–HCl (pH 6.8), 2% (w/v) SDS, 1% (v/v) 2-mercaptoethanol, 10% (v/v) glycerol, a trace of bromophenol blue

Method

1. Co-cultivate 10 ml of virus-producing cells with 40 ml of uninfected cells for 2–3 days. On the day before labelling, add 50 ml of tissue culture medium.
2. Pellet the cells, resuspend them in 10 ml of cysteine- and/or methionine-free medium, and incubate at 37°C for 30 min. Pellet the cells and resuspend them in 5 ml of cysteine- and/or methionine-free medium to which 1–2 mCi of [^{35}S]methionine and/or [^{35}S]cysteine have been added. Label the cells for 4–6 h (or overnight) in a 5 cm plate at 37°C in the presence of 5% (v/v) CO_2.
3. Pellet the cells and retain the supernatant, which can be used to label a second batch of cells or from which nascent virus may be immunoprecipitated. Wash the cells once in 10 ml of PBS and resuspend them in 2 ml of ice-cold lysis buffer. Incubate on ice for 10 min, pellet the cells in a bench microcentrifuge, and retain the supernatant. Re-extract the pellet with

1 ml of lysis buffer. Centrifuge the combined supernatants at 12 000 *g* for 10 min at 4°C. Freeze aliquots of the cleared extracts at −20°C.

4. Add 20–50 μl of labelled cell extract to 50 μl of lysis buffer diluted 1:3 in PBS and add BSA to 200 μg/ml. Mix on ice and add 1–3 μl of antiserum, patient's serum, monoclonal ascites, or 5 μg of purified antibody. Incubate on ice for at least 4 h. Add 50 μl of a 20% suspension of protein A–Sepharose in lysis buffer using a cut-off pipette tip and incubate for 1 h on a rotating wheel at 4°C.
5. Pellet the Sepharose beads for 2 min in a microcentrifuge and wash them three times with 0.5 ml of buffer A and twice with 0.5 ml of buffer B. Elute the beads with 30 μl of SDS–PAGE loading buffer by heating at 95°C for 5 min.
6. Subject samples to SDS–PAGE and stain the gel with Coomassie brilliant blue. Destain, enhance (Amplify, Amersham International Ltd), dry, and autoradiograph the gel.

An ELISA to measure the amount of viral envelope protein produced in culture can be performed in a similar manner to the p24 ELISA described in *Protocol 1*, except that antibody D7324 (to HIV-1 gp120) or D7335 (to HIV-2 gp105) should substitute for D7320 and the appropriate patients' sera should be used (40). Both antibodies are available from Aalto Bioreagents. A further variant of this ELISA can be used to measure the interaction between the HIV envelope proteins and their receptor, CD4 (33).

6. Transfecting full-length proviral DNA and generating virus from cloned DNA

A replication-competent (infectious) molecular clone of an immunodeficiency virus can be transfected into a variety of cells to generate virus. We often use HeLa or COS cells because they are easy to transfect and they replicate human immunodeficiency viruses very well, although they do not express CD4. We use the calcium phosphate precipitation or DEAE–dextran methods (*Protocol 16*). We achieve the highest transfection efficiency with a commercially available liposome preparation, lipofectin (Gibco-BRL).

Although COS and HeLa cells are easily transfectable, the level of virus production is rather low compared to the virus titre obtained from standard laboratory strains grown in established T-cell lines. There is, however, enough virus present after transfection to allow detection in a p24 assay for HIV-1 or in an infectivity assay. In order to obtain higher virus titres, it is necessary to co-cultivate the transfected COS or HeLa cells with an appropriate susceptible cell line. We usually start co-cultivation 2–3 days after transfection and, for transfected HIV-1 or HIV-2 genomes, use H9, 8166, MOLT4, or CEM cell lines. After co-cultivating for 2–3 days, we passage the

T-cell until syncytia become apparent or a p24 antigen or RT assay becomes strongly positive.

Protocol 16. DNA transfection techniques

Materials

- adherent cells (for example COS, HeLa, 3T3) seeded on the day prior to the transfection at approximately 5×10^5 cells/28 cm^2 flask in Dulbecco's minimal essential medium (DMEM) containing 10% (v/v) fetal calf serum (FCS)
- 25 mg/ml DEAE–dextran
- plasmid DNA: ethanol-precipitated, washed twice with 70% ethanol, and dissolved in sterile water
- DMSO solution: DMEM containing 50 mM Hepes–NaOH (pH 7.4), 10% (v/v) DMSO

Method

A. *Transfection with DEAE–dextran*

1. Wash the subconfluent cells twice with DMEM. Mix 2 ml of DMEM, 30 μl of DEAE–dextran, and 5–10 μg of plasmid DNA. Add the mixture to the drained cells and incubate for 4 h at 37°C.
2. Replace the mixture with DMEM containing 10% FCS. Alternatively, the cells can be DMSO-boosted after removing the DEAE–dextran mixture in order to enhance the transfection efficiency. To do this, replace the medium with 2 ml of DMSO solution. After 2 min, wash twice with DMEM and add DMEM containing 10% FCS.
3. Incubate the cells for 1–2 days. If secreted proteins are to be analysed or if the culture is to be labelled with radioactive amino acids, the medium can be replaced with DMEM or labelling medium on the day after transfection and the cultures incubated for another day.
4. Harvest the cells 2 days after transfection if expressed proteins are to be analysed. If the aim is to recover virus from the transfection, the transfected cells can be co-cultivated with suitable T-cell lines or phytohaemaglutinin-stimulated peripheral blood lymphocytes at this point.

B. *Transfection using calcium phosphate precipitation*

1. Precipitate 10–20 μg of plasmid DNA with calcium phosphate (see Chapter 8, *Protocol 6*).
2. Replace the culture medium with the precipitated DNA. After 10 min, add 8 ml of prewarmed DMEM containing 10% FCS, and incubate for 4–16 h at 37°C. The precipitated DNA suspension has a more adverse

effect on certain cell lines and should be replaced with DMEM containing 10% FCS after 4 h. Other cell lines (for example HeLa, 3T3) will tolerate the precipitate for 16 h.

3. Subject the transfected cells to DMSO boost if required (see step A2).
4. Carry out steps A3 and A4.

Replication-competent molecular clones can also be transfected directly into appropriate T-cell lines (for example H9, MOLT4, or CEM). For this we use the DEAE–dextran (*Protocol 16*) or lipofectin methods. T-cell lines transfected in this way with HIV-1 and HIV-2 genomes may have to be cultured for about 1–3 weeks before virus production, as assessed by syncytia formation or positive p24 antigen or RT assays, is detectable. This time span is quite variable, however, as some infectious molecular clones will lead to detectable virus production only a few days after transfection.

7. Safety considerations

Since many immunodeficiency viruses are human pathogens, safety precautions must be introduced for some experiments covered in this chapter. The cloning of subgenomic proviral DNA frgments in non-expression vectors and the expression of individual viral genes in prokaryotic as well as eukaryotic vectors can be carried out in ACGM category 1 laboratories, and require standard precautions appropriate for experiments with recombinant DNA. Since full-length proviral genomes can result in the production of infectious virions when transfected into mammalian cells, however, transfection of such genomes must be done under ACGM and ACDP category 3 conditions, as must the initial extraction of chromosomal or Hirt DNA from infected cells. Cloning of full-length viral genomes can be done under ACGM category 2 conditions. As it is possible to induce infection with SIV_{mac} in macaques by intramuscular injection of several hundred micrograms of proviral DNA (41), it is advisable to avoid the use of sharp instruments when dealing with large amounts of cloned viral DNA (for example when preparing large amounts of plasmids). We, therefore, use commercially available disposable affinity columns (Quiagen) for plasmid preparations instead of carrying out conventional purification by caesium chloride gradient centrifugation.

References

1. Maniatis, T., Fritsch, E. F., and Sambrook, J. (ed.) (1989). *Molecular cloning: a laboratory manual*. Cold Spring Harbor Press, Cold Spring Harbor, NY.
2. Ausubel, F., Brent, R., Kingston, R. E., Moore, D. D., Seidman, J. G., Smith, J. A., and Struhl, K. (ed.) (1987). *Current protocols in molecular biology*. John Wiley & Sons, NY.

3. Vartanian, J. P., Meyerhans, A., Åsjö, B., and Wain-Hobson, S. (1991). *J. Virol.,* **65,** 1779.
4. Guyader, M., Emerman, M., Sonigo, P., Clavel, F., Montagnier, L., and Alizon, M. (1987). *Nature,* **326,** 662.
5. Chakrabarti, L., Guyader, M., Alizon, M., Daniel, M., Desrosiers, R. C., Tiollais, P., and Sonigo, P. (1987). *Nature,* **328,** 543.
6. Hirsch, V. M., Edmondson, P., Murphey-Corb, M., Arbeille, B., Johnson, P. R., and Mullins, J. I. (1989). *Nature,* **341,** 573.
7. Mann, D. L., O'Brien, S., Gilbert, D. A., Reid, Y., Popovic, M., Read-Connole, E., Gallo, R. C., and Gazdar, A. F. (1989). *AIDS Res. Hum. Retroviruses,* **5,** 253.
8. Moore, J. P., McKeating, J., Weiss, R. A., and Sattentau, Q. (1990). *Science,* **250,** 1139.
9. Ferns, R. B., Partridge, J. C., Spence, R. P., Hunt, N., and Tedder, R. S. (1989). *AIDS,* **3,** 829.
10. Goff, S. P. (1990). *J. Acqu. Immunodef. Syndromes,* **3,** 817.
11. Alizon, M., Sonigo, P., Barré-Sinoussi, F., Chermann, J. C., Wain-Hobson, S. (1984). *Nature,* **312,** 757.
12. Hattori, N., Michaels, F., Fargnoli, K., Marcon, L., Gallo, R. C., and Franchini, G. (1990). *Proc. Natl Acad. Sci. USA,* **87,** 8080.
13. Saiki, R. K., Gelfand, D. H., Stoffel, S., Scharf, S. J., Higuchi, R., Horn, G. T., Mullis, K. B., and Erlich, H. A. (1988). *Science,* **239,** 487.
14. Simmonds, P., Balfe, P., Ludlam, C. A., Bishop, J. O., and Brown, A. J. (1990). *J. Virol.,* **64,** 5840.
15. Balfe, P., Simmonds, P., Ludlam, C. A., Bishop, J. O., and Brown, A. J. (1990). *J. Virol.,* **64,** 6221.
16. Laure, F., Courgnaud, V., Rouzioux, C., Blanche, S., Veber, F., Burgard, M., Jacomet, C., Griscelli, C., and Brechot, C. (1988). *Lancet,* **ii,** 538.
17. Loche, M. and Mach, B. (1988). *Lancet,* **ii,** 418.
18. Simmonds, P., Balfe, P., Peutherer, J. F., Ludlam, C. A., Bishop, J. O., and Brown, A. J. L. (1990). *J. Virol.,* **64,** 864.
19. Ou, C.-Y., Kwok, S., Mitchell, S. W., Mack, D. H., Sninsky, J. J., Krebs, J. W., Feorino, P., Warfield, D., and Schochetman, G. (1988). *Science,* **239,** 295.
20. Hart, C., Schochetman, G., Spira, T., Lifson, A., Moore, J., Galphin, J., Sninsky, J., and Ou, C.-Y. (1988). *Lancet,* **ii,** 596.
21. Rayfield, M., De-Cock, K., Heyward, W., Goldstein, L., Krebs, J., Kwok, S., Lee, S., McCormick, J., Moreau, J. M., Odehouri, K., Schochetman, G., Sninsky, J., and Ou, C.-Y. (1988). *J. Infect. Dis.,* **158,** 1170.
22. Schulz, T. F., Whitby, D., Hoad, J. G., Corrah, T., Whittle, H., and Weiss, R. A. (1990). *J. Virol.,* **64,** 5177.
23. Marchuk, D., Drumm, M., Saulino, A., and Collins, F. S. (1991). *Nucl. Acids Res.,* **19,** 1154.
24. Winship, P. R. (1989). *Nucl. Acids Res.,* **17,** 1266.
25. Gyllensten, U. B. and Erlich, H. A. (1988). *Proc. Natl Acad. Sci. USA,* **85,** 7652.
26. Gheysen, D., Jacobs, E., de Foresta, F., Thiriart, C., Francotte, M., Thines, D., and de Wilde, M. (1989). *Cell,* **59,** 103.
27. Gowda, S. D., Stein, B. S., Engleman, E. G. (1989). *J. Biol. Chem.,* **264,** 8459.
28. Shioda, T. and Shibuta, H. (1990). *Virology,* **175,** 139.
29. Larder, B., Purifoy, D., Powell, K., and Darby, G. (1987). *EMBO J.,* **6,** 3133.

30. Smith, D. B. and Johnson, K. S. (1988). *Gene,* **67,** 31.
31. Berman, P., Riddle, L., Nakamura, G., Haffar, O. K., Nunes, W. M., Skehel, P., Byrn, R., Groopman, J., Matthews, T., and Gregory, T. (1989). *J. Virol.,* **63,** 3489.
32. Stephens, P. E. and Cockett, M. I. (1989). *Nucl. Acids Res.,* **17,** 7110.
33. Moore, J. P., McKeating, J. A., Jones, I. M., Stephens, P. E., Clements, G., Thomson, S., and Weiss, R. A. (1990). *AIDS,* **4,** 307.
34. Morikawa, Y., Overton, H. A., Moore, J. P., Wilkinson, A. J., Brady, R. L., Lewis, S. J., and Jones, I. M. (1990). *AIDS Res. Hum. Retroviruses,* **6,** 765.
35. Morikawa, Y., Moore, J. P., Wilkinson, A. J., and Jones, I. M. (1991). *Virology,* **180,** 853.
36. Chakrabarti, S., Robert-Guroff, M., Wong-Staal, F., Gallo, R. C., and Moss, B. (1986). *Nature,* **320,** 535.
37. Goudsmit, J. (1988). *AIDS,* **2** (suppl. 1), 41.
38. Schulz, T. F., Oberhuber, W., Hofbauer, J. M., Hengster, P., Larcher, C., Gürtler, L., Tedder, R. S., Wachter, H., and Dierich, M. P. (1989). *AIDS,* **3,** 165.
39. Huang, M. L., Essex, M., and Lee, T.-H. (1991). *J. Virol.,* **65,** 5073.
40. Moore, J. P. and Jarrett, R. F. (1988). *AIDS Res. Hum. Retroviruses,* **4,** 369.
41. Letvin, N. L., Lord, C. I., King, N. W., Wyand, M. S., Myrick, K. V., and Haseltine, W. A. (1991). *Nature,* **349,** 573.

6

Retroviral vectors

ANDREW W. STOKER

1. Introduction

Retroviruses can transfer genetic material efficiently into vertebrate cells, and incorporate it stably into the cellular genome. Many differentiated and undifferentiated cell types of rodents, birds, and primates can be infected, and high viral expression levels can be achieved. Furthermore, retroviral gene transfer is effective *in vivo* from early embryonic stages onwards, as well as in cultured cells. Given this versatility, retroviruses are excellent tools for gene transfer in experimental animal systems, and are potential vehicles for gene therapy in humans. Engineered derivatives of naturally-occurring retroviruses, the retroviral vectors, are being used increasingly, and with great success, for the transfer and expression of foreign genes (1–3). This chapter describes the structures of retroviral vectors, their varied applications, and methods for their construction and use.

2. Retroviruses and derived vectors

The principal types of retroviruses that have been used in vector technology are the murine leukaemia viruses (MLV), the avian sarcoma and leukosis viruses (ASLV), and the reticuloendotheliosis viruses (REV). Harvey murine sarcoma virus, mouse mammary tumour virus, and murine myeloproliferative sarcoma virus have also been used. The detailed genetics and biology of these viruses cannot be described here, and readers should consult refs 3 and 4 for more details.

Retroviruses pass through their life cycle as single-stranded (ss)RNA, unintegrated circular or linear double-stranded DNA, and an integrated double-stranded DNA provirus (*Figure 1*). The viral genome contains a number of essential *cis*-acting sequences (*Figure 2*), including transcriptional elements, reverse transcription signals, and packaging sequences which direct the encapsidation of viral RNA into virus particles. The transcriptional control sequences of a retrovirus reside in the long terminal repeat (LTR) elements flanking the genome, which contain enhancer elements, TATA boxes, and polyadenylation signals. One attraction of retroviruses as vectors is the high activity of LTRs in diverse tissues.

Three viral genes, *gag*, *pol*, and *env*, are required in forming virus particles. The genes are carried in *cis* in replication-competent retroviruses and helper-independent vectors (Section 2.1). Viral RNA can, however, equally well use viral proteins supplied in *trans* to assemble virus. This has been pivotal in the development and application of replication-defective (rd) retroviral vectors which lack viral genes (Section 2.2).

The viral genes encode capsid proteins (*gag*), reverse transcriptase (*pol*), and the viral surface glycoprotein (*env*). *Env* determines the species and subspecies specificity of virus infection. In brief, murine ecotropic *env* (of Moloney MLV (Mo–MLV)) allows infection of rodent cells, and amphotropic *env* (of MLV 4070A) can target rodent, primate, and some avian cells. REV virus of *env* subgroup A will infect rodent, avian, and some primate cells. With ASLV there are several *env* subgroups: A is commonly used for retroviral vectors, allowing infection of avian cells only; subgroups D and C (strain Bratislava 77) permit infection of avian and some rodent cells. For more information see ref. 4.

Each retroviral vector described below, whether it retains viral genes or not, has the capacity to carry foreign genes within its genome and to deliver and express them in host cells. These vectors are commonly available, and readers should refer to the references given in the text and in *Figures 3* and *5* for the appropriate sources.

2.1 Replication-competent vectors

Helper-independent vectors RCAN and RCAS (*Figure 3*) are replication-competent (5). They were derived principally from Rous sarcoma virus (RSV), which is unique among ASLV in carrying not only a complete set of

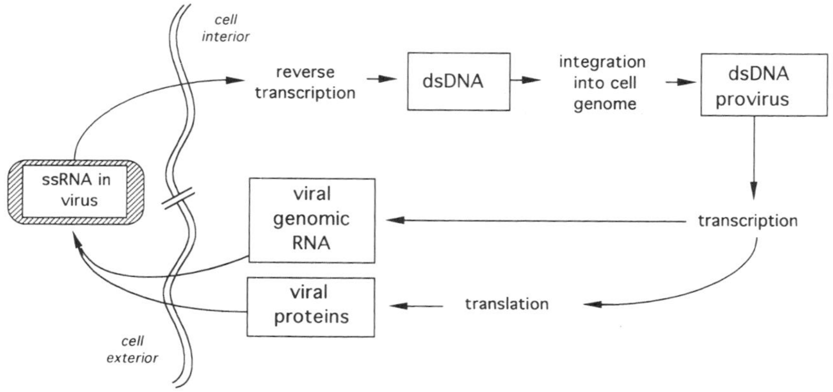

Figure 1. The life cycle of a retrovirus.

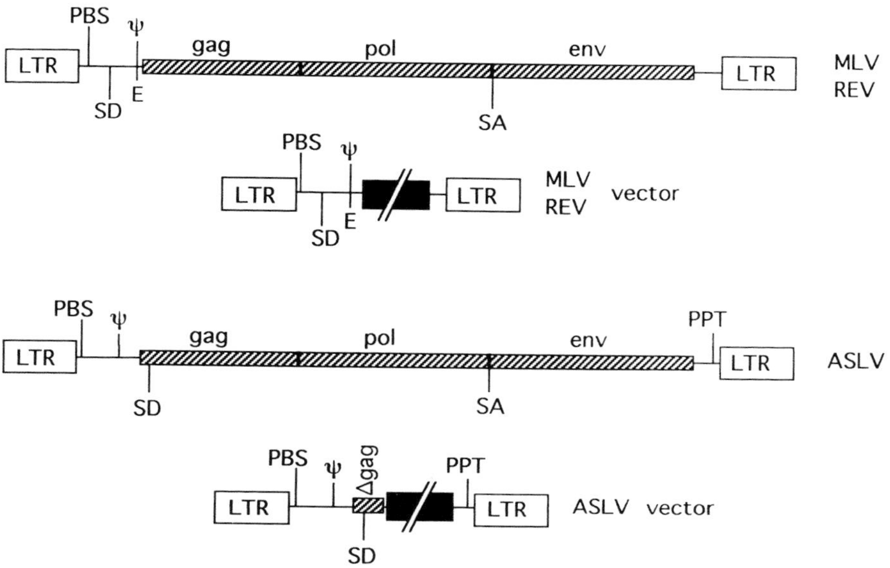

Figure 2. Schematic diagram of retroviruses from the MLV, REV, and ASLV families. Essential *cis*-acting sequences are: LTR, long terminal repeat; PBS, tRNA primer binding site (to prime reverse transcription); ψ, the viral RNA packaging signal (known as E in REV); PPT, polypurine tract. Hatched boxes represent viral genes *gag*, *pol*, and *env*. Black boxes represent foreign gene sequences inserted into replication-defective vectors retaining necessary *cis*-acting regions.

viral genes but also a non-essential 3′ gene, the oncogene v-*src*. V-*src* has been excised and replaced with a *Cla*I insertion site to create vectors which can carry foreign genes of up to 1.5 kbp. Adaptor plasmids facilitate the insertion of foreign gene cassettes into the vector backbones. Small internal promoters can be used in these gene cassettes. RCAN and RCAS are available in *env* subgroups, A, B, and D.

2.2 Replication-defective vectors

Rd vectors are the primary choice for many experimental applications. They have had their viral genes almost entirely removed, offering a large insertion space for non-viral DNA (see Section 5.1). Because rd vectors cannot replicate autonomously, viral proteins must be supplied in *trans*, and methods for achieving this are discussed in Sections 3 and 7. When rd vectors reach a target cell, they are no longer supplied with viral proteins and become captive transgenes in that cell. The benefits of this are discussed in Section 4.

Schematic diagrams of rd vectors and examples of existing vectors are shown in *Figures 4* and *5*. For the purposes of this chapter, the vectors have been classified as type 1–6. Vectors of types 1 and 2 are basic structures with

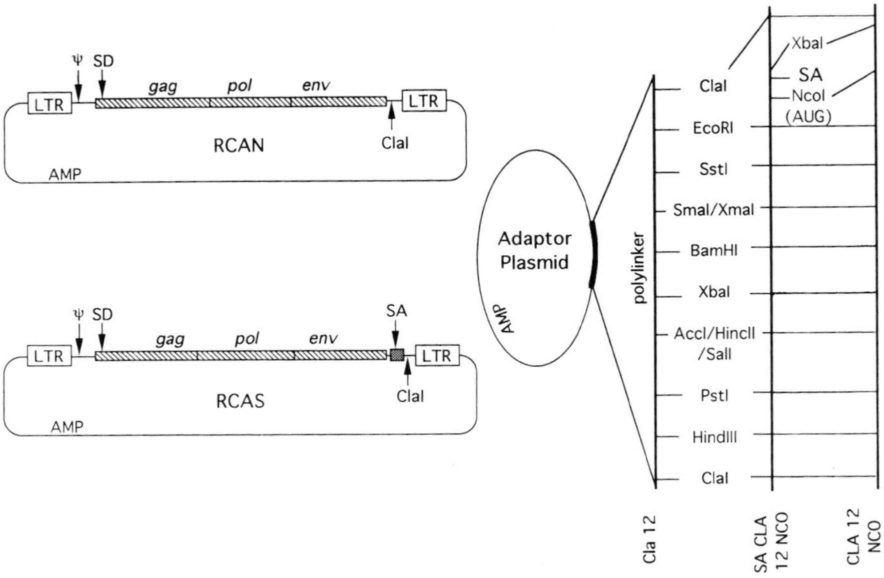

Figure 3. Schematic diagram of helper-independent vectors RCAS and RCAN. Also shown is an adaptor plasmid with three polylinker variations. The indicated sequences are the viral RNA packaging signal (ψ) and the splice donor and acceptor sites (SD and SA). SA is also present in adaptor SA CLA 12 NCO. Adapted from ref. 5.

gene expression under LTR or internal promoter control, respectively. Of particular value are dual expression vectors which can transfer and express two genes, the second of which is expressed from an internal promoter (type 3) or a spliced RNA (type 4). Examples of internal promoters which have been used include the thymidine kinase promoter of herpes simplex virus, the early promoter of SV40, the immediate early gene promoter of human cytomegalovirus, the β-actin promoter and the β-globin promoter. A third kind of dual expression vector has recently been developed (type 5). These express a single mRNA transcript containing two consecutive open reading frames. Both frames are translated from the *same* mRNA owing to the presence of picornavirus sequences that facilitate initiation of translation from internal AUG codons. Type 5 vectors have been successfully developed for Mo-MLV, REV, and ASLV (*Figure 5*, vector LZ1C1 and LNPOZ) (6–8).

The type 6, or self-inactivation (SIN) vectors, lose their LTR enhancer functions upon infection of a host cell (*Figure 5*, vector MT-N) (9). This is achieved by introducing a deletion into the 3′ LTR of the cloned vector. During viral replication in a host cell, the 3′ deletion is duplicated into the 5′ LTR, leaving the vector with no LTR enhancers (*Figure 4*). Thereafter, these vectors rely on internal promoters to express their genes. The principal purpose of SIN vectors is to prevent insertional activation of cellular genes,

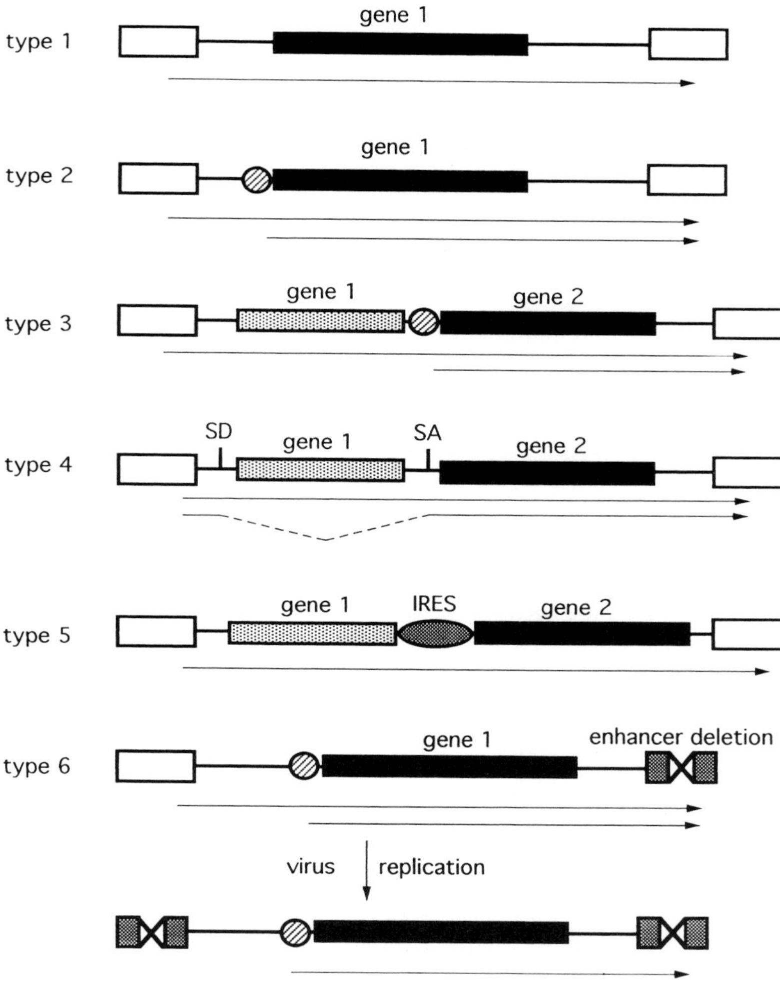

Figure 4. Schematic representation of replication-defective retroviral vectors. Each vector is classified as type 1 through 6, depending upon the number of genes carried and the transcriptional and translational context of each gene. Open boxes indicate the LTRs, the hatched circle the internal promoter, and the oval the picornavirus sequence (IRES) facilitating internal translational initiation of gene 2. Arrowed lines denote RNA transcripts, the dashed line removal of the intron by RNA splicing, and SD and SA the splice donor and acceptor sites.

particularly oncogenes, by LTR enhancers. In addition, SIN vectors can avoid competition between the LTR and internal promoter/enhancers, and the vectors cannot be rescued easily by helper virus. Effective SIN vectors are yet to be developed for ASLV, since removal of the 3′ enhancer apparently debilitates 5′ LTR activity.

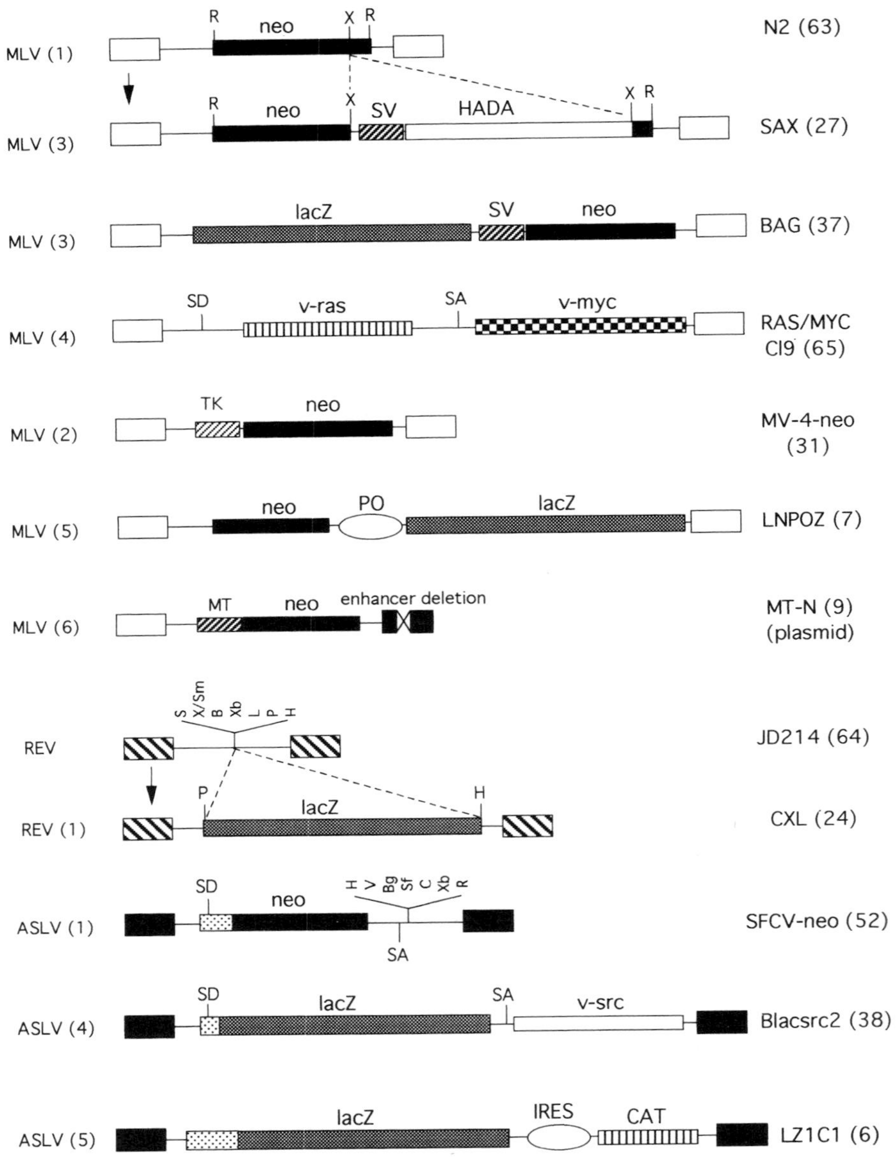

Figure 5. Examples of replication-defective retroviral vectors (not to scale). To the left of each vector, its source (MLV, REV, or ASLV) and type (1–6; see Section 2.2, and *Figure 4*) are indicated. The name of each vector and the relevant reference are given on the right. Restriction enzyme sites shown are: B, *Bam*HI; Bg, *Bgl*I; C, *Cla*I; R, *Eco*RI; H, *Hin*dIII; L, *Sal*I; P, *Pst*I; S, *Sac*I; Sf, *Sfi*I; V, *Eco*RV; X, *Xho*I; X/Sm, *Xma*I/*Sma*I; Xb, *Xba*I. The genes represented encode the following products: *neo*, neomycin phosphotransferase; *HADA*, human adenosine deaminase; *lacZ*, β-galactosidase; v-*ras*, v-*myc*, and v-*src*, viral onco-

3. Packaging cell lines

The success of rd retroviral vectors has depended largely on the development of specialized packaging cell lines to produce infectious virus. These cells are engineered to supply viral proteins in *trans* for rd vectors without the cells releasing replication-competent virus (*Figure 6*; *Table 1*). This is achieved by expressing the viral proteins from RNAs lacking the packaging signal ψ (also called E in REV). An attendant problem with simpler packaging cell designs (ψ2, ψAM, Q2bn, PA12, D17-C3) has been the eventual evolution of replication-competent helper viruses. More recent designs include *gag*/*pol* and *env* genes as separate transcriptional units and LTRs replaced with non-viral promoters and polyadenylation signals. In the cell lines DSN, DAN, PA317, Clone 32, GP+E-86, ψCRIP/ψCRE, and Isolde, therefore, recombination is a very minimal hazard. Titres of 10^5–10^6 infectious viruses/ml can be achieved with all the packaging cell lines. The factor which limits titre is usually the vector and not the cell. See references in *Table 1* for sources of these cell lines.

4. Applications of retroviral vectors

4.1 General considerations

There are several prerequisites when using retroviral vectors for gene transfer. First, the virus must be able to gain physical access to cells in culture or in target tissues. This is straightforward for cells in culture. It can be achieved for cells in tissues by bathing the tissue or organism in a virus suspension or by microinjecting virus at the site of interest (see Section 9). Second, cells must contain surface receptor sites for the viral env glycoprotein (Section 2). There appears to be little cell-type specificity of receptor expression within individual organisms, but there are differences between species and within species (see also Section 2). As examples, receptors for amphotropic MLV env occur in many species, but receptors for ASLV env A occur only in avian species. There are, however, some chicken strains which lack receptors for env subgroup A. Finally, the host cell must undergo at least one round of DNA replication to ensure stable infection. If this does not occur within a few hours of virus adsorption a provirus will not be generated in the host DNA (23). Post-mitotic cells (for example mature neurons) or cells in stationary phase are not stably infectable.

genes; *CAT*, chloramphenicol acetyl transferase. The internal promoters are: SV, SV40 early promoter; TK, thymidine kinase promoter of herpes simplex virus; MT, metallothionine promoter. White ovals represent 5′-nontranslated regions of picornaviruses: PO, poliovirus; IRES, encephalomyocarditis virus. SD and SA indicate the splice donor and acceptor sites. The stippled box in ASLV vectors is residual *gag* gene sequence fused in frame with *lacZ*. Downward-oriented arrows indicate precursor/product relationships.

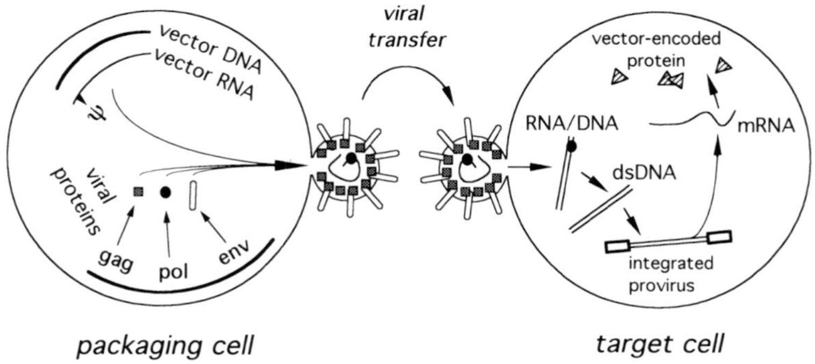

Figure 6. The concept behind retroviral vector packaging cells. The mRNAs encoding gag, pol, and env do not contain the packaging signal ψ and are not incorporated into viruses from the packaging cell. The vector RNA is packaged and transmitted to a target cell, where replication to the proviral stage is completed. The target cell contains no viral proteins and the vector is trapped thereafter.

4.2 Stability of vector expression

Proviral LTRs are active in many cell types. In cultured primary fibroblasts (24) and epithelial cells (25), vector expression can be very stable. Long-term expression of vectors *in vivo* is also documented (2, 25, 26), but there are some exceptions (27). A specific problem is encountered when early mouse embryo cells are infected: the LTRs of MLV-derived viruses are stably inactivated when introduced into cells *in vivo* before day 9 of embryogenesis or into undifferentiated embryonic stem cells in culture (28). Vector expression can be achieved in these cells, however, if an internal promoter is present (vector MV-4 neo, *Figure 5*) (29–32). ASLV vectors will infect and express in chick embryo cells from the blastoderm stage onwards (33). In summary, stable expression of vectors *in vivo* must not be taken for granted, but should be examined on a vector-by-vector and tissue-by-tissue basis.

4.3 Cell lineage marking

Once integrated into a host cell, rd vectors are transmitted as integrated proviruses to all descendent cells. Retroviruses have therefore been used very effectively as genetic markers of cell lineages *in vivo* (34–40). A marker gene, commonly *Escherichia coli lacZ*, facilitates the identification of expressing cells (post fixation) within whole mounted tissues or in cell culture. The BAG vector expresses *lacZ* and has been used widely for such lineage marking (*Figure 5*). *LacZ* encodes β-galactosidase, which can be detected histochemically (*Protocol 4*) and immunologically. The alkaline phosphatase gene has also been used recently as a histochemical marker in vectors (41).

Two advantages of this marking technique are its genetic stability and lack of cellular side effects. Lineage marking has been achieved in most embryo

Table 1. Packaging cell lines

Packaging cell	Type of vector packaged	Permissive host cells for vectors		
		Rodent	Avian	Human
ψ2 (10)[a]	MLV	+	−	−
ψAM (11)	MLV	+	+	+
PA12 (12)	MLV	+	+	+
PA317 (13)	MLV	+	+	+
Clone 32 (14)	MLV	+	−	−
GP + E − 86 (15)	MLV	+	−	−
ψCRIP (16)	MLV	+	+	+
ψCRE (16)	MLV	+	−	−
D17–C3 (17)	REV, MLV	+[b]	+	+/−[c]
DSN (18)	REV, MLV	+[b]	+	+/−[c]
DAN (18)	REV, MLV	+	+	+
pHF–G[d] (19)	ASLV	−	+	−
Isolde[d] (20)	ASLV	−	+	−
Q2bn/Q4dh[d] (21)	ASLV	−	+	−

[a] References are given in parentheses.
[b] Mouse cells are infected poorly.
[c] Some primate cells are permissive for infection and expression, but not productive replication, of REV and its vectors (22).
[d] ASLV packaging cells are currently limited to subgroup A *env*.

tissue types. It is not possible, however, to target retroviral vectors to individual and specific progenitor cells; the resolution of targeting is within groups of cells only. Because of this, care has been taken to ensure that groups of marked cells are true genetic clones resulting from single infection events. The number of marked clones in a tissue must first be kept low in order to minimize clonal overlap. After this, it can be shown by co-infecting tissues with mixtures of vectors (separate vectors expressing cytoplasmic β-galactosidase, nuclear β-galactosidase, or alkaline phosphatase) that mixed clones are very rare (39–41).

Finally, as suggested in Section 4.2, the stability of vector expression may be a concern, especially in lineage studies. It is known that transgene expression in mice can be affected by genomic position effects at DNA integration sites; both up- and down-regulation have been seen. It is conceivable, although it has not been demonstrated clearly, that retrovirus expression may also be affected by position effects. If down-regulation of LTR function should occur, the marker gene may not be expressed reliably in all cells arising from a single progenitor; the true lineage patterns may then be misrepresented. Although inactivation of a strong LTR is unlikely to be common, a different lineage marking system may be needed to confirm the fidelity of the retroviral approach.

4.4 Genetic manipulation of cells

Retroviral vectors offer a most effective mechanism for transferring genes into vertebrate cells *in vivo* and in culture, and have been used successfully with many cellular genes, including oncogenes. Retroviral vectors also offer a particularly efficient way of introducing genes into suspension cells (for example haematopoetic cells) which are refractory to several other gene transfer methods. When transferring foreign genes into cells using vectors, a histochemical or selectable marker gene is often co-expressed from a dual expression vector (vector types 3–5, *Figure 4*). This aids in identifying the targeted cells in tissues and facilitates their direct selection in culture. Fluorescein-conjugated substrates for β-galactosidase have also been used for fluorescence-activated cell sorting of live cells (25).

Another potential application of retroviral expression vectors is in human gene therapy. Mammalian tissues including liver, skin fibroblasts, blood, and endothelium have been successfully targeted with retroviral vectors, resulting in stable gene expression in culture (27, 42, reviewed in ref. 2). There has been some success with this approach in animals, but it remains to be seen whether it will be practicable or acceptable in humans.

Retroviral vectors have also been used for introducing and expressing genes in transgenic strains of animals. Primordial germ cells of pre- and post-implantation stage mouse embryos (43), and blastoderm stage chick embryos (44, 45), can be infected with retroviruses (both rd and helper-independent). Mouse embryonic stem cells can also be infected in culture and used to generate chimeric embryos and ultimately transgenic mouse strains (46). The stable inactivation of MLV vectors in embryonic stem cells and preimplantation embryos is a problem which can be overcome by careful vector design (Sections 4.2 and 5.1). Whereas there are other established methods available for creating transgenic mice, avian transgenesis currently relies on the retroviral approach.

4.5 Expression of antisense RNA

Some attempts have been made to express antisense sequences from retroviral vectors. Although antisense segments of *src* and *myc* have been expressed in cells at high levels from RSV-derived vectors, there is little apparent interference with the cellular mRNAs (47, 48). A major finding is that cells expressing high levels of sense RNA (of cellular or viral origin) are not infectable with viruses containing the complementary antisense sequence. This block to infection does not occur in cells expressing low levels of sense RNA, suggesting that infection of cells before the onset of target gene expression (for example before cell differentiation) may still be a viable approach.

In recent work, an apparently successful vector expressing antisense integrin sequences was derived from an REV vector (49). This vector, LZ15, reduced the expression of integrin in cultured cells and altered the migration of optic tectal cells *in vivo*.

The success of the retroviral antisense approach has therefore been variable, but its potential applications will doubtless drive researchers to develop more reliable vectors.

5. Benefits and limitations of vector designs

5.1 Replication-defective vectors

The upper size limits for viral RNA packaging are approximately 10 kb for ASLV vectors, 10.5 kb for MLV vectors, and 13 kb for REV vectors. Two to three kilobases must be set aside for essential *cis*-acting sequences, leaving 7–10 kb free for foreign sequences. If a vector is larger than the maximal size it will not be incorporated efficiently into a virus, it may suffer deletions, and virus titres will fall. Structural instability will also occur when large direct repeats are incorporated: deletions will frequently remove the inter-repeat region (50). Similarly, large inverted repeats are unstable owing to the formation of hairpin structures.

The LTR is highly active in most cell types but may be adversely affected if an internal promoter/enhancer is present (vectors type 2 and 3, *Figure 4*). LTRs and internal promoters often compete negatively (51). If the internal promoter is too strong, too little full-length viral RNA is made and virus titres will be low. Conversely, if the LTR dominates, then internal promotion may be too low to be useful. Promoter competition may also vary with cell type if the internal promoter is tissue-specific. It has proven difficult to predict the transcriptional behaviour of dual-promoter vectors, but promoter competition may not always be a problem. For example, the *neo* gene does not need high levels of expression to confer drug resistance and can remain useful with a low efficiency promoter.

Type 4 vectors which use mRNA splicing for gene 2 may avoid the above problem, but ASLV type 4 vectors suffer from a different one: spliced mRNAs may be packaged and transmitted as vectors themselves (38). This could be a problem if the spliced gene 2 is a co-expressed marker for gene 1, since independent transmission of the marker is possible. This problem does not affect MLV or REV vectors owing to their genome organization (*Figure 2*): MLV/REV spliced mRNAs do not retain ψ and are not incorporated efficiently into viruses. The packaging signal ψ is included in subgenomic mRNAs of ASLV, and when ALSV vectors are expressed in packaging cell lines (where helper virus genomes do not compete for encapsidation) the subgenomic mRNAs can be transmitted at reasonable frequencies (in excess of 10^4 particles/ml).

The relative positions of the splice donor site and the AUG codon for *gag* also differ between MLV, REV, and ASLV (*Figure 2*). The AUG is not incorporated into subgenomic RNAs of MLV or REV, but is included with six codons of *gag* in ASLV. *Gag* translation must therefore be terminated in spliced ASLV RNAs. A stop codon is already present after the splice acceptor site in v-*src* mRNA of RSV, but with other acceptors there may be a need to engineer artificial stop codons (52). Note that interfering with the *gag* AUG itself is not an option in vectors where residual *gag* sequence is fused in frame with *lacZ* (vectors SFCV-neo, Blacsrc2, LZ1C1, *Figure 5*).

Type 5 vectors (*Figure 4*) use internal translational initiation in bi-cistronic mRNAs to avoid some of the problems discussed above. If the successes reported recently with these vectors hold true (6, 7), then vector types 3 and 4 may become obsolete in many applications.

5.2 Replication-competent vectors

High virus titres and levels of expression can be achieved with these vectors. These vectors also facilitate the widespread infection of cells in culture and in tissues. While current ASLV packaging cells are limited to *env* A (infecting only avian cells), the RCAS/RCAN vectors have the added flexibility of *env* A, B, and D: infection of rodent cells and *env* A-resistant chicken cells is therefore possible. Potential drawbacks with replication-competent vectors are the 1.5 kb size limit for inserted genes, and the capacity to carry only single gene cassettes.

When using these vectors for transgenic animal studies (Section 4.4), there is a risk of activating cellular genes through LTR enhancer action. To minimize this risk, additional vectors have been constructed using an LTR with low enhancer activity (vectors RCOS and RCON; 53).

6. Designing and constructing retroviral vectors

The following questions should be considered:

- What species-specificity is required?
- What size gene cassette(s) is to be included?
- Which vector is more appropriate: rd or helper-independent?
- Is a marker gene necessary?
- If a dual expression vector is needed, which type is most appropriate?
- Should an internal promoter be tissue-specific?

For many experimental applications, a suitable vector, or the backbone of one which can be modified easily, may already be available. *Figures 3* and *5* show some vectors which are in use, several of which serve as recipients for new gene cassettes (N2, JD214, SFCV-neo, RCAS, and RCAN). The advantages of using existing vectors are that they have been tested and their potential uses and possible drawbacks are documented.

New vectors can be constructed *in vitro* from existing plasmid vectors and the cloned gene of choice. To construct a new vector, basic techniques for recombinant DNA manipulation are required. Since these techniques are adequately described elsewhere (54, 55), they will only be referred to in general terms here. Specifically, the equipment and knowledge necessary to do the following are needed:

- Make large-scale plasmid preparations.
- Digest plasmids with restriction enzymes and perform phenol:chloroform extraction and ethanol precipitation.
- Perform agarose gel electrophoresis and purify DNA fragments.
- Ligate DNA fragments to plasmids and transform the products into bacterial hosts.
- Screen bacterial colonies by performing restriction enzyme analysis of plasmid minipreparations.

To construct a new vector, you will initially need 50–100 μg each of the target vector plasmid and the plasmid containing the required gene cassette. Using standard methods, the gene cassette should be excised from its plasmid, gel purified, and ligated into a unique cloning site in the vector. Examples of this are shown in *Figure 5*: SAX was made by inserting an *Xho*I fragment containing the human adenosine deaminase gene under the control of the SV40 promoter into a unique *Xho*I site in N2; CXL was derived by inserting a *Pst*I/*Hin*dIII cassette into JD214.

Several steps may be needed if more than one cassette is to be inserted, or if an internal promoter must be inserted separately. Check that the inserts are present in only one copy and oriented correctly. Take note of the upper size limits for all retroviruses (Sections 5.1 and 5.2). Finally, do not delete or interrupt any of the essential *cis*-acting regions of the retrovirus backbone (Section 2 and *Figure 2*).

If you are constructing helper-independent vectors from RCAN and RCAS, then consider the source of the splice acceptor (i.e. whether it is to be supplied by RCAS or the insert cassette) and the AUG codon for the inserted gene (*Figure 3*). These will determine the kind of adaptor plasmid which should be used. The DNA inserts must have *Cla*I cohesive termini which can be supplied in three ways:

- The insert may already have *Cla*I termini.
- Non-*Cla*I termini can be made compatible with the *Cla*I cloning site using commercially available oligonucleotide adaptors or linkers.
- Inserts can be cloned into an adaptor plasmid (*Figure 3*) and then excised using *Cla*I.

If the insert contains internal *Cla*I sites, then use of *Cla*I adaptors is the easiest option.

7. Obtaining infectious virus from recombinant plasmids

To obtain infectious virus, retroviral DNA must be introduced into eukaryotic cells expressing the viral proteins gag, pol, and env. These proteins can be supplied in three ways:

- Packaging cell lines (for use with rd vectors) (Section 3, *Table 1*, and *Figure 6*).
- Helper retroviruses (for use with rd vectors) such as Mo-MLV, Rous-associated virus-1 (ALV), or REV. These retroviruses are replication-competent and express all viral proteins. Co-transfection of vector with helper virus DNA produces a mixed population of helper virus and vector.
- Helper-independent vectors. These encode their own viral proteins.

For the purpose of the methods described in this chapter, it is assumed that the reader is familiar with cell culture techniques, the details of which may depend on the cell line used.

7.1 Transfecting cells with vector DNA

There are several methods for transfecting vector DNA into eukaryotic cells (55). The calcium phosphate precipitation method works well for introducing retroviral vector DNA into adherent packaging cells (*Protocol 1*).

Protocol 1. Calcium phosphate-mediated DNA transfection

Materials

- cells plated subconfluently 12–24 h in advance on 60 mm Petri dishes ($1–5 \times 10^5$ cells/cm^2)
- plasmid DNA[a] dissolved in sterile distilled water or 1 mM Tris–HCl (pH 8.0), 0.1 mM EDTA
- 2 × Hepes-buffered saline (HBS): dissolve 1.0 g Hepes, 1.6 g NaCl, 74 mg KCl, 25 mg Na_2HPO_4 in 90 ml of water, adjust the pH carefully to 7.05 with 0.5 M NaOH, adjust final volume to 100 ml, sterile-filter and store at −20 °C

Method

1. Mix 1–20 μg of plasmid DNA with 62.5 μl of 2 M $CaCl_2$ and adjust the final volume to 500 μl with water. Add the solution dropwise with gentle agitation to 500 μl of 2 × HBS and incubate at room temperature for 15–30 min until visibly milky.
2. Add the mixture dropwise to the culture medium over the cells and swirl to mix. Incubate the cells at 37 °C for 4–8 h.

3. Remove the medium and gently add 5 ml of 15% (v/v) glycerol in 1 × HBS or culture medium. After 30–100 sec (cells vary in their sensitivity to this step[b]), aspirate the medium and rinse the cells with 1 × HBS or medium. Add culture medium and return the dishes to the incubator.

[a] DNA purified by caesium chloride gradient centrifugation gives good results, as does DNA purified by polyethylene glycol precipitation.
[b] A straightforward way to gauge transfection efficiencies is to transfect *lacZ* expression vectors, then fix the cells after 20 h and stain with X-gal, as described in *Protocol 4*. The percentage of transfected cells is determined by light microscopy.

Within a few hours of transfection, viral RNA and proteins are synthesized and virus particles are released. Circularized plasmids provide good transient release of viruses for 20–50 h after transfection. To establish long-term production of virus, use a selectable marker (either in the vector or co-transfected in a plasmid) to isolate stably transfected cell clones. Drug treatment can be started 24 h after transfection. A higher frequency of stable transfectants is obtained if the vector DNA is linearized at a unique restriction site before it is used in *Protocol 1*. Several drug-resistant clones should be assayed for virus titre since considerable variation is found between individual transfectants. Also monitor virus titre over time, since extended passaging can result in reduced titre.

If a helper virus is to be used to provide viral proteins, then cloned helper virus DNA must be co-transfected with the vector. A ten-fold excess of vector over helper DNA ((w/w) as uncut plasmids) should be used when preparing DNA for *Protocol 1*. Co-transfection of these DNAs will lead to stable virus production without further selection.

7.2 Harvesting virus from transfected cells

Virus may be harvested from transfected cells using *Protocol 2*.

Protocol 2. Collecting virus from transfected cells

Materials

- 60 mm dishes of cells transfected with circular plasmids 20–50 h previously, or subconfluent stably-transfected and selected cells
- Centricon-30 microconcentrator devices (Amicon)

Method

1. Place fresh medium on the cells, 1–2 ml/dish. If the virus is to be concentrated using microfiltration (step 4), reduce the serum levels to reduce filter blockage. Use serum-free medium if necessary.

Protocol 2. *Continued*

2. Incubate the dishes at 37°C for 2–4 h,[a] then transfer the medium to sterile 1.5 or 5 ml polypropylene tubes. Repeat steps 1 and 2 if more virus is needed.
3. Centrifuge the virus suspension at 3000 *g* for 5 min (or 1500 *g* for 15 min) at 4°C to pellet cell debris; retain the virus-containing supernatant.
4. Several alternative steps may now be performed.
   - Use the virus immediately to infect cells or tissues.
   - Concentrate the virus by centrifugation in microconcentrators (5000 *g* for 30–60 min at 4°C). The virus concentrate remains in the upper chamber in a final volume of 50 μl.
   - Concentrate the virus by centrifugation at 15 000 r.p.m. in a Beckman SW28 or J14 rotor for 12–18 h at 2°C. Carefully remove the supernatant and resuspend the pellet in 0.2–1% of the original volume of fresh culture medium or fetal calf serum. Resuspend the pellet by shaking rather than pipetting.
   - Freeze the virus at −70°C. It can be stored for over a year.[b]

[a] The half-life of viable virus is approximately 2–4 h at 37°C. Extension of virus collection times increases the proportion of inactivated virus, which will compete with active virus for cell surface binding.
[b] The virus titre is reduced two- to five-fold upon thawing at 37°C (see Section 9.1).

8. Determining virus titres

8.1 Titrating replication-defective vectors

Rd virus titres are calculated by infecting cells in culture and visualizing infectious centres histo- or immuno-chemically or counting drug-resistant colonies (*Protocol 3*). NIH/3T3 or canine D17 cells are commonly used for MLV and REV vector assays, and primary chick embryo fibroblasts or the quail line QT6 are used for ASLV vector assays.

Protocol 3. Infection of cells with virus and calculation of titres

Materials

- 60 mm dishes of subconfluent cell monolayers prepared 24 h before they are required
- phosphate-buffered saline containing divalent cations (PBS^+): 140 mM NaCl, 2.7 mM KCl, 8 mM Na_2HPO_4, 1.4 mM KH_2PO_4, 0.7 mM $CaCl_2$, 0.5 mM $MgCl_2$ (pH 7.2)

- 10 mg/ml polybrene (hexadimethrine bromide) in PBS^+; store at 4°C
- an appropriate drug for selection of resistant colonies

Method

1. Add polybrene to the virus suspension to a final concentration of 5–10 μg/ml.[a]
2. Place fresh medium containing appropriate dilutions of virus on to the dishes.[b] Use a small inoculum volume (for example 0.25–0.5 ml/dish), incubate at 37°C for 15–30 min and then top up with fresh medium and incubate overnight. Alternatively, add the virus to 2–4 ml of fresh medium in the dish and incubate overnight at 37°C.
3. If drug selection is to be used go to step 4, otherwise proceed to step 5.
4. Start the drug selection 24 h after infection. For G418 or hygromycin B, use 100–500 μg/ml. Replace the culture medium every 2–3 days. Proceed to step 6.
5. After 3–4 days in culture, fix the cells and stain with X-gal (*Protocol 4*).
6. The virus titre can be calculated from the number of surviving drug-resistant colonies or from the number of X-gal-stained foci.[c]

[a] Polybrene is a polycation which increases the efficiency of virus infection. It acts, in part, by neutralizing net negative surface charges on the cells and virus.

[b] As a guide, vectors are usually obtained in a range of 10^4–10^6 infectious units/ml.

[c] X-gal-stained foci may not be entirely coherent if motile fibroblasts are used. Accurate counts are achieved only when foci are well separated, that is, at greater virus dilutions.

β-Galactosidase expression can be detected histochemically in fixed cells (*Protocol 4*). The assay produces an intense insoluble blue product in expressing cells, and is excellent for detecting marked cells within whole mounted tissues. Cells and tissues should be infected with a *lacZ*-expressing vector for at least 24 h, but preferably 2–3 days, before staining. The cells should be stained as soon as possible after fixation.

Protocol 4. X-gal assay for β-galactosidase expression in fixed cells and tissues

Materials

- phosphate-buffered saline (PBS): 170 mM NaCl, 3.4 mM KCl, 10 mM Na_2HPO_4, 1.8 mM KH_2PO_4 (pH 7.2)
- 2 or 4% (w/v) paraformaldehyde in PBS
- rinsing solution: 0.1% (w/v) bovine serum albumin, 0.25% (v/v) Triton X-100 in PBS

Protocol 4. *Continued*

- X-gal buffer: 5 mM potassium ferricyanide, 5 mM potassium ferrocyanide, 2 mM $MgCl_2$ in PBS
- 50 mg/ml X-gal (5-bromo-4-chloro-3-indolyl β-D-galactopyranoside) in dimethyl sulphoxide or dimethylformamide: store in a glass container at 4°C in the dark
- stain solution: 1 μl of 50 mg/ml X-gal in 1 ml of X-gal buffer

Method

1. Fix cultured cells in paraformaldehyde solution for 15 min at room temperature.[a] Fix tissues in 4% paraformaldehyde solution, incubating tissues 1–2 mm thick for 30 min and tissues of 5 mm or more for 1–2 h.
2. Wash the cells for 5 min in rinsing solution. Wash tissues for the same period as that used for fixation.
3. Rinse the cells or tissues in X-gal buffer for 5 min or 15–30 min, respectively.
4. Place the samples into stain solution.[b] Incubate in the dark overnight at room temperature or for 2–6 h at 37°C. Tissues should be incubated overnight to ensure penetration of the stain. It is vital to include negative control tissue in order to monitor the extent of endogenous background staining.
5. Rinse the cells or tissues with water or PBS to remove the stain buffer. Store the samples in water or PBS.[c]

[a] 2 mM $MgCl_2$ and 1 mM EGTA can be included in the paraformaldehyde solutions to reduce endogenous background staining.

[b] Steel forceps should not be used to manipulate tissues in the stain solution, since the metal reacts to form a blue precipitate.

[c] The X-gal product is resistant to most chemicals, including those used during preparation for tissue sectioning. Immunohistochemical or immunofluorescent detection of other cellular proteins is difficult in cells containing the X-gal product. *In situ* hybridization of DNA or RNA is also impractical after this staining.

8.2 Titrating replication-competent virus

There are several quite different assays for replication-competent viruses, each of which can be used to screen for contaminants in rd virus stocks or to titrate helper-independent vectors.

- End-point dilution and reverse transcriptase (RT) assay. Cells are infected with serial dilutions of virus (*Protocol 3*), allowed to grow for several days, and then the supernatant is collected and screened for viral RT (56). The titre can be extrapolated from the lowest virus dilution at which RT is detected.

- As a variation, the infected cells are overlaid with agar 6–12 h after infection, and an immunocytochemical assay is performed 4–10 days later using gag antibodies (the EFU assay; 57).
- When starting with rd vectors, the supernatant collected in the first assay above can be screened for secondary transmission of rd virus (use *Protocol 3*) or for RT. If either is detected, then helper virus must also be present.
- There are cellular syncytial assays for establishing MLV virus titres. These include the S+L-(58) and XC assays (59).

9. Practical applications of retroviral vectors

9.1 General comments

Virus which has been frozen should be thawed rapidly at 37°C, shaken briefly, and placed on ice. A brief centrifugation step at 3000 *g* may help to clear any debris. Once inoculated, either in culture or *in vivo*, the virus adsorbs quickly to cells, and stable proviral integration occurs within a few hours in dividing cells. In rapidly-growing embryo tissue, vector expression can be detected within 10 h of inoculation (38).

If vectors expressing *lacZ* are used *in vivo*, the targeted cells can be visualized using the X-gal histochemical assay after fixing the tissues (*Protocol 4*). Fixed tissues can also be processed for immunofluoresence analysis or immunohistochemistry to detect β-galactosidase or any other vector gene product for which an antibody is available.

9.2 Injection of virus into tissues *in vivo*

Finely-drawn glass capillaries are required for microinjection of virus into tissues. Capillaries with a length of 10 cm, an internal diameter 0.75–0.8 mm and an external diameter 1.0–1.1 mm are suitable. They should be drawn out using an electronically-controlled pulling device (Narishige Scientific Instruments) to an external tip diameter of 25–50 μm. If the tip clogs during injection, it can often be cleared by gently tapping with a sterile scalpel or steel needle or rinsing with sterile saline; leaving a small droplet of expelled virus suspension on the tip after each use can prevent clogging. Since injection of small volumes of virus is often necessary (see below), use a pneumatic or hydraulic microinjection device in order to control the injection volume. A micromanipulator is also required for precise injections.

For many experiments, the target tissue will be embryonic and consequently very small (for example limb buds, otocysts, or somites), and the volume of injected fluid is highly restricted. Because of this, the titre must be sufficient to ensure injection of adequate virus in volumes of 10–1000 nl. Methods for concentrating virus are given in *Protocol 2*. Polybrene can be added to virus solutions before injection into tissues to increase the efficiency of infection; final concentrations of 50–100 μg/ml are commonly used. Adequate infection can be obtained without polybrene, and its necessity should be judged at the time. Note that the effective virus titre *in vivo* can be significantly lower than that judged by tissue culture assays. Consequently, the efficiency of infection *in vivo* should be estimated each time a new vector or target site is used.

9.2.1 Infection of chick embryos

Injection of chicken embryos with virus can be performed readily *in ovo* (*Protocol 5*).

Protocol 5. Infection of chick embryos

Materials

- fertile chicken eggs at an appropriate developmental stage
- virus at an appropriate titre
- 0.05% (w/v) trypan blue

Method

1. Sterilize the egg shell by wiping with 70% (v/v) ethanol. Remove a 1–2 cm^2 'window' of egg shell and shell membrane above the embryo using sharp scissors and forceps. A syringe and 19 gauge needle may be used optionally to puncture the shell at the narrow end and to remove 4–5 ml of albumin prior to opening the shell window; this provides more space for working within the egg.
2. Inoculate blastoderms by injecting virus directly beneath the epiblast or even through the epiblast itself.[a,b] Relatively large volumes (5–20 μl) can be injected in this way. When injecting at later embryonic stages (after day 4), carefully tear a hole in the amniotic membrane over the embryo before injection, since the membrane obscures visibility and may be difficult to inject through.
3. Inject the virus directly into the target tissue, controlling the volume as necessary. Mixing the virus with trypan blue before injection can help to visualize the inoculum during and after injection.
4. Seal the egg shell with adhesive tape or Parafilm and incubate it for an appropriate time.
5. Remove the embryo from the egg and prepare it for histology, histochemistry, etc.

[a] Suspensions of virus in culture medium or PBS^+ are usually less dense than albumin, and can float away from the embryo.

[b] Up to 50% of embryos survive this procedure and develop normally. Initial development can, however, be delayed by 6–12 h.

9.2.2 Infection of mouse and rat embryos

Isolated mouse embryos up to 10-days-old can be infected by incubating them directly in virus suspensions or by co-culturing them with packaging cells releasing virus (32, 60). Injection of virus or virus-producing packaging cells can also be performed *in utero* after suitable surgical preparation (34). Ref. 61 describes comprehensive methods for manipulating mouse embryos. Injection of neonatal rat retinal tissue has also been successful (37).

10. Troubleshooting

Several potential problems associated with retroviral vectors have been described in Section 5. If a problem arises, check first that the vectors have been constructed with all the necessary *cis*-acting sequences (Section 2). Low virus titres and poor expression of vector genes can be caused by a vector being too large, containing deleterious repeat structures, or having an imbalance between internal promoters and viral LTRs. Replication-competent virus in rd vector stocks is caused by vector sequences recombining with the homologous viral sequences of a packaging cell (Section 3). Such contaminants usually occur at very low levels (frequencies of 10^{-4} or less), and are least likely to occur with the more stringent cell lines DSN, DAN, PA317, clone 32, GP+E-86, ψCRIP/ψCRE, and Isolde (*Table 1*).

10.1 Examining vector expression

To check whether a new vector is expressing a gene correctly, examine the RNA and protein expression patterns in transfected cells (*Protocols 6* and *7*). If possible, use a control vector which is structurally stable and expresses well in the appropriate cells (several of the vectors in *Figure 5* are suitable). Standard procedures for isolating and characterizing RNA and DNA are given in refs 54 and 55, and should be used in conjunction with *Protocol 6*.

10.1.1 RNA expression

Northern hybridization may be used to gauge the efficiency of vector expression relative to a control vector (*Protocol 6*). This method also indicates the relative levels of full-length and subgenomic RNAs expressed by a dual-expression vector. RNase protection assays should be performed if more accurate quantitation of RNA levels is required; this approach can also be used to locate the exact sites of splicing and internal transcription.

Protocol 6. Analysing vector RNA expression by Northern hybridization

Materials

- vector plasmid
- monolayers of the appropriate cells, or tissue from a transgenic animal

Protocol 6. *Continued*

Method

1. Transfect 10 μg of vector plasmid into the cells and incubate for 24–48 h (*Protocol 1*). Alternatively, prepare fresh dishes of cultured cells which have been selected for stable vector expression or obtain tissue from a transgenic animal.
2. Extract total cellular RNA from the above cells.
3. Electrophorese 5–10 μg of RNA on a denaturing agarose gel, and transfer it to a nylon membrane.
4. Prepare a suitable [^{32}P]DNA probe from the vector. Choose a 500–1000 bp segment whose sequence is present in all vector-derived RNAs[a,b] and label it by the random priming method using [α-^{32}P]dCTP.
5. Hybridize the probe to the RNA immobilized on the filter. Visualize the hybridized probe by autoradiography.

[a] For example, a probe from gene 2 of the vectors in *Figure 4*.
[b] The probe will also hybridize to helper virus RNAs if it contains retroviral structural sequences.

10.1.2 Protein expression

Quantitative assays for marker gene products, such as β-galactosidase and chloramphenicol acetyltransferase (62), are well established. A solution assay for β-galactosidase is given in *Protocol 7*, using the substrate *o*-nitrophenyl β-D-galactopyranoside (ONPG). *LacZ* expression can also be detected histochemically in fixed cells and whole mounted tissues in a qualitative assay (*Protocol 4*). Again, use a control vector if available.

Protocol 7. Quantitative ONPG assay for β-galactosidase in transfected cells

Materials

- cell monolayers prepared at least 18 h in advance and freshly transfected with plasmid DNA or stably expressing a retroviral vector
- PM-2 buffer: 33 mM NaH_2PO_4, 66 mM Na_2HPO_4, 0.1 mM $MnCl_2$, 2 mM $MgSO_4$, 40 mM 2-mercaptoethanol
- 4 mg/ml ONPG in PM-2 buffer

Method

1. Wash the cells once with PBS and place the plates on ice.
2. For 35 mm plates, add 0.2 ml of 0.1% (w/v) sodium dodecyl sulphate in PBS. Incubate on ice for 5 min to allow cell lysis.

3. Add 0.8 ml of PM-2 buffer, mix well, and pipette several times to fully suspend the lysate. Place the lysate in a 1.5 ml microcentrifuge tube on ice.
4. Aliquot 0.5 ml of lysate into a fresh tube and add 100 μl of 4 mg/ml ONPG to all samples. Mix the samples and incubate them at 37°C. Allow the reaction to proceed until the yellow colour is well developed—an OD_{405} value of 0.1–1.0 is appropriate. If the colour is allowed to over-develop linearity will be lost. Note the total reaction time for each sample.
5. Stop each reaction by placing the tube on ice and adding 0.25 ml 1 M Na_2CO_3.
6. Freeze the samples briefly, thaw and centrifuge them at 12 000 *g* for 10 min.
7. Determine the OD_{405} value for each supernatant.

The reaction rates are calculated by plotting OD_{405} values against time. The rate correlates with the amount of enzyme during the linear reaction phase. Normalize the rates for protein concentration in each lysate, or alternatively equalize the protein concentrations using PM-2 buffer before starting the reactions.

10.2 Structural stability of vectors

A structurally unstable vector will suffer proviral rearrangements. A symptom of this may be aberrant RNA patterns (*Protocol 6*). However, given that there may be other explanations for aberrant RNAs (for example cryptic promoters or splicing sites), it is advisable to examine proviral DNA directly (*Protocol 8*).

Protocol 8. Examining proviral DNA

Materials

- appropriate cell monolayers (for example 3T3, QT6, chick embryo fibroblasts, or stable cell clones expressing a vector)
- virus suspension

Method

1. Infect cells with virus at a multiplicity of 1–10 infectious units/cell.[a] Do this at least 20 h prior to step 2. Alternatively, use stable cell clones.[b]
2. Isolate high molecular weight genomic DNA from 10^7 cells (54, 55; see Chapter 5, *Protocol 3*).
3. Digest 10 μg of DNA with a restriction enzyme that cuts at least twice within the vector sequences.

Protocol 8. *Continued*

4. Separate the fragments by agarose gel electrophoresis and transfer them to a nylon membrane (54, 55).
5. Prepare a DNA probe and hybridize it to the immobilized DNA (*Protocol 6*).
6. Visualize the hybridized probe by autoradiography.

[a] Since thousands of integration events are examined, only commonly-occurring rearrangements will be seen.
[b] Examine several independent cell clones.

Using this method it is possible to determine, to a resolution of about 100 bp, whether the proviral structure corresponds to the theoretical vector map. Vector instability will be evidenced by aberrant sizes of hybridized fragments.

If a problem is encountered with a vector, try to pinpoint its source. Because too little is known about the mechanisms controlling retroviral splicing and interference between promoters, it is difficult to adjust the vector structure appropriately in an attempt to overcome problems arising from these factors. It is usually necessary to redesign a vector significantly in order to obtain a better one. This could entail using a different internal promoter or switching to a different vector type (for example from type 3 to 5 for a dual expression vector).

11. Safety considerations

A major health and safety consideration concerns the ability of retroviruses to infect human cells. ASLV systems are intrinsically safe since these viruses do not infect human cells. The use of ASLV packaging cells and vectors, therefore, does not require the use of high containment laboratory facilities. Although REV systems have also been seen as safe, there is a recent report clearly showing that REV-based vectors can infect primate cells, including human ones, in culture (66). Infection is followed by efficient expression in the cells, but a productive infection is not seen—that is, primate cells do not synthesize and further transmit infectious virus. Thus, the hazards inherent with replication-competent contaminants in vector stocks, or their evolution *in vivo*, *may not* be a problem with REV vectors in humans. Nevertheless, precautions should be taken to prevent accidental infection of researchers with the primary rd vectors, especially if vectors can express harmful genes such as oncogenes. Containment facilities appropriate for human pathogens should be considered.

MLV ecotropic vectors are considered safe to use under conditions similar to those used for ASLV or REV. Ecotropic cell lines include ψ2, clone 32, and GP+E-16. Amphotropic MLV systems, including packaging cell lines ψAM, PA12, PA317, and ψCRIP, produce viruses which infect and will replicate in human cells. These should be used under strict containment facilities for human pathogens. Although the more stringent cell lines such as ψCRIP apparently do not give rise to replication-competent virus, the hazards from a primary infection with an rd vector should be borne in mind.

Acknowledgements

I thank Jack Price and Paul Martin for their comments on the manuscript. I am also indebted to Mina Bissell (Berkeley, California) in whose laboratory I developed the ASLV packaging cells and vectors. I am currently funded by a Royal Society University Research Fellowship.

References

1. Gilboa, E., Eglitis, M. A., Kantoff, P. W., and Anderson, W. F. (1986). *Biotechniques,* **4,** 504.
2. Eglitis, M. A. and Anderson, W. F. (1988). *Biotechniques,* **6,** 608.
3. Weiss, R. A., Teich, N. M., Varmus, H. E., and Coffin, J. M. (ed.) (1985). *RNA tumor viruses, 2/supplements and appendices.* Cold Spring Harbor Press, Cold Spring Harbor, NY.
4. Weiss, R. A., Teich, N. M., Varmus, H. E., and Coffin, J. M. (ed.) (1984). *RNA tumor viruses, 1/text.* Cold Spring Harbor Press, Cold Spring Harbor, NY.
5. Hughes, S. H., Greenhouse, J. J., Petropoulos, C. J., and Sutrave, P. (1987). *J. Virol.,* **61,** 3004.
6. Ghattas, I. R., Sanes, J. R., and Majors, J. E. (1991). *Mol. Cell. Biol.,* **11,** 5848.
7. Adam, M. A., Ramesh, N., Miller, A. D., and Osborne, W. R. A. (1991). *J. Virol.,* **65,** 4985.
8. Koo, H.-M., Brown, A. M. C., Kaufman, R. J., Provock, C. M., Ron, Y., and Dougherty, J. P. (1992). *Virology,* **186,** 669.
9. Yu, S.-F., von Rden, T., Kantoff, P. W., Garber, C., Seiberg, M., Rüther, U., Anderson, W. F., Wagner, E. F., and Gilboa, E. (1986). *Proc. Natl Acad. Sci. USA,* **83,** 3194.
10. Mann, R., Mulligan, R. C., and Baltimore, D. (1983). *Cell,* **33,** 153.
11. Cone, R. D. and Mulligan, R. C. (1984). *Proc. Natl Acad. Sci. USA,* **81,** 6349.
12. Miller, A. D., Law, M.-F., and Verma, I. M. (1985). *Mol. Cell. Biol.,* **5,** 431.
13. Miller, A. D. and Buttimore, C. (1986). *Mol. Cell. Biol.,* **6,** 2895.
14. Bosselman, R. A., Hsu, R.-Y., Bruszewski, J., Hu, S., Martin, F., and Nicolson, M. (1987). *Mol. Cell. Biol.,* **7,** 1797.
15. Markowitz, D., Goff, S., and Bank, A. (1988). *J. Virol.,* **62,** 1120.
16. Danos, O. and Mulligan, R. C. (1988). *Proc. Natl Acad. Sci. USA,* **85,** 6460.
17. Watanabe, S. and Temin, H. M. (1983). *Mol. Cell. Biol.,* **3,** 2241.
18. Dougherty, J. P., Wisniewski, R., Yang, S., Rhode, B. W., and Temin, H. M. (1989). *J. Virol.,* **63,** 3209.
19. Savatier, P., Bagnis, C., Thoraval, P., Poncet, D., Belakebi, M., Mallet, F., Legras, C., Cossett, F.-L., Thomas, J.-L., Chebloune, Y., Faure, C., Verdier, G., Samarut, J., and Nigon, V. (1989). *J. Virol.,* **63,** 513.

20. Cosset, F.-L., Legras, C., Chebloune, Y., Savatier, P., Thoraval, P., Thomas, J. L., Samarut, J., Nigon, V. M., and Verdier, G. (1990). *J. Virol.,* **64,** 1070.
21. Stoker, A. W. and Bissell, M. J. (1988). *J. Virol.,* **62,** 1008.
22. Koo, H.-M., Brown, A. M. C., Ron, Y., and Dougherty, J. P. (1991). *J. Virol.,* **65,** 4769.
23. Miller, D. G., Adam, M. A., and Miller, A. D. (1990). *Mol. Cell. Biol.,* **10,** 4239.
24. Milkawa, T., Fischman, D. A., Dougherty, J. P., and Brown, A. M. C. (1991). *Exp. Cell Res.,* **195,** 516.
25. Smith, G. H., Gallahan, D., Zwiebel, J. A., Freeman, S. M., Bassin, R. H., and Callahan, R. (1991). *J. Virol.,* **65,** 6365.
26. Keller, G. and Wagner, E. F. (1989). *Genes and Development,* **3,** 827.
27. Kantoff, P. W., Kohn, D. B., Mitsuya, H., Armentano, D., Sieberg, M., Zwiebel, J. A., Eglitis, M. A., McLachlin, J. R., Wiginton, D. A., Hutton, J. J., Horiwitz, S. D., Gilboa, E., Blaese, R. M., and Anderson, W. F. (1986). *Proc. Natl Acad. Sci. USA,* **83,** 6563.
28. Jaenisch, R., Fan, H., and Croker, B. (1975). *Proc. Natl Acad. Sci. USA,* **72,** 4008.
29. Soriano, P., Cone, R. D., Mulligan, R. C., and Jaenisch, R. (1986). *Science,* **234,** 1409.
30. Stewart, C., Schuetze, S., Vanek, M., and Wagner, E. (1987). *EMBO J.,* **6,** 383.
31. Wagner, E. F., Vanek, M., and Vennström, B. (1985). *EMBO J.,* **4,** 663.
32. Savatier, P., Morgenstern, J., and Beddington, R. S. P. (1990). *Development,* **109,** 655.
33. Reddy, S. T., Stoker, A. W., and Bissell, M. J. (1991). *Proc. Natl Acad. Sci. USA,* **88,** 10505.
34. Austin, C. P. and Cepko, C. L. (1990). *Development,* **110,** 713.
35. Frank, E. and Sanes, J. R. (1991). *Development,* **111,** 985.
36. Price, J. (1987). *Development,* **101,** 409.
37. Price, J., Turner, D., and Cepko, C. (1987). *Proc. Natl Acad. Sci. USA,* **84,** 156.
38. Stoker, A. W., Hatier, C., and Bissell, M. J. (1990). *J. Cell Biol.,* **111,** 217.
39. Bonnerot, C., Rocancourt, D., Briand, P., Grimber, G., and Nicolas, J.-F. (1987). *Proc. Natl Acad. Sci. USA,* **84,** 6795.
40. Galileo, D. S., Gray, G. E., Owens, G. C., Majors, J., and Sanes, J. R. (1990). *Proc. Natl Acad. Sci. USA,* **87,** 458.
41. Fields-Berry, S. C., Halliday, A. L., and Cepko, C. L. (1992). *Development,* (In press.)
42. Zwiebel, J. A., Freeman, S. M., Kantoff, P. W., Cornetta, K., Ryan, U. S., and Anderson, W. F. (1989). *Science,* **243,** 220.
43. Jaenisch, R. (1988). *Science,* **240,** 1468.
44. Bosselman, R. A., Hsu, R.-Y., Boggs, T., Hu, S., Bruszewski, J., Ou, S., Souza, L., Kozar, L., Martin, F., Nicolson, M., Rishell, W., Schultz, J. A., Semon, K. M., and Stewart, R. G. (1989). *J. Virol.,* **63,** 2680.
45. Hughes, S. H., Kosik, E., Fadly, A. M., Salter, D. W., and Crittenden, L. B. (1986). *Poult. Sci.,* **65,** 1459.
46. Robertson, E., Bradley, A., Kuehn, M., and Evans, M. (1986). *Nature,* **323,** 445.
47. To, R. Y.-L., Booth, S. C., and Neiman, P. E. (1986). *Mol. Cell. Biol.,* **6,** 4758.
48. Hopper, P. and Coffin, J. M. (1988). *Current communications in molecular biology: viral vectors,* pp. 139–45. Cold Spring Harbor Press, Cold Spring Harbor, NY.

49. Galileo, D. S., Majors, J., Horwitz, A. F., and Sanes, J. R. (1991). *Neuron*, **9,** 117.
50. Rhode, B. W., Emerman, M., and Temin, H. M. (1987). *J. Virol.*, **61,** 925.
51. Emerman, M. and Temin, H. M. (1984). *Cell,* **39,** 459.
52. Fuerstenberg, S., Beug, H., Introna, M., Khazaie, K., Munoz, A., Ness, S., Nordström, K., Sap, J., Stanley, I., Zenke, M., and Vennström, B. (1990). *J. Virol.,* **64,** 5891.
53. Petropoulos, C. J. and Hughes, S. H. (1991). *J. Virol.,* **65,** 3728.
54. Sambrook, J., Fritsch, E. F., and Maniatis, T. (1989). *Molecular cloning: a laboratory manual.* Cold Spring Harbor Press, Cold Spring Harbor, NY.
55. Perbal, B. (1988). *A practical guide to molecular cloning.* John Wiley, NY.
56. Goff, S., Traktman, P., and Baltimore, D. (1981). *J. Virol.,* **38,** 239.
57. Stoker, A. W. and Bissell, M. J. (1987). *J. Gen. Virol.,* **68,** 2481.
58. Bassin, R. H., Tuttle, N., and Fischinger, P. J. (1971). *Nature,* **229,** 564.
59. Klement, V., Rowe, W. P., Hartley, J. W., and Pugh, W. E. (1969). *Proc. Natl Acad. Sci. USA,* **63,** 753.
60. Jahner, D. and Jaenisch, R. (1980). *Nature,* **287,** 456.
61. Copp, A. J. and Cockkroft, D. L. (ed.) (1990). *Postimplantation mammalian embryos: a practical approach.* IRL Press, Oxford.
62. Gorman, C. M., Moffat, L. F., and Howard, B. (1982). *Mol. Cell. Biol.,* **2,** 1044.
63. Keller, G., Paige, C., Gilboa, E., and Wagner, E. F. (1985). *Nature,* **318,** 149.
64. Dougherty, J. P. and Temin, H. M. (1986). *Mol. Cell. Biol.,* **6,** 4387.
65. Thompson, T. C., Southgate, J., Kitchener, G., and Land, H. (1989). *Cell,* **56,** 917.
66. Koo, H.-M., Brown, A. M. C., Ron, Y., and Dougherty, J. P. (1991). *J. Virol.,* **65,** 4769.

7

Poliovirus antigen chimeras

DAVID J. EVANS

1. Introduction

The term chimera is derived from the Greek word χιμαροσ which described the mythological fire-breathing monster, slain by Bellerophon, which had a lion's head, a goat's body, and a serpent's tail. Although far less exotic than Bellerophon's monster, the term is also applied to hybrid poliovirus particles in which one or more of the well-characterized antigenic sites on the virus capsid have been replaced—using recombinant DNA technology—with defined amino acid sequences from another source. Poliovirus antigen chimeras have several potential applications for antigen presentation.

- Generation of monoclonal antibodies against defined peptide sequences, presented in a known immunogenic location on the poliovirus particle.
- Use in diagnostic assays to present defined antigenic sequences.
- Design and development of novel vaccines.

This chapter addresses the construction and characterization of chimeric picornaviruses, with particular emphasis on the most widely exploited example, poliovirus. The reader is referred to the recent review by Rose and Evans for a more detailed overview of the applications of chimeric poliovirus particles (1).

2. Poliovirus and poliovirus vaccines

Poliovirus is the causative agent of paralytic poliomyelitis, a disease largely eradicated in the developed world by the widespread use of the killed (Salk) and live attenuated (Sabin) vaccines. The oral poliovirus vaccine (OPV) developed by Albert Sabin in the 1950s contains derivatives of each of the three serotypes (1, 2, and 3) of poliovirus generated by multiple passage in culture of wild type (WT) strains of the virus (2). The resulting live attenuated strains exhibit the following characteristics that have resulted in their establishment as extremely effective vaccines.

- They induce good, long-lasting levels of immunity in recipients.
- They are relatively inexpensive to manufacture.
- They are easy to administer—on a sugar lump or as a drop on the tongue.

Extensive use of the Sabin vaccine strains has demonstrated that they are extremely safe and efficacious. However, there still exists a small risk of vaccine-associated paralytic poliomyelitis for vaccinees or their close contacts, with approximately 1 case of paralysis per 1.2 million recipients, which has largely been attributed to the reversion of the type 2 or type 3 strains to neurovirulence (3). A series of detailed studies from this and other laboratories have determined the molecular basis for the attenuation of the Sabin vaccine strains (4), and are currently elucidating the mechanism by which the attenuation mutations modify the neurovirulent phenotype of the WT virus. These studies have suggested a number of ways in which the current strains could be improved to provide safer vaccines which exhibit a reduced incidence of reversion.

Parallel studies on the three-dimensional and antigenic structure of the poliovirus particle have suggested an alternative method in which safer poliovaccines could be engineered. Since serotypes 2 and 3 were implicated in the majority of vaccine-associated cases of paralytic poliomyelitis, it seemed reasonable to modify the antigenicity of the type 1 strain so that it resembled type 2 or type 3, whilst retaining the stable attenuation phenotype of type 1. Nomoto *et al.* achieved this by replacing the entire capsid region of Sabin type 1 with the corresponding region of Sabin type 3 (5). Other laboratories undertook to modify the antigenicity in stages, by exchanging individual antigenic sites of type 1 with the analogous regions of type 2 or type 3 (6–8). The latter studies involve the construction of poliovirus antigen chimeras, or hybrids, which are providing new insights into picornavirus antigenicity. Although the original reasoning behind the construction of chimeric virus particles—the development of improved poliovaccines—now appears ambitious in the light of results obtained from the construction of intertypic antigen chimeras, the studies have resulted in the development of poliovirus as an 'epitope presentation system' for expression of defined amino acid sequences.

2.1 The poliovirus genome and the three-dimensional and antigenic structures of poliovirus particles

Poliovirus possesses a single-stranded positive sense RNA genome of approximately 7500 nucleotides which encodes a single polypeptide that is subsequently processed by virally-encoded proteases to generate the proteins required for virus replication and capsid formation. The single open reading frame is flanked by 5′ and 3′ non-coding regions containing *cis* acting functions required for virus replication. The seminal observation by Racaniello and Baltimore (9) that a cloned poliovirus cDNA is infectious for mammalian cells in culture (i.e. transfection of cells permissive for poliovirus replication enables recovery of infectious virus particles), has allowed dissection of the structure and function of the virus using recombinant DNA technology.

The complete cDNA sequences of attenuated and neurovirulent representatives of all three serotypes of poliovirus have been determined, allowing the identification of highly conserved and variable regions of the virus genome (10). Subsequent determination of the three-dimensional structure of the poliovirus particle at near-atomic resolution (11) demonstrated that some of the regions exhibiting significant amino acid sequence heterogeneity form the surface-protruding loops of the capsid proteins. The icosahedral poliovirus particle is composed of 60 copies of each of the four capsid proteins VP1–VP4, the latter being entirely internal. VP1, VP2, and VP3 possess a conserved core structure composed of an eight-stranded anti-parallel β-pleated sheet, flanked by two short α-helices. The turns between the relatively conserved β strand exhibit the amino acid heterogeneity observed in sequence alignments, and contribute towards the antigenicity of the poliovirus particle.

A variety of strategies have been employed to identify regions of the virus capsid recognized by neutralizing antibodies, and a total of four antigenic sites have been located (12, 13). Antigenic sites 2, 3, and 4 are conformational in nature, being composed of discontinuous regions of the capsid proteins. In contrast, the region of VP1 that lies between the βB and βC strands and protrudes at the pentameric apex of the virus particle is largely linear in nature. This region is designated antigenic site 1 (see *Table 1*). The relative ease with which this region of the capsid can be genetically engineered has made it the focus for studies involving the construction of poliovirus antigen chimeras.

2.2 An overview of the design and construction of poliovirus chimeras

The design, construction, and characterization of poliovirus antigen chimeras involves a number of stages. The examples given in this chapter all involve poliovirus, which is the picornavirus used most extensively for the construction of chimeras. However, the structural and functional similarity between

Table 1. Antigenic determinants of poliovirus type 1 located by selection of mAb neutralization-resistant escape mutants[a]

Antigenic site	Capsid protein		
	VP1	VP2	VP3
1	97, 99–101,[b] 144	ND[c]	ND
2	221–224, 226	164, 165, 168–170, 270	ND
3	ND	ND	58–60, 71, 73
4	ND	72	76

[a] Adapted from the data reviewed by Minor (13).
[b] Hyphenated residue numbers indicate that all inclusive amino acids have been implicated in contributing to the antigenic site (e.g. 99–101 = 99, 100, and 101).
[c] No changes to this capsid protein have been detected.

poliovirus and other picornaviruses implies that the overall approach will be widely applicable. The stages involved in the design and construction of poliovirus antigen chimeras are as follows:

(a) Selection of a suitable region of the capsid for modification, and preliminary experiments to demonstrate suitability of the chosen region.
(b) Design and construction of a cassette vector to facilitate extensive modification of the selected region.
(c) Generation of recombinant cDNA by ligation of the prepared cassette vector with annealed complementary oligonucleotides encoding the selected epitope.
(d) Transfection of cells permissive for virus replication with RNA generated *in vitro* from the recombinant cDNA.
(e) Recovery and purification of the chimeric virus (if viable).
(f) Antigenic and immunogenic characterization of the chimeric virus particle.

The current poliovirus cassette vectors are described in the following section as they illustrate the broad principles involved. There follows more generalized details on the design of cassette systems, epitope selection, and the construction, recovery, and characterization of chimeric virus particles.

3. Poliovirus cassette vectors

The efficient and extensive modification of a defined region of DNA is facilitated by construction of a cassette system, in which convenient restriction sites are used to replace the unwanted region with DNA from another source. In the most basic form, a poliovirus cassette vector would consist of a full-length poliovirus cDNA, cloned in a suitable plasmid vector, in which restriction sites unique to the entire construct flank the region of the poliovirus genome to be replaced. The careful selection of sites unique to the plasmid vector obviates the requirement for additional subcloning and rebuilding procedures, or for the use of partial restriction digests.

A number of additional features could be included in the design of the vector to facilitate the construction and recovery of antigen chimeras. The first is the ability to generate RNA runoff transcripts of the poliovirus genome, by the addition of a promoter for the DNA-dependent RNA polymerase of bacteriophages T7 or SP6 preceding the poliovirus cDNA. This approach is preferable to DNA transfections, as runoff RNA from T7 (or SP6) promoters is approximately 100-fold more infectious than the cDNA template (14). Two further additions, that with careful design can conveniently be included simultaneously, include modifications which prevent recovery of viable poliovirus from the vector unless the cassette region has been modified, and which facilitate the screening of recombinant poliovirus cDNAs.

3.1 The vectors pCAS1 and pCAS7

The majority of the poliovirus antigen chimeras constructed in this laboratory have been generated from two cassette vectors based upon a Sabin type 1 cDNA cloned in a pBR322-derived plasmid. The original vector pCAS1, and more particularly the improved pCAS7, exhibit the majority of the desirable features described in the preceding section. The construction of pCAS1 has been previously described in detail (15). Briefly, the vector consists of a full length Sabin 1 cDNA, preceded by a promoter for bacteriophage T7 RNA polymerase, cloned in the pBR322-derived plasmid vector pFB1[2] (Pharmacia), which confers tetracycline resistance. The sequences flanking antigenic site 1 of the cDNA were modified by the introduction of unique *Sal*I and *Dra*I restriction sites to allow rapid replacement of the intervening region with pairs of annealed complementary oligonucleotides.

Although pCAS1 has been used successfully for the construction of a large number of antigen chimeras, extensive use of the vector brought to light a number of problems inherent in its design. The most serious of these was the use of the pFB1[2]-derived plasmid vector carrying a tetracycline resistance marker, which resulted in slow growth of *Escherichia coli* strains harbouring the vector and exhibited an additional unacceptably high level of plasmid instability. Furthermore, pCAS1 could give rise to viable virus without modification of the cassette region; it was therefore critical to confirm that each potential cDNA was a recombinant before proceeding to the transfection step.

To overcome these problems an improved vector, designated pCAS7, was designed and constructed. This vector included the following modifications:

(a) Use of a different pBR322-derived vector (pJM1) allowing selection using kanamycin rather than tetracycline.
(b) Replacement of antigenic site 1 to prevent the cDNA from yielding infectious poliovirus unless the cassette region has been modified, thereby ensuring that any viable virus recovered must be a recombinant.
(c) Introduction of two restriction sites to aid in the screening and recovery of recombinant cDNAs.

The modifications made to the poliovirus cDNA during construction of pCAS1 and pCAS7 are illustrated in *Figure 1*. pJM1 is a 3.7 kbp pAT153-based vector that was generated by replacing the entire tetracycline resistance gene of pAT153 (base pairs 1–1369 of the parental pBR322) with a gene cartridge encoding resistance to the antibiotic kanamycin derived from Tn903 (J. Meredith and D. J. Evans, unpublished results). *E. coli* strains harbouring recombinants constructed in the pJM1-based pCAS7 grow well, and the plasmid exhibits minimal instability problems. Since the cassette region of pCAS1 was, by definition, replaced during the construction of recombinants, it was modified to facilitate the screening and recovery of hybrid cDNAs. The a)

```
                93                                                   103
                 |                                                    |
          T   V   D   N   S   A   S   T   K   N   K   D   K   L   F
Sabin 1  ACC GTG GAT AAC TCA GCT TCC ACC AAG AAT AAG GAT AAG CTA TTT

          T   V   D   N   S   A   S   T   K   N   K   F   K   L   F
pCAS1    ACC GTC GAC AAC TCA GCT TCC ACC AAG AAT AAG TTT AAA CTA TTT
              SalI                                     DraI

          T   V   D   R   G   R   V   *
pCAS7    ACC GTC GAC CGC GGA CGC GTT TAA  A   CTA
                     SstII    MluI
```

b)

```
                                         ---------T7 runoff transcript---------->
5' T7    GAATTCTAATACGACTCACTATAGGGTTAAAACAGCTCTGGGG
           EcoRI                           1
```

c)

```
pCAS7 3'      AAAAAAAAAAAAAAAAAAAAAAAAAA↓CGCGTCCCT
                                        MluI
```

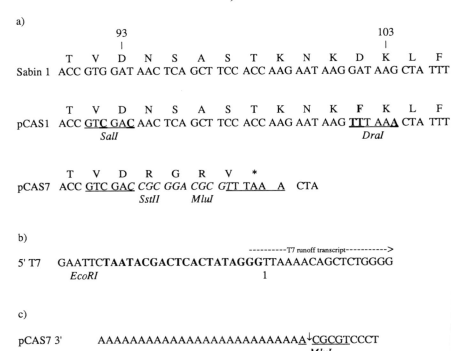

Figure 1. (a) The cDNA and amino acid sequence of antigenic site 1 of Sabin 1, showing the modifications made during the construction of pCAS1 and pCAS7. (b) The extreme 5′ end of the Sabin 1 cDNA in the pCAS1/7 cassette vectors, indicating the point of transcriptional initiation from the introduced T7 promoter (bold) and the first nucleotide of the poliovirus cDNA. (c) The extreme 3′ end of the Sabin 1 cDNA in pCAS7, indicating the introduced *Mlu*I restriction site used to linearize the template for use in a T7 reaction.

original *Sal*I and *Dra*I sites were retained, but the intervening region was replaced with an oligonucleotide spacer containing two restriction sites, *Sst*II and *Mlu*I. The spacer region was designed to introduce a frameshift in the poliovirus polyprotein coding sequence, resulting in premature termination of translation within the recognition sequence for *Dra*I (TT*TAA*A) and thereby making the vector incapable of generating infectious poliovirus without further modification (see *Figure 1a*). The inclusion of a *Mlu*I site within the cassette region was coupled with replacement of the GC-tails at the 3′ extreme of the poliovirus cDNA (which remained after the original cloning of the cDNA) with an additional *Mlu*I site. Inclusion of this site serves two purposes: (a) it allows rapid screening of potential recombinants by *Mlu*I digestion (unmodified vector generates two bands of 5.4 and 4.7 kbp, whereas recombinants give a single band of 10.1 kbp), and (b) it allows linearization of the cDNA template prior to T7 transcription and excludes all non-virus sequences from the 3′ end of the transcript (see *Figure 1c*), since this has previously been reported to increase the infectivity of T7 transcripts (16). Full details of the construction of pCAS7 have been reported recently (17).

4. *De novo* design and construction of picornavirus cassette vectors

4.1 Choice of position to make insertion

The region of the capsid chosen for the construction of antigen chimeras is of fundamental importance. It is critical that the chosen region can be modified in the desired manner without abrogating virus viability. The extensive knowledge of the antigenic and three-dimensional structure of the poliovirus particle made the selection of antigenic site 1 a relatively straightforward choice. It should be noted that attempts to modify the antigenic characteristics of site 2b (which, although possessing certain linear characteristics, forms one of the discontinuous components of antigenic site 2—see *Table 1*) have been far less successful (18). This region of the capsid is clearly subject to certain constraints that do not apply to the free-standing antigenic site 1. In the absence of the detailed structural and antigenic information available for poliovirus, use must be made of picornavirus capsid sequence alignments (10) to locate regions that are potentially exposed on the surface, and, by analogy with the polio- and rhinoviruses, antigenic in nature.

Before proceeding with the construction of a cassette system to facilitate the generation of a large number of antigen chimeras, the sequence of the selected region of the capsid should be modified in a limited manner to confirm that it can accommodate amino acid variation. Since different serotypes of a virus are likely to be closely related structurally, a suitable replacement would be the amino acid sequence of the analogous region of a different serotype. The chosen sequence can be introduced to the virus backbone using any standard procedures for site-directed mutagenesis. Prior to the construction of the pCAS1 cassette system, Burke *et al.* (6) had demonstrated that an intertypic poliovirus antigen chimera could be generated in which eight amino acids of antigenic site 1 of the Sabin 1 strain of poliovirus were replaced with the corresponding region of a type 3 strain by site-directed mutagenesis. The resulting viable chimeric poliovirus, designated S1/3.10, exhibited composite antigenicity and dual immunogenicity. Similarly, Reimann *et al.* (19) demonstrated that it is possible to substitute the VP1 βB–βC loop of coxsackievirus (CV) B3 with the analogous region of CVB4, and generate a viable antigen chimera exhibiting composite CVB3/4 antigenicity and immunogenicity.

Having demonstrated that the chosen region of the capsid can accommodate some sequence variation without losing virus viability, the cassette system can be designed and constructed. This involves the selection and introduction of suitable restriction sites into the virus cDNA together with any necessary modifications of the plasmid vector.

The availability of suitable computer software makes the design of the cassette vector (i.e. selection of unique restriction sites) straightforward. The University of Wisconsin Genetics Computing Group (UWGCG) suite of sequence analysis programs (20) is particularly suited to this task, though other software may be equally applicable. The UWGCG program MAPSORT, run with the command line qualifer/SILENT (it may be necessary to first alter the default settings for command line qualifiers using COMCHECK), will locate suitable restriction sites within the region of interest, including those that can be introduced by modification of the nucleotide sequence without consequent alteration of the translation product. It is important to remember that the two limitations placed upon the choice of restriction sites are: (a) that they are unique to the final vector (it may therefore be necessary to alter or substitute the plasmid vector), and (b) that they should not result in an alteration to the coding region after ligation with oligonucleotides encoding the introduced epitope. For example, the *Dra*I site used in the construction of pCAS1, in addition to necessitating the use of a vector lacking *Dra*I sites, introduced a coding change to the polyprotein—Asp102Phe (see *Figure 1a*). Although this change was subsequently shown not to affect virus viability, the substituted amino acid (Phe102) is always removed upon *Dra*I digestion, leaving the flanking region of the polyprotein unaltered. The addition of all the blunt restriction endonuclease 'half-sites' (e.g. the 5′ 'half-sites' of *Sma*I and *Pvu*II are, respectively, CCC and CAG, whereas the 3′ 'half-sites' of the same enzymes are GGG and CTG) to the associated file ENZYME.DAT which is read by MAPSORT (and can be copied to the user's directory by typing FETCH ENZYME.DAT at the $ prompt) can be used to realize the full potential of this approach. The use of blunt restriction sites considerably increases the likelihood of locating two restriction sites close to the required region. Subsequent use of the cassette vector is facilitated if the two restriction sites chosen are not compatible, and so do not ligate during the construction of recombinant cDNAs. This also circumvents any problems in determining the orientation of the inserted oligonucleotides. However, in certain cases the use of two enzymes leaving blunt ends represents the most convenient modification of the cassette region or cannot be avoided. Self-ligation of the vector can be reduced by phosphatase treatment (though the replacement oligonucleotides must be phosphorylated with polynucleotide kinase to replace the phosphate groups required for ligation), and it is sometimes possible to include other subtle modifications that enables the orientation of the insert to be determined (as described in the following section).

4.2 Cassette vector construction by PCR mutagenesis

A large variety of site-directed mutagenesis strategies have been developed, each with particular advantages and disadvantages. We routinely use the Bio-

Rad Muta-Gene system based upon the technique devised by Kunkel *et al.* (21) to introduce single, or more extensive, mutations into a region of DNA. However, the recent explosion of novel amplification and cloning techniques based upon the polymerase chain reaction (PCR) allows rapid generation of cassette vectors without recourse to standard methods of site-directed mutagenesis (22). The advantage of this approach is the speed with which the completed recombinant cassette vector can be constructed, since the time-consuming steps of subcloning and preparation of template for mutagenesis (for example generation of uracil-containing M13 DNA) are no longer necessary.

Protocol 1 describes the construction of a cassette system that allows the subsequent modification of residues 142–148 of poliovirus VP1, which form the βD–βE loop, by the use of overlapping PCR-mediated mutagenesis. This vector, designated pCAS8, is based upon pCAS7 and illustrates several of the concepts described in the preceding paragraphs. The sequence of this region is shown in *Figure 2a*.

a)

```
                     142                  148
                      |                    |
 M  E  F  T  F  V  V  T  A  N  F  T  E  T  N  N  G  H  A  L  N  Q  V
ATGGAATTTACCTTTGTGGTTACTGCAAATTTCACTGAGACTAACAATGGGCATGCCTTAAATCAAGTG
TACCTTAAATGGAAACACCAATGACGTTTAAAGTGACTCTGATTGTTACCCGTACGGAATTTAGTTCAC
```

b)

```
  3'-TAAATGGAAACACCAATTGCCGGCGCCTCCGGA-5'      Primer P3
     |||||||||||||||||
ATGGAATTTACCTTTGTGGTTACTGCAAATTTCACTGAGACTAACAATGGGCATGCCTTAAATCAAGTG
TACCTTAAATGGAAACACCAATGACGTTTAAAGTGACTCTGATTGTTACCCGTACGGAATTTAGTTCAC
                                              || |||||||||||||||||||
            Primer P2     5'-GGCCGCGGAGGCCTGTTAACGGGCATGCCTTAAATCAAG-3'
```

c)

```
 M  E  F  T  F  V  V  N  G  R  G  G  L  F  T  G  M  P  *
ATGGAATTTACCTTTGTGGTTAACGGCCGCGGAGGCCTGTTAACGGGCATGCCTTAAATCAAGTG
TACCTTAAATGGAAACACCAATTGCCGGCGCCTCCGGACAATTGCCCGTACGGAATTTAGTTCAC
```

d)

```
 M  E  F  T  F  V  V  T  A  N  F  T  E  T  N  N  G  H  A  L  N  Q  V
ATGGAATTTACCTTTGTGGTTACCGCAAACTTCACCGAAACCAACAACGGGCATGCCTTAAATCAAGTG
TACCTTAAATGGAAACACCAATGGCGTTTGAAGTGGCTTTGGTTGTTGCCCGTACGGAATTTAGTTCAC
```

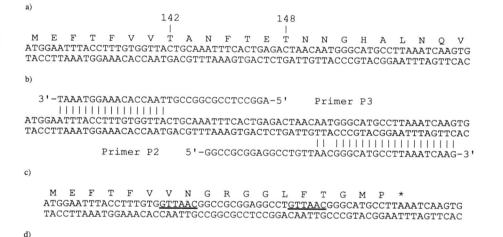

Figure 2. (a) The cDNA and amino acid sequence of the βD–βE loop of Sabin 1 that forms the basis of the cassette vector pCAS8. (b) The primers (P3 and P2) described in *Protocol 1* and used for PCR mutagenesis, showing regions of homology with the cDNA template. (c) The cDNA sequence of the cassette vector following PCR mutagenesis and rebuilding. The two *Hpa*I sites that define the cassette region are doubly underlined, and flank the intervening region that contains *Sfi*I, *Sst*II, and *Stu*I restriction sites. The point at which translation of the polyprotein is prematurely terminated is indicated. (d) Reconstruction of the original βD–βE loop of Sabin 1 using pCAS8 showing the use of the introduced *Bst*EII site (doubly underlined) to determine the orientation of the oligonucleotide insert.

Protocol 1. Cassette construction by PCR mutagenesis

Materials

- 10 × *Taq* polymerase buffer: 500 mM KCl, 100 mM Tris–HCl (pH 8.4), 15 mM $MgCl_2$, 1 mg/ml gelatin (Perkin–Elmer)
- *Taq* polymerase (Amplitaq; Cetus)
- 2 mM dNTPs: 2 mM each of dATP, dTTP, dCTP, and dGTP
- four oligonucleotide primers (P1–4) at 100 ng/μl[a]

Method

1. Prepare the reagents for two separate PCR reactions in 500 μl microcentrifuge tubes; using 1 μg of template reduces the number of cycles required.

	Reaction 1	**Reaction 2**
• pCAS7 DNA template (1 μg)	10 μl	10 μl
• primer P1	5 μl	—
• primer P2	5 μl	—
• primer P3	—	5 μl
• primer P4	—	5 μl
• 2 mM dNTPs	10 μl	10 μl
• 10 × *Taq* polymerase buffer	10 μl	10 μl
• sterile water	60 μl	60 μl
• *Taq* polymerase	2 U	2 U

2. Cover both reactions with 60 μl of sterile mineral oil and amplify in a thermal cycler by completing 10 cycles of 95 °C for 1 min, 55 °C for 2 min, and 72 °C for 2 min, followed by a single cycle of 95 °C for 1 min, 55 °C for 2 min, and 72 °C for 10 min.
3. Separate the amplification products of both reactions from the Sabin 1 cDNA template by electrophoresis through a 2.5% (w/v) agarose gel, and recover DNA fragments of 284 bp (Reaction 1) and 120 bp (Reaction 2) using standard procedures. It is important that the purified fragments are not contaminated with the template DNA or the amplification primers.
4. Anneal the two reaction products from step 3 by performing a PCR reaction *in the absence of primers*. Add the following to a microcentrifuge tube:

• purified 284 bp reaction 1 product	20–50 ng
• purified 120 bp reaction 2 product	20–50 ng
• 2 mM dNTPs	10 μl

- 10 × *Taq* polymerase buffer 10 μl
- sterile water to a total volume of 90 μl
- *Taq* polymerase 2 U

Cover the reaction with 60 μl of sterile mineral oil and amplify for 5 cycles using identical conditions to those described in step 2. Omit the final 10 min extension cycle.

5. Add 5 μl of each of primers P1 and P4 to the tube and complete a further 20 cycles using the reaction conditions used in step 2.
6. Remove the mineral oil overlay by ether extraction and precipitate the products by adding 50 μl of 7.5 M ammonium acetate and 220 μl of ice-cold ethanol. Purify the 394 bp PCR product on a 2.5% agarose gel, and extract using standard procedures.
7. Digest the purified fragment with *Sal*I and *Sna*B1, gel-purify the 203 bp fragment using standard procedures and ligate it to pCAS7 cut with *Sal*I and *Sna*B1.
8. Confirm the presence of the cassette region by restriction enzyme digestion and DNA sequencing.

[a] Primers P1 (sense) and P4 (antisense) are complementary to nucleotides 2627–2649 and 3021–3003 of the Sabin 1 cDNA respectively, and primers P2 (sense) and P3 (antisense) are mutagenic oligonucleotides, as illustrated in *Figure 2b*.

Following construction of the cassette vector, the ability to generate viable virus must be confirmed. This is achieved by replacement of the cassette region with oligonucleotides encoding the original sequence of the epitope (note that since pCAS8 is based upon pCAS7, both cassette regions must be replaced to generate a viable virus). The construction of recombinant cDNAs and RNA transfections are covered in Sections 5.3 and 5.4, respectively. pCAS8 exhibits several design features that facilitate the construction and screening of recombinant cDNAs. The cassette region includes three restriction sites *Sst*II, *Stu*I, and *Sfi*I (GGCCnnnnnCCGG), and introduces a frameshift to the poliovirus polyprotein which results in premature termination of translation after residue 152 of VP1 (see *Figure 2c*). The *Hpa*I sites selected for the construction of the cassette are unique to the full-length cDNA. Phosphatase treatment of the vector prior to ligation is unnecessary; self-ligation of the vector re-creates a *Hpa*I site and this population of molecules can therefore be excluded from subsequent transformation by *Hpa*I digestion (the oligonucleotides synthesized for insertion into pCAS8 are designed to destroy both *Hpa*I sites upon ligation). A problem associated with the use of two blunt restriction sites for cassette construction is the inability to 'force' the insert into a particular orientation. To facilitate the screening of recombinants, oligonucleotides for insertion into pCAS8 are designed so that the codon for Thr143 is always 5′-ACC-3′. Oligonucleotide pairs that have ligated in the correct orientation will create a *Bst*EII restriction site (see *Figure 2d*) that can be identified rapidly by restriction digestion. *Bst*EII cuts at nucleotides 3291 and 3911 of pCAS8, generating fragments of 0.7 and 9.4 kbp; the gain of an additional *Bst*EII site due to the correct orientation of ligated oligonucleotides generates fragments of approximately 0.4, 0.7, and 9.0 kbp. The additional 0.4 kbp fragment can be readily visualized by gel electrophoresis in 2% agarose.

5. Construction and recovery of antigen chimeras

5.1 Choice of epitope for presentation

The present design of poliovirus antigen chimeras is based upon replacement of a defined linear epitope (antigenic site 1) with a contiguous peptide sequence from another protein. Manipulation of epitopes that are largely conformational in nature (portions of antigenic site 2) have been less successful (see Section 4.1). The selection of suitable linear epitopes for presentation on the surface of the virus particle involves two factors: (a) identification of an antigenic determinant on the native protein, and (b) possible modification of the epitope to maximize the chances of formation of a viable antigen chimera. The precise identification of suitable epitopes for expression is beyond the scope of this chapter, but would typically require information from the characterization of monoclonal antibody (mAb) escape mutants, the screening of immune sera with defined overlapping synthetic peptides (Pepscan analysis), or computer modelling studies. The computer prediction of linear epitopes in proteins has recently been thoroughly reviewed (23).

The modification of the surface of the poliovirus particle can result in structural changes to the capsid that are incompatible with virus viability—approximately 30% of the chimeras generated in this laboratory are non-viable. Careful selection of the precise sequence to be engineered into the virus will maximize the chances of recovering a viable virus, and can also affect whether the peptide adopts an antigenic conformation. Preliminary analysis of the extensive panel of viable and non-viable poliovirus-based chimeras generated in this laboratory suggest that one of the major determinants of viability may be the charge distribution within the engineered loop, particularly at the N-terminus of the peptide (H. Stirk, D. Evans and J. Thornton, unpublished results). A significant proportion of non-viable chimeras have an imbalance of charged residues (aspartic/glutamic acid or lysine/arginine) in the first four residues following the Val–Asp dipeptide that forms the N-terminus of the cassette vector (see *Figure 3a*). Presumably an incorrect charge distribution could adversely affect the conformation adopted by the virus polypeptide during proteolytic processing, capsid assembly, or interaction with the cellular receptor—all of which must be completed for the generation of viable chimeric virus particles.

a) V D **1 2 3 4** 5 6 7 8 9 K L F

b) V D [G] **1 2 3 4** 5 6 7 8 9 [G] K L F

V D [A] **1 2 3 4** 5 6 7 8 9 [A] K L F

c) V D N S A S **1 2 3 4** 5 6 7 8 9 T K N K D K L F

Figure 3. (a) Schematic representation of antigenic site 1 of a poliovirus chimera, indicating the four residues (bold) that must exhibit a balanced charge for virus viability. (b) Positions for the insertion of poly-glycine [G] or poly-alanine [A] to retain virus viability. (c) Insertion of a sequence at the natural turn of poliovirus antigenic site 1 (underlined).

One way in which the charge distribution—particularly at the critical N- and C-termini of the sequence—of the introduced epitope can be controlled is to flank the foreign residues with a series of spacing amino acids (see *Figure 3b*). This could be in the form of short stretches of poly-glycine or poly-alanine (typically 4–5 residues at either end of the insert). Alternatively, and used most successfully in this laboratory, the introduced sequence can be flanked by the amino acids that naturally occupy antigenic site 1 of poliovirus. In this case, the inserted sequence should be located at the natural turn of antigenic site 1 (see *Figures 1a* and *3c*), which, by examination of the three-dimensional structure of the poliovirus particle (11), lies between residues Ser97 and Thr98 of VP1.

5.2 Oligonucleotide design and purification

Two complementary oligonucleotides are designed for insertion into the cassette vector, which in the case of pCAS1 or pCAS7 must, when annealed, generate a 5′ site compatible with *Sal*I-digested vector (a four nucleotide overhang of TCGA), and a blunt 3′ site compatible with *Dra*I-digested vector. Oligonucleotides for use with pCAS1/7 are routinely designed to retain the *Sal*I site and destroy the *Dra*I site, as this facilitates both the screening of recombinants and the further manipulation of the inserted sequences. Since we have not observed any consistent differences in the growth or viability of chimeras containing codons, within the introduced epitope, that are used infrequently in the poliovirus polyprotein coding region (J. Meredith and D. J. Evans, unpublished results), screening and manipulation of recombinant cDNAs is also aided by including restriction enzyme sites within the oligonucleotides, taking advantage of the degeneracy of the genetic code. This can be rapidly achieved by using the UWGCG program BACKTRANSLATE to generate an oligonucleotide sequence from the desired amino acid sequence of the epitope, and then using the output from this as the input for the a)

```
ACHIKLQRST
```

b)

```
 A   C   H   I   K   L   Q   R   S   T
GCT TGC CAC ATC AAA CTG CAG CGT TCC ACC

GCA TGC CAC ATA AAA CTA CAA CGA TCA ACA
GCC TGT CAT ATC AAG CTC CAG CGC TCC ACC
GCG         ATT     CTG     CGG TCG ACG
GCT                 CTT     CGT TCT ACT
                    TTA     AGA AGC
                    TTG     AGG AGT
```

c)

```
GCATGCCACATCAAGCTTCAGAGATCTACC
  SphI         HindIII    BglII
```

d)

```
5'-TCGAGCATGCCACATCAAGCTTCAGAGATCTACC-3'
      3'-CGTACGGTGTAGTTCGAAGTCTCTAGATGG-5'
```

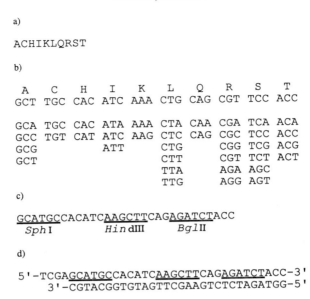

Figure 4. (a) Hypothetical epitope for insertion into pCAS7. (b) Output from the UWGCG program BACKTRANSLATE, indicating the 'best' oligonucleotide (based upon codon usage in the poliovirus polyprotein) immediately below the single letter amino acids, together with possible variation due to the degeneracy of the genetic code. (c) Output from the UWGCG program MAPSORT/SILENT, indicating the restriction sites that can be introduced into the sequence without modification of encoded amino acids. (d) The two oligonucleotides to be synthesized, including the addition of nucleotides necessary for 'in frame' ligation with the *Sal*I–*Dra*I cut vector.

program MAPSORT (with the command line qualifer/SILENT) as shown in *Figure 4*.

Oligonucleotides, synthesized using any suitable chemistry, should be purified using a method that efficiently separates full-length products from prematurely terminated (n-1) products. We routinely purify 20- to 60-mers using 15–20% polyacrylamide gels, elute the product into water and de-salt by centrifugation at low speed through a Sephadex G-25 column. Oligonucleotides are stored frozen at a concentration of 100 ng/μl. Synthesis of oligonucleotides longer than approximately 60 nucleotides, although routinely possible on many DNA synthesizers, generally becomes increasingly inefficient. Furthermore, subsequent modification of the introduced epitope is impossible without re-synthesis of the entire sequence. To avoid unnecessary expense it is often more efficient to design the epitope as two pairs of complementary oligonucleotides, that overlap in the central region of the epitope. Modification of any region within the introduced sequence can then be achieved by re-synthesizing only one pair of oligonucleotides. The inclusion of suitable restriction sites within the introduced oligonucleotides that are unique to the full-length cDNA can also facilitate further manipulation of the engineered epitope.

5.3 Recombinant cDNA construction and screening

The efficient design of the cassette vector ensures that construction and screening of recombinant cDNAs is a straightforward undertaking. In practice it is usually most efficient to prepare a stock supply of pre-digested vector that can be used for ligation with annealed oligonucleotides and for parallel control ligations. In the case of pCAS7, 1–2 μg of plasmid DNA are digested with *Dra*I, and a small aliquot is run on a gel to confirm that all the material is linearized, prior to digestion with *Sal*I. The double-digested vector requires no purification other than phenol extraction and ethanol precipitation.

Approximately 400 ng (4 μl of the 100 ng/μl oligonucleotide stock) of each of the two complementary oligonucleotides are added to a microcentrifuge tube and heated in a boiling-water bath for 3 min. The water bath is removed from the heater and allowed to cool slowly to room temperature. (In the case of four overlapping oligonucleotides encoding the epitope, it is often more efficient to anneal each complementary pair together and then perform a three-fragment ligation, rather than anneal all four oligonucleotides together. However, results may vary depending upon the specific sequence and degree of overlap of the oligonucleotides.) Two to three microlitres of the annealed oligonucleotides are ligated with 20–50 ng of *Sal*I–*Dra*I digested pCAS7 in a total volume of 10 μl. Following ligation, the reaction is digested with one or more of the restriction enzymes whose sites should be absent from the recombinant cDNA (i.e. those present within the replaced cassette region). Depending upon the design of the oligonucleotides for insertion into pCAS7, this is generally *Dra*I or *Sst*II. A control ligation, in which the oligonucleotides are omitted, should be performed in parallel.

Half of the digested ligation reactions is used to transform a suitable host strain of *E. coli* (for example DH5α or MC1061) made competent using standard procedures. Plasmid DNA extracted from the resulting colonies is screened with one or more restriction enzymes to confirm its recombinant nature. In the case of pCAS7 we initially screen potential recombinants with *Mlu*I (which yields a single band of 10.1 kbp for recombinant constructs, as described in Section 3.1), and confirm the presence of the oligonucleotides by the ability to linearize the construct with *Sal*I and any restriction site included within the oligonucleotides, and the inability to digest the construct with *Dra*I or *Sst*II. The nucleotide sequence of the modified region of the cDNA encoding the capsid must also be determined to confirm the authenticity of the oligonucleotides used. The method of Murphy and Kavanagh (24) allows rapid verification of the sequence directly from the full-length double-stranded cDNA recombinant using a primer complementary to nucleotides 2627–2649 of the cDNA.

5.4 Recovery of viable virus chimeras

Recovery of a viable virus from the cDNA construct requires transfection of a cell line permissive for virus replication. The specific cell line chosen depends upon the virus vector; Hep2c, HeLa, or Ohio HeLa lines are routinely used for the recovery of poliovirus chimeras. Although calcium phosphate-mediated DNA transfection of cells with the virus cDNA was initially used for recovery of infectious virus, it is more efficient to utilize a promoter for a DNA-dependent RNA polymerase to synthesize a runoff RNA which is subsequently used for transfection (14). RNA produced in this manner is intermediate in transfection efficiency between cDNA and virus RNA and offers distinct advantages for recovery of chimeras that may grow less well than the parental virus. The promoter for bacteriophage T7 or SP6 RNA polymerase must be engineered immediately preceding the extreme 5′ end of the virus cDNA (see *Figure 1b*) to ensure that the number of non-virus synthesized is kept to a minimum.

The significant polymerase error rate of picornaviruses necessitates plaque purification of the chimera prior to further characterization. In certain cases it is possible to generate plaques directly from a chimeric virus by transfection of permissive cells with RNA produced *in vitro* using T7 RNA polymerase. This approach allows the plaque phenotype to be determined along with an initial round of plaque purification (which should in any event be followed by a second round). However, not all recombinant cDNAs can be recovered in this manner, some require initial passage by transfection of a flask of cells, followed by two rounds of plaque purification. The observed requirement for a balanced charge distribution within the engineered loop of a chimera (Section 5.1), is, in part, based upon adaptive changes observed following the recovery of a series of related chimeras expressing an immunodominant epitope from the envelope glycoprotein of HIV-1. The cDNA constructs in question did not generate plaques directly following transfection of the T7 reaction, but did allow the recovery of viable virus after initial transfection of a flask of cells. Comparison of the cDNA sequence and the sequence of the virus RNA identified as amino acid substitution introducing an altered charge within the modified antigenic site 1. Presumably, the original cDNA sequence, although incapable of generating a virus which had the ability to form plaques, allowed sufficient initial translation and replication to occur to generate an adapted virus capable of forming plaques.

The T7 transcription reaction is described in *Protocol 2*, and is followed by protocols for transfection of a flask (*Protocol 3*) or a six-well plate of permissive cells for the direct generation of virus plaques (*Protocol 4*). In practice, it is possible to avoid repetition of the T7 reaction by dividing the products of a single reaction between both transfections.

Protocol 2. T7 transcription reaction

Materials

- RNase-free water (treated with diethyl pyrocarbonate (DEPC))
- linearized plasmid DNA,[a] resuspended at 20 ng/μl in water treated with DEPC
- 10 × T7 RNA polymerase buffer: 400 mM Tris–HCl (pH 8.0), 80 mM $MgCl_2$, 500 mM NaCl (Stratagene)
- 5 mM NTPs: 5 mM each of ATP, UTP, CTP, and GTP
- RNase inhibitor (e.g. RNAguard; Pharmacia)
- T7 RNA polymerase (100 U/μl; Stratagene)
- 2 × Hepes-buffered saline (HBS): dissolve 1.6 g NaCl, 74 mg KCl, 27 mg $Na_2HPO_4{\cdot}2H_2O$, 200 mg dextrose, and 1 g Hepes in 90 ml water, adjust to pH 7.05 with 500 mM NaOH, make up to 100 ml with water, filter through a 0.22 μm filter and store in aliquots at −20 °C
- 20 mg/ml DEAE–dextran

Method

1. Add the following to a microcentrifuge tube and incubate at 37 °C for 1 h.

• linearized plasmid DNA (approximately 100 ng)	5 μl
• 10 × T7 RNA polymerase buffer	2.5 μl
• 20 mM spermidine	2.5 μl
• 5 mM NTPs	2.5 μl
• 100 mM DTT	7.5 μl
• RNase inhibitor	0.5 μl
• DEPC-treated water	5 μl
• T7 RNA polymerase	0.5 μl

2. Electrophorese 13 μl of the T7 reaction from step 1 on a 1% agarose gel to confirm the presence of a transcript. Transcripts derived from pCAS7 linearized with *Mlu*I have an apparent size of 2.1 kb. If the product is extensively smeared, the reaction may have been contaminated with RNAses.
3. To the remainder of the T7 reaction (13 μl from step 1), add 500 μl of 2 × HBS, 467 μl of DEPC-treated water, and 20 μl of DEAE–dextran. Incubate on ice for 25 min.
4. Transfect cells in flasks (*Protocol 3*) or six-well plates (*Protocol 4*).

[a] In the case of pCAS7, *Mlu*I or *Xho*I is used. The enzyme site must lie 3′ to the poly-A tail of the virus cDNA, and must generate either a 5′ overhang or blunt end. Plasmid DNA prepared by alkaline lysis procedures generally produces better T7 transcripts than DNA produced by methods involving the use of caesium chloride.

Protocol 3. Transfection of cells in a flask with T7 transcripts

Materials

- 75 cm^2 flask of subconfluent (approximately 80–90% confluent) cells permissive for virus infection (e.g. Hep2c or HeLa cells)
- growth medium: Eagle's minimal essential medium (EMEM) containing 2–5% (v/v) fetal calf serum (FCS), 20 mM glutamine, antibiotics (penicillin and streptomycin at 50 μg/ml), 0.22% (w/v) sodium bicarbonate
- phosphate-buffered saline (PBS): 170 mM NaCl, 3.4 mM KCl, 10 mM Na_2HPO_4, 1.8 mM KH_2PO_4 (pH 7.2)
- T7 reaction in HBS/DEAE–dextran from step 3 of *Protocol 2*

Method

1. Prepare the flask of cells for transfection by washing twice with 10 ml of pre-warmed PBS. During this period allow the T7 reaction to warm up to room temperature.
2. Add the T7 reaction to the cell sheet, rock the flask gently to ensure even distribution, and incubate it at room temperature for 30 min.
3. Wash the cell sheet with PBS and add 10 ml of growth medium.
4. Incubate the flask at 34°C in an atmosphere containing 5% (v/v) CO_2.

Protocol 4. Transfection of cells in a six-well plate for plaque formation

Materials

- six-well plate of subconfluent permissive cells
- transfection buffer: 500 μl 2 × HBS, 480 μl DEPC-treated water, 20 μl DEAE–dextran
- agar overlay: 2% (w/v) agar autoclaved in PBS, cooled to 42°C, added to an equal volume of pre-warmed growth medium, and maintained at 42°C until required
- T7 reaction in HBS/DEAE–dextran from step 3 of *Protocol 2*

Method

1. Prepare the six-well plate of cells for transfection by washing each well twice with 3 ml of pre-warmed PBS. During this period allow the T7 reaction to warm up to room temperature.
2. Prepare at least 250 μl of two serial ten-fold dilutions of the T7 reaction in transfection buffer.

3. Add the undiluted or diluted T7 reaction to two wells each, rock the plates gently, and incubate at room temperature for 30 min.
4. Add 3 ml/well of agar overlay, allow the agar to set at room temperature, and incubate the plates inverted at 34°C in an atmosphere containing 5% CO_2.

It is important to perform parallel positive and negative transfection controls, as chimeric virus cDNAs show considerable variability in the time taken for a cytopathic effect (c.p.e.) to develop. We routinely use *Xho*I-linearized pT7Sabin1 (a full-length Sabin 1 cDNA preceded by a T7 promoter) and pCAS7 linearized with *Mlu*I as positive and negative controls, respectively. Comparison of the recombinant cDNA with the controls helps considerably in monitoring progress of the transfections. In addition, individual T7 reactions vary in quality and can affect the time that elapses before c.p.e. becomes apparent. It is therefore good practice to prepare a common mixture of ingredients for the T7 reaction, rather than adding reagents separately to the test and control templates.

6. Characterization of poliovirus antigen chimeras

6.1 Safety considerations

The construction of poliovirus antigen chimeras does not involve modification of a region of the capsid that has been implicated in the attenuation phenotype of the virus. Furthermore, the studies conducted in this laboratory have involved the attenuated Sabin 1 vaccine strain of poliovirus. It is therefore considered highly unlikely that modified viruses will exhibit an alteration of the attenuation phenotype of the parental Sabin 1 virus, but this has not yet been formally tested.

Evidence is accumulating, however, that antigenic site 1 may play a role in determining the tropism of the virus particle. Mouse neurovirulence of the type 2 Lansing strain of poliovirus (a mouse-adapted strain) is dependent upon sequences within site 1 (25). Furthermore, the normally mouse-avirulent P1/Mahoney strain can be converted to a strain that is fully virulent for mice by modification of sequences within site 1, so that they resemble those in the analogous position of the P2/Lansing virus (7, 26). In contrast to these results, the 'canyon hypothesis' proposed by Rossman *et al.* has implicated the cleft surrounding the pentameric apex of the rhinovirus (and closely related poliovirus) particle as the region of the virus that interacts with, and determines the specificity for, the cellular receptor (27). Although there is supporting evidence for the canyon hypothesis from site-directed mutagenesis studies on the rhinovirus canyon (28), the results obtained from analysis of mouse neurovirulence of the P1/Mahoney–Lansing chimeras suggests that antigenic site 1 could affect the tropism of the hybrid viruses. It is therefore possible that certain chimeras may acquire novel cell tropisms and thereby constitute a safety hazard.

Prior to the preparation and purification of large amounts of a virus chimera we routinely confirm that the virus is neutralized by pooled human serum containing antibodies against poliovirus. In addition, the tropism of the virus is investigated by confirming that infection of a suitable permissive cell line is blocked by a mAb directed against the cellular receptor for poliovirus (29). A virus that had acquired an additional tropism might be expected to infect cells in the presence of an anti-receptor mAb. None of the extensive panel of poliovirus antigen chimeras constructed and tested to date have either escaped neutralization by pooled human anti-poliovirus serum or overcome the block imposed by the anti-receptor mAb. The majority of chimeras are therefore handled under containment conditions appropriate for poliovirus. However, in certain circumstances, for example involving chimeras in which a modification of tropism is intended or where the potential exists for induction of an anti-self immune response, higher containment conditions are considered appropriate.

6.2 Preparation of virus stocks for further characterization

Master stocks of poliovirus chimeras recovered following transfection of suitable cell lines must be prepared before further characterization (*Protocol 5*). These are used as inocula for all sub-stocks prepared for RNA sequencing, assays of antigenicity, and as immunogens. One or two rounds of plaque purification ensure homogeneity of the virus stock.

Protocol 5. Preparation of master stocks

Materials

- cell monolayers showing plaques from transfection experiments (see *Protocols 3* and *4*)
- serum-free EMEM

Method

1. Pick off individual plaques with a Pasteur pipette and elute the virus from the agar plug by storage at 4°C in 500 μl of PBS.
2. Remove the growth medium from a confluent 75 cm^2 flask of cells permissive for virus growth and wash the cell sheet twice with PBS.
3. Seed the cell sheet with 250 μl of virus eluted from a plaque. Allow the virus to adsorb for 30 min at room temperature with intermittent rocking of the flask to ensure even coverage of the monolayer.

4. Add sufficient serum-free EMEM to cover the cell sheet (5–10 ml) and incubate at 34°C or 37°C (for Sabin- or Mahoney-based chimeras, respectively) until complete c.p.e. is observed.
5. Freeze the flask and its contents at −20°C and thaw at room temperature. Repeat this step twice. Decant the contents into a sterile plastic tube and centrifuge at 4500 *g* to remove cell debris. Store aliquots of the virus-containing supernatant at −20°C.

Sub-stocks are raised in a similar manner as described in *Protocol 5* by inoculating the cell monolayer with 100–250 μl of the master stock, and are used for all subsequent procedures. To enable standardization and quantification of results obtained during assays of antigenicity, the titre of the sub-stock must be determined. Two methods are used routinely—the $TCID_{50}$ (50% tissue culture infectious dose) assay (*Protocol 6*) and the plaque assay (*Protocol 7*).

Minor (30) compared the advantages and disadvantages of the two methods. The $TCID_{50}$ assay is less accurate, but its ease generally outweighs the advantages of the plaque assay. The accuracy of the $TCID_{50}$ assay is usually sufficient for most subsequent procedures, and the assay can readily be performed in parallel with assays of antigenicity (see Section 6.4 and *Protocol 10*).

Protocol 6. $TCID_{50}$ assay for poliovirus antigen chimeras

Materials

- sub-stock of chimera from *Protocol 5*
- 96-well flat-bottomed tissue culture plates
- Ohio HeLa or Hep2c cells diluted to 1×10^5 cells/ml in EMEM
- formol–saline: 4% (v/v) formaldehyde in 0.9% (w/v) NaCl
- crystal violet solution: 0.5 g crystal violet dissolved in 20 ml of ethanol and made up to 1 litre with formol–saline

Method

1. Prepare ten-fold serial dilutions of the virus stock in serum-free EMEM or PBS.
2. Inoculate each of (a minimum of) four wells of a 96-well tissue culture plate with 40 μl of diluted virus.
3. Add 100 μl of cell suspension. Add cells to virus-free wells as a negative control. Replace the plate cover and incubate the plate at 34°C in a humidified atmosphere containing 5% CO_2.

Protocol 6. *Continued*

4. Observe the appearance of c.p.e. in the virus-containing wells by microscopy, and stain the cell sheet after 5 days with crystal violet solution. The virus titre is expressed as the dilution required to infect 50% of the cultures, and is usually expressed as $TCID_{50}$/ml.

Protocol 7. Plaque assay of poliovirus antigen chimeras

Materials

- sub-stock of chimera from *Protocol 5*
- six-well tissue culture plates seeded 24 h in advance with 1×10^6 Ohio HeLa or Hep2c cells/well in 3 ml of medium and incubated at 37°C in a humidified atmosphere containing 5% CO_2

Method

1. Prepare ten-fold serial dilutions of virus stock in serum-free EMEM or PBS.
2. Remove the growth medium from each well by aspiration and wash the cell sheet twice with PBS at 37°C.
3. Add a measured amount of virus dilution to each well (100–250 μl) and adsorb at 34°C for 30 min. Ensure that the cell sheet remains covered with the diluted virus stock by rocking the plate periodically, since drying of the monolayer will kill the cells.
4. Add 3 ml of agar overlay to each well and rock the plate to ensure even distribution. When the overlay has solidified, incubate the plate inverted in a humidified atmosphere containing 5% CO_2 at 34°C (for Sabin-derived constructs) or 37°C (for Mahoney-based chimeras).
5. Monitor the appearance of plaques at intervals by light microscopy. After 3–5 days remove the agar overlay using a small sterile spatula and stain the cell sheet with crystal violet solution. Express the titre of the virus chimera in p.f.u./ml.

6.3 Virus genome sequencing

The sequence of the chimeric virus RNA must be determined to confirm that the authentic introduced sequence is present, and that no adaptive changes have occurred during recovery of viable virus. The sequence can be determined directly from the virus RNA or indirectly by PCR amplification of a region of the genome spanning antigenic site 1. The former approach involves virus purification and depends upon the chimera growing to a satisfactory titre. In contrast, the sensitivity of PCR amplification facilitates rapid screening of small amounts of virus-containing media, but requires direct DNA sequencing or subsequent cloning of the PCR fragment. The method used for purification and extraction of virus RNA is given in *Protocol 8*, and is adapted from that reported by Rico-Hesse *et al.* (31).

Protocol 8. Virus purification for RNA sequencing

Materials

- sub-stock of chimera from *Protocol 5*
- TN buffer: 50 mM Tris–HCl (pH 7.8), 50 mM NaCl
- 30% sucrose solution: 30% (w/v) sucrose in 1 M NaCl, 20 mM Tris–acetate (pH 7.5), 0.1% (w/v) bovine serum albumin
- TSE: 10 mM Tris–HCl (pH 8), 50 mM NaCl, 1 mM EDTA
- 10% (w/v) sodium dodecyl sulphate (SDS)

Method

1. Add 250 μl of the sub-stock to a 75 cm^2 flask of permissive cells and adsorb for 30 min at 37°C. Add 10 ml of growth medium and incubate at 34°C until complete c.p.e. is observed.
2. Freeze–thaw the flask twice, transfer the contents to a sterile plastic centrifuge tube and remove cell debris by centrifuging at 10 000 *g* for 10 min. Transfer the supernatant to a sterile round-bottomed centrifuge tube.
3. Precipitate virions by adding polyethylene glycol 6000 to 7% (w/v) and NaCl to 2.3% (w/v). Incubate overnight at 4°C with constant stirring using a magnetic stirring bar.
4. Pellet precipitated virions by centrifuging at 10 000 *g* for 10 min, discard the supernatant, and resuspend the pellet in 1.5 ml of TN buffer. Insoluble material can be removed by further centrifugation at 4500 *g* for 10 min.
5. Pellet virions by centrifugation at 100 000 *g* (e.g. 40 000 r.p.m. in a Beckman TLS55 rotor in a TL100 centrifuge) at 4°C for 2 h through a 0.5 ml cushion of 30% sucrose solution.
6. Resuspend the pellet in 400 μl of TSE, and add 40 μl of 10% SDS and 4 μl of 2-mercaptoethanol. Dissociate the virions by incubating at 60°C for 2 min.
7. Extract the disrupted virions three times with an equal volume of 1:1 (v/v) phenol:chloroform and precipitate the virus RNA by adding 5 M LiCl to 0.2 M and 2.5 volumes of ice-cold ethanol to the aqueous phase.
8. Centrifuge at 13 000 *g* for 10 min and resuspend the RNA pellet in 5 μl of DEPC-treated sterile water.

Virus RNA extracted according to the method described in *Protocol 18* is routinely sequenced using the modification of the chain termination method originally described by Rico-Hesse *et al.* (31). An alternative strategy (32) for direct sequencing of virus RNA has been reported to generate more consistently labelled products with less cross-banding.

Direct sequencing of poliovirus RNA requires relatively time-consuming purification of virus and template, and can generate ambiguous results due to template impurity or poor growth of the chimera (and therefore limited amount of the RNA template). An alternative strategy for determining the sequence of the modified region of the capsid is to exploit the sensitivity of PCR. This requires very small amounts of virus-containing tissue culture supernatant, and can be readily completed within the time taken to purify the RNA template in *Protocol 8*. The RNA extraction, reverse transcription, and PCR amplification of poliovirus chimeras modified in antigenic site 1 is described in *Protocol 9*.

Protocol 9. RNA extraction, reverse transcription, and PCR amplification

Materials

- positive and negative primers (complementary to nucleotides 2627–2649 and 2924–2905 of Sabin 1, respectively) at 100 ng/μl
- vanadyl ribonucleoside complexes (VRC; Sigma)
- TE: 10 mM Tris–HCl (pH 8.0), 1 mM EDTA
- Moloney murine leukaemia virus (Mo-MLV) reverse transcriptase (RT; Life Technologies)
- 5 × RT buffer: 250 mM Tris–HCl (pH 8.3), 375 mM KCl, 15 mM $MgCl_2$ (supplied with RT)

Method

1. To prevent RNA degradation, add 25 μl of VRC to 250 μl of clarified tissue culture supernatant containing the viable antigen chimera.
2. Lyse the virus particles by adding 14 μl of 10% SDS and incubating for 10 min on ice.
3. Add 20 μl of 20 mg/ml proteinase K and incubate for 20 min at 37°C.
4. Extract the sample twice with an equal volume of phenol:chloroform, taking care to avoid material at the interface.
5. Precipitate the extracted RNA by adding 4 μl of 2 M sodium acetate and 750 μl of ice-cold ethanol. Centrifuge the sample at 12 000 *g* for 10 min. Resuspend the RNA-containing pellet (coloured black owing to the presence of VRC) in 40 μl of TE. Store frozen until required.

6. Prepare the following reagents in a 500 μl microcentrifuge tube for the reverse transcription reaction and incubate at 37°C for 5 min.
   - RNA (from step 5) 2 μl
   - negative sense primer 1 μl
   - 2 mM dNTPs 2.5 μl
   - sterile water 1.5 μl
   - 5 × RT buffer 2 μl
   - RT (200 U) 1 μl
7. Add the following (see *Protocol 1* for buffer components).
   - 2 mM dNTPs 10 μl
   - 10 × *Taq* polymerase buffer 10 μl
   - negative sense primer 500 ng
   - positive sense primer 500 ng
   - sterile water to a total volume of 100 μl
   - *Taq* polymerase 2 U

   Cover the reaction with 60 μl of sterile mineral oil and amplify by completing 30 cycles of 95°C for 1 min, 55°C for 2 min, and 72°C for 2 min.
8. Confirm successful amplification of a 300 bp fragment by agarose gel electrophoresis of 10 μl of the PCR reaction.

PCR amplifies the population of molecules present after reverse transcription of the RNA template. Rather than cloning and sequencing individually amplified molecules, it is preferable to determine directly the sequence of the population present after amplification. This reduces the significance of the majority of *Taq* polymerase errors when the sequence is interpreted. Several methods exist for direct sequencing of PCR products—in our experience all are significantly affected by the quality of the template. The sequence is often of better quality if the primer lies within those used in the original PCR. The methods of Winship (33) and the *fmol* system (Promega) yield consistently good results.

6.4 Antigenic characterization

The antigenic characteristics of a chimeric virus particle define the authenticity with which a foreign epitope is presented on the surface of the poliovirus capsid. Chimeras that react well with antibodies generated against the foreign protein are likely to present the foreign epitope in a conformation closely resembling that in the native context. However, the conclusion that can be drawn from this type of analysis depend very much upon the nature of the epitopes recognized by the antibodies used in the assay. Those that are known to recognize a linear epitope, and that also exhibit reactivity with peptides derived from the epitope, are most likely to react with the peptide sequence presented on the surface of the poliovirus particle. In contrast, antibodies that are unreactive with linear peptides are far less likely to bind to the chimeric virus particle. Nonetheless, the results depend largely upon the antibodies used in the assays, and do not preclude antibodies induced by the chimera from recognizing the native foreign protein. These comments apply to both mAbs and polyclonal antisera, though the latter are far more likely to contain a range of reactivities, including a proportion that recognizes linear epitopes.

The interaction of the chimeric virus with antibodies can be readily monitored in two ways: by observing the ability of the antibody to neutralize the infectivity of the virus particle, resulting in a reduction in plaque numbers or $TCID_{50}$, or by the restriction of virus diffusion through a semi-solid support. The ability to bind to the virus in the latter assay is generally reflected by the neutralization of the virus in the former. Minor (30) has previously described the generation of purified radiolabelled virus for use in a radio-immunodiffusion assay, and the method for performing the 'antigen blocking' assay. *Protocol 10* describes the neutralization assay used to qualify the antigenic characteristics of a chimeric virus.

Protocol 10. Neutralization assay by reduction of $TCID_{50}$

Materials

- suitable antibody
- virus stock of known titre (*Protocol 6*)

Method

1. Prepare two- or ten-fold dilutions of antibody in serum-free EMEM or PBS.
2. Prepare a diluted virus stock at 2.5 $TCID_{50}$/μl in serum-free EMEM.
3. Dispense 10 μl aliquots of diluted antibody in each of (a minimum of) two wells of a sterile 96-well tissue culture plate and add 40 μl of diluted virus to each well. Set up control wells containing no antibody, or virus alone. Incubate for 1–2 h at 34°C.
4. Add 100 μl of permissive cells (i.e. 1×10^4 cells) to each well containing virus and/or antibody, and incubate at 34°C for 5 days in an atmosphere containing 5% CO_2.
5. Stain the wells with crystal violet and score for the presence or absence of c.p.e.
6. Compare the control wells with antibody-containing wells and, after confirmation of virus titre, express the neutralization titre of the antibody in terms of the reciprocal titre that will neutralize 100 $TCID_{50}$ units of virus.

7. Summary

The design, construction and characterization of poliovirus antigen chimeras described in this chapter involve the application of a number of techniques from the fields of virology and molecular biology. These research areas are advancing rapidly and are affected significantly by the development of new technologies. However, the underlying principles involved are relatively simple and allow the approach to be applied to any of the closely related members of the picornavirus family. The potential applications of picornavirus antigen chimeras in vaccine design and development, antigen presentation and characterization, and the generation of valuable diagnostic reagents, have yet to be fully realized.

Acknowledgements

I acknowledge the significant contributions made to these studies by colleagues at the University of Reading (Jeffrey Almond, Karen Burke, Janet Meredith, Charlotte Rose, and Sean Whelan), the National Institute for Biological Standards and Control (Philip Minor, Morag Ferguson, and Cherylyn Vella) and University College London (Helen Stirk and Janet Thornton). This work was funded by the Medical Research Council and the MRC AIDS Directed Programme.

Note added in proof

Whilst this chapter was in preparation Dedieu *et al.* (34) published results describing the use of sequences flanking the introduced foreign epitope to ensure the viability of the chimera. These results, and the importance of critical charged residues within the βB-βC loop, are similar to those presented in Section 5.1.

References

1. Rose, C. S. P. and Evans, D. J. (1991). *Tibtech,* **9,** 415.
2. Sabin, A. B. and Boulger, L. R. (1973). *J. Biol. Stand.,* **1,** 115.
3. Assaad, F. and Cockburn, W. C. (1982). *Bull. WHO,* **60,** 231.
4. Almond, J. W. (1987). *Annu. Rev. Microbiol.,* **41,** 153.
5. Kohara, M., Abe, S., Komatsu, T., Tago, K., Arita, M., and Nomoto, A. (1988). *J. Virol.,* **62,** 2828.
6. Burke, K. L., Dunn, G., Ferguson, M., Minor, P. D., and Almond, J. W. (1988). *Nature,* **332,** 81.
7. Martin, A., Wychowski, C., Couderc, T., Crainic, R., Hogle, J., and Girard, M. (1988). *EMBO J.,* **7,** 2839.
8. Murray, M. G., Kuhn, R. J., Arita, M., Kawamura, N., Nomoto, A., and Wimmer, E. (1988). *Proc. Natl Acad. Sci. USA,* **85,** 3203.

9. Racaniello, V. R. and Baltimore, D. (1981). *Science,* **214,** 916.
10. Palmenberg, A. C. (1989). In *Molecular aspects of picornavirus infection and detection* (ed. B. L. Semler and E. Ehrenfeld), pp. 211–241. ASM, Washington.
11. Hogle, J. M., Chow, M., and Filman, D. J. (1985). *Science,* **229,** 1358.
12. Minor, P. D., Ferguson, M., Evans, D. M., Almond, J. W., and Icenogle, J. P. (1986). *J. Gen. Virol.,* **67,** 1283.
13. Minor, P. D. (1990). In *Current topics in microbiology and immunology 161: picornaviruses* (ed. V. R. Racaniello), pp. 122–154. Springer-Verlag, Berlin.
14. Van der Werf, S., Bradley, J., Wimmer, E., Studier, F. W., and Dunn, J. J. (1986). *Proc. Natl Acad. Sci. USA,* **83,** 2330.
15. Burke, K. L., Evans, D. J., Jenkins, O., Meredith, J., D'Souza, E. D. A., and Almond, J. W. (1989). *J. Gen. Virol.,* **70,** 2475.
16. Sarnow, P. (1989). *J. Virol.,* **63,** 467.
17. Evans, D. J. and Almond, J. W. (1991). *Methods Enzymol.,* **203,** 386.
18. Murdin, A. D. and Wimmer, E. (1989). *J. Virol.,* **63,** 5251.
19. Reimann, B. Y., Zell, R., and Kandolf, R. (1991). *J. Virol.,* **65,** 3475.
20. Devereux, J., Haeberli, P., and Smithies, O. (1984). *Nucleic Acids Res.,* **12,** 387.
21. Kunkel, T. (1985). *Proc. Natl Acad. Sci. USA,* **82,** 488.
22. Higuchi, R. (1990). In *PCR protocols,* (ed. M. A. Innis, D. H. Gelfand, J. J. Sninsky and T. J. White), pp. 177. Academic Press, San Diego.
23. Pellequer, J. L., Westhof, E., and Van Regenmortel, M. H. (1991). *Methods Enzymol.,* **203,** 176.
24. Murphy, G. and Kavanagh, T. (1988). *Nucleic Acids Res.,* **16,** 5198.
25. La Monica, N., Kupsky, W. J., and Racaniello, V. R. (1987). *Virology,* **161,** 429.
26. Murray, M. G., Bradley, J., Yang, X. F., Wimmer, E., Moss, E. G., and Racaniello, V. R. (1988). *Science,* **241,** 213.
27. Rossmann, M. G., Arnold, E., Erickson, J. W., Frankenberger, E. A., Griffith, J. P., Hecht, H. J., Johnson, J. E., Kamer, G., Luo, M., Mosser, A. G., Rueckert, R. R., Sherry, B., and Vriend, G. (1985). *Nature,* **317,** 145.
28. Colonno, R. J., Condra, J. H., Mizutani, S., Callahan, P. L., Davies, M. E., and Murcko, M. A. (1988). *Proc. Natl Acad. Sci. USA,* **85,** 5449.
29. Minor, P. D., Pipkin, P. A., Hockley, D., Schild, G. C., and Almond, J. W. (1984). *Virus Res.,* **1,** 203.
30. Minor, P. D. (1985). In *Virology: a practical approach,* (ed. B. W. J. Mahy), pp. 25–41. IRL Press, Oxford.
31. Rico-Hesse, R., Pallansch, M. A., Nottay, B. K., and Kew, O. M. (1987). *Virology,* **160,** 311.
32. Fichot, O. and Girard, M. (1990). *Nucleic Acids Res.,* **18,** 6162.
33. Winship, P. (1989). *Nucleic Acids Res.,* **17,** 1266.
34. Dedieu, J. F., Ronco, J., Van der werf, S., Hogle, J. M., Henin, Y., and Girard, M. (1992). *J. Virol.* **66,** 3161.

8

Baculovirus expression vectors

LORNA M. D. STEWART and ROBERT D. POSSEE

1. Introduction

Progress in molecular virology has given us sufficient understanding of genetic organization and regulation in certain viruses of eukaryotes to enable us to utilize these systems as expression vectors of foreign genes. Analysis of the baculovirus genome has identified genes which are redundant when the virus is grown in cell culture. Furthermore, the promoters associated with these genes are exceptionally active, leading to the production of considerable quantities of protein within virus-infected cells. Consequently, baculoviruses have been used as a vehicle for the production of a variety of foreign gene products. In most cases the yields are very high and provide material suitable for studies of protein structure–function, therapeutic applications, or as vaccines. Foreign genes have also been inserted into the baculovirus genome to enhance the effectiveness of these agents as virus insecticides.

Viruses of the Baculoviridae have a large, double-stranded circular genome (80–120 kbp) (1) and are infectious only to invertebrates (primarily insects) (2). An individual virus is usually named after the species from which it was originally isolated. Baculoviruses are classified, according to structural considerations, as nuclear polyhedrosis viruses (NPVs), granulosis viruses (GVs), or non-occluded viruses (NOVs). The NPVs have rod-shaped nucleocapsids, containing the DNA, which are enveloped by a lipoprotein membrane to form virus particles. These are further embedded within an occlusion body comprising the polyhedrin protein. The occlusion bodies or polyhedra contain several virus particles which may have a single nucleocapsid (SNPV) or multiple nucleocapsids (MNPV). Each baculovirus is genetically programmed to produce one or the other type. The NPVs have been shown to infect larvae of *Lepidoptera, Hymenoptera, Diptera, Coleoptera*, and *Trichoptera*. The GVs have a similar structure but only one virion, containing a single nucleocapsid, is embedded in granulin protein. These viruses are considered to infect lepidopteran insects exclusively. NOVs are not occluded at any stage in their life cycle and are confined mainly to coleopteran species.

The virus most widely used as an expression vector is *Autographa californica* MNPV (AcMNPV), which was derived from the larvae of the alfalfa looper.

It grows readily in cell culture systems and has become the prototype baculovirus for genetic and molecular studies. These data permitted the development of the expression vector system. AcMNPV will be the major topic of discussion in this chapter.

2. The life cycle of baculoviruses

In the natural environment, baculoviruses are able to persist for long periods (years) by virtue of the protective protein coat comprising the polyhedrin protein. These structures are relatively resistant to degradation and facilitate the spread of infection between individual larvae (3). The polyhedra are ingested by larvae and dissolve in the alkaline pH of the midgut to release infectious virus particles. In susceptible larvae the virion envelope fuses with the microvilli of midgut epithelial cells and the nucleocapsids are transported to the nucleus where they uncoat and initiate virus replication. Cell culture studies have recognized four periods of transcription during a baculovirus infection cycle (reviewed in ref. 4): immediate early, delayed early, late, and very late phases. Immediate early transcription (0–3 h post infection (p.i.)) involves genes which require no viral products for their expression; host cell factors are presumed to be sufficient for their activation. The product of one of these genes (IE-1) has been shown to work as a trans-activator of other early genes (5). Transcription of delayed early genes (3–6 h p.i.) requires the presence of the immediate early virus gene products. Late (6–18 h p.i.) and very late (18–72 h p.i.) genes are transcribed at the same time as viral DNA synthesis and the formation of progeny virus. Examples of very late genes are those coding for the structural protein polyhedrin (6) and the p10 protein, which has been implicated as a cytoskeletal structure important in cell lysis (7). The polyhedrin and p10 proteins are synthesized in virus-infected cells in abundant quantities and may account for 25–40% of total protein late in infection.

In the course of the baculovirus infection, two genetically identical but morphologically distinct types of virus structure are produced. Nucleocapsids (from about 12 h p.i.) may traverse the nuclear membrane and bud through the plasma membrane to produce budded virus which spreads infection to other cells within the insect host or in an *in vitro* system. Alternatively, the nucleocapsid may be enveloped and occluded in the nucleus to form polyhedra (observed at 24 h p.i.), where they are retained and released subsequent to the death of the larva to act as infectious agents for other larvae. In cell culture, polyhedra have no role in propagation of the virus because the alkaline conditions required for their dissolution would be lethal for the cells. Polyhedra are clearly visible by light microscopy of virus-infected cultured cells.

3. Insect cell culture

A prerequisite for any expression system is the ability to propagate the virus in cell culture. The insect cells used for maintenance of AcMNPV were derived from pupal ovarian tissue of the fall army worm (*Spodoptera frugiperda*). The cell lines which are routinely used are *S. frugiperda* (IPLB-Sf-21) (8) and *S. frugiperda* (IPLB-Sf-9) (ATCC CRL 1711) (9). Recently, a cell line has become available (BTI-TN-5B1-4, from Invitrogen) which is reported to give a ten-fold increase in protein yields compared with *S. frugiperda* cells. However, it is not clear whether this increased yield will apply to all foreign gene products.

In addition to routine good laboratory practice, the following guidelines should be followed:

- All glassware should be clean and detergent-free.
- Use of a laminar flow cabinet is recommended for all uninfected cell work, although virus infections may be carried out on the bench.
- It is suggested that cells are used over a defined number of passages after liquid nitrogen storage to ensure reproducibility of results.
- Virus-infected material must be autoclaved or disinfected prior to disposal.

Cell lines are maintained in a complex medium, TC100 (Gibco), which is supplemented with 5 or 10% (v/v) fetal calf serum (FCS), penicillin (500 U/ml) and streptomycin (500 μg/ml). The use of antibiotics is optional; we prefer not to include them when growing cells routinely. Stocks of cells are grown at room temperature or 28°C in monolayer culture in glass or plastic flasks, or maintained in suspension in spinner flasks. Apparatus for the spinner cultures can be produced simply by sterilizing a flat-bottomed, round glass flask containing a magnetic stirring bar and sealed with a foil cap. The medium and cells are subsequently dispensed into the vessel and stirred to facilitate growth. When monolayer cultures become confluent the medium is removed from the bottles, replaced with a small amount of fresh medium, and the cell layer is agitated or scraped into suspension. Cells are then counted and dispensed to new flasks as required. A seeding density of 5×10^4 cells/cm^2 is normally used for routine passage. Suspension cultures are diluted 1 in 20 when the cell density reaches 2×10^6 cells/ml. These cells double in number approximately every 24 h.

For long-term storage, stocks of *S. frugiperda* cells are maintained in liquid nitrogen. Cells in the logarithmic phase of growth are harvested from the monolayer or spinner cultures. Their concentration is adjusted to 2×10^6 cells/ml with TC100 + 10% (v/v) FCS containing 10% (v/v) dimethylsulphoxide (DMSO) (culture grade) and 1 ml aliquots are then placed in cryogenic tubes and chilled immediately on ice. Subsequently, the tubes are transferred to the vapour-phase chamber of the liquid nitrogen container overnight before being placed into the liquid nitrogen for storage. To revive the cells, the frozen suspension is thawed rapidly and seeded into a 25 cm^2 flask containing 5 ml of TC100 + 10% (v/v) FCS and incubated at 28°C. After 1–2 h, viable cells will have adhered to the flask and the medium can be decanted and replaced.

S. frugiperda cells may be used to propagate AcMNPV. For small-scale virus production (for example to amplify the virus from a plaque isolate) the work may be conveniently performed in Petri dishes (tissue culture grade; 35 mm diameter). This procedure is described in *Protocol 1*. Titration of the virus produced is by plaque assay (see *Protocol 2*). If larger quantities of virus are needed, the cells may be grown in 60 mm diameter dishes or 25 or 75 cm^2 flasks. Seeding densities for these vessels are given in *Table 1*. The use of cells in suspension culture for preparing larger stocks of virus is described in *Protocol 3*.

Protocol 1. Preparation of virus stocks in *S. frugiperda* cells

Materials

- sandwich box lined with moist tissue paper
- incubator at 28°C
- 35 mm diameter Petri dishes (tissue culture grade)
- logarithmic phase *S. frugiperda* cells
- TC100 growth medium: TC100 supplemented with 5 or 10% (v/v) FCS, 500 U/ml penicillin, 500 μg/ml streptomycin
- virus inoculum

Method

1. Seed 35 mm Petri dishes each with 10^6 cells in 1.5–2.0 ml of TC100 growth medium and incubate overnight at 28°C.
2. Remove the medium from the cells and add the virus inoculum (100 μl) at a multiplicity of infection (m.o.i.) of 0.1 plaque-forming units (p.f.u.)/cell. If using virus isolated from a single plaque (see Section 7.3), inoculate the cells with 100 μl of the 500 μl plaque stock.
3. Incubate the cells for 1 h at room temperature on a level surface. It is important to ensure the surface is level so that the virus suspension covers the cell monolayer evenly. If preferred, gently rock the dish(es) every 15–20 min to bathe the cell monolayer.
4. Remove the inoculum with a Pasteur pipette and discard this safely, according to local rules. Replace the inoculum with 1.5–2.0 ml of TC100 growth medium.
5. Incubate the cells in an humidified sandwich box and observe daily for signs of infection. In general, virus stocks will need to be harvested after 4–5 days.
6. Harvest the cells and medium together using a Pasteur pipette to dislodge cells from the dish. Virus-infected cells tend to detach from the substratum and float in the medium.

7. Remove the cells by low speed centrifugation (2500 *g* for 10 min) and store the clarified medium at 4°C (days–weeks) or −70°C (months–years). Titrate as described in *Protocol 2*.
8. To amplify the virus stock further, simply repeat the procedure with larger flasks.

Table 1. Seeding densities for *S. frugiperda* cells

Dish/flask size	Seeding density[a]	Volume of medium (ml)
35 mm dish	1×10^6	1.5–2.0
60 mm dish	2×10^6	3.0–4.0
25 cm² flask	1×10^6	3.0–4.0
75 cm² flask	$0.5–1.0 \times 10^7$	10
Spinner cultures	$1–2 \times 10^5$/ml[b]	40–500

[a] Cells are incubated at 28°C overnight before use.
[b] Grow cells to a density of 5×10^5 ml prior to infection for production of high-titre virus stocks and $1–2 \times 10^6$ ml for the production of recombinant protein.

Protocol 2. Titration of virus by plaque assay (ref. 10)

Materials

- supplemented TC100 growth medium (*Protocol 1*)
- Logarithmic phase *S. frugiperda* cells in 35 mm Petri dishes, tissue culture grade
- 3% (w/v) low-gelling temperature agarose (FMC Sea-plaque) dissolved and sterilized by autoclaving in distilled water
- 0.5% (w/v) neutral red stain in water, filter sterilized and stored in the dark
- phosphate-buffered saline (PBS): 140 mM NaCl, 2.7 mM KCl, 8 mM Na_2HPO_4, 1.4 mM KH_2PO_4 (pH 7.2)

Method

1. Seed logarithmic phase *S. frugiperda* cells (1.5×10^6 cells in 2 ml of TC100 (5% FCS) into a 35 mm diameter Petri dish and incubate at 28°C for 2–4 h.
2. Prepare ten-fold dilutions of the virus in TC100 (5% FCS). The dilutions used are determined by the source of the virus. For example, to titrate virus from an individual plaque isolate, use undiluted, 10^{-1}, 10^{-2}, and 10^{-3} dilutions. To titrate other virus stocks, assay 10^{-4}, 10^{-5}, 10^{-6}, and 10^{-7} dilutions.

Protocol 2. *Continued*

3. Remove the medium from the *S. frugiperda* cells and inoculate with 100 μl of each dilution (2–3 plates) or with TC100 growth medium alone as a control. Incubate the dishes at room temperature for 1 h to allow the virus to attach to the cells.
4. Remove the virus inoculum carefully and cover the cells with 2 ml of overlay medium (3% (w/v)) low gelling temperature agarose at 37°C, diluted 1:1 with TC100 (10% FCS).
5. When the overlay has set, add 1 ml TC100 (5% FCS) to each Petri dish and incubate the dishes at 28°C for 3–4 days.
6. After incubation, stain the cells with neutral red. Dilute the stock neutral red 1:20 with PBS and add 1 ml of this to each dish. After 2–4 h at 28°C remove the stain and invert the dish to allow the plaques to clear for 12 h in the dark.[a]
7. Examine the plates and count the plaques.

[a] The virus plaques may be stained with X-gal to allow detection of β-galactosidase (see Section 7).

Protocol 3. Purification of virus particles and preparation of viral DNA

Material

- supplemented TC100 growth medium (*Protocol 1*)
- titrated virus stock (*Protocol 2*)
- TE: 10 mM Tris–HCl (pH 8.0), 1 mM EDTA
- 10% (w/v) and 50% (w/v) sucrose in TE (sterile)
- one-step sucrose gradient: prepare this in a 15 ml ultraclear swing-out centrifuge tube (e.g. for Beckman SW41 rotor) by layering 5 ml of 10% (w/v) sucrose on to 5 ml of 50% (w/v) sucrose
- Sarkosyl solution: 20% (w/v) *N*-lauryl sarcosine in 10 mM Tris–HCl (pH 8.0), 10 mM EDTA
- CsCl gradient: 5 ml of 50% (w/w) CsCl solution containing ethidium bromide (25 μg/ml) in a 15 ml ultraclear swing-out centrifuge tube (e.g. for Beckman SW41 rotor); an angle or vertical rotor may be used in place of the swing-out rotor

Method

1. Set up a 400–500 ml spinner culture seeded with *S. frugiperda* cells at 1–2 $\times 10^5$ cells/ml in TC100 (5% FCS). Incubate at 28°C until the density has reached 5×10^5 cells/ml (2–3 days), then inoculate the spinner with virus at a m.o.i. of 0.1–0.2 p.f.u./cell.

2. At 4–5 days p.i., when all cells are infected, decant the culture into bottles and remove cell debris by centrifugation at low speed (2500–3000 *g*) for 10 min. Transfer the clarified supernatant to ultracentrifuge tubes and pellet the virus at 100 000 *g* for 1 h at 4°C.
3. Decant the medium and soak the pellet in a small volume (1–2 ml) of TE overnight at 4°C. Gently resuspend the pellet. A low speed centrifugation step (see step 2) may be introduced at this stage to remove the remaining cellular debris. Layer the virus on top of a one-step sucrose gradient. Centrifuge at 100 000 *g* for 1 h at 4°C.
4. The virus will band at the interface between the two solutions. Harvest the virus from above using a wide-bore Pasteur pipette, or from below by puncturing the bottom and letting the sucrose drip through the hole.
5. Dilute the virus 1:5 in TE and pellet at 100 000 *g*.
6. Soak the pellet overnight in a small volume of TE (1–2 ml). Resuspend gently and store at 4°C. Do not freeze the virus since this causes lysis of the virus particles and subsequent difficulties in purifying the DNA.
7. Transfer 400 μl of the purified virus suspension into a sterile microcentrifuge tube and add 100 μl of Sarkosyl solution. Incubate at 60 °C for 30 min.
8. Layer the lysed virus suspension on to the CsCl gradient. Top up the tube with liquid paraffin and spin at approximately 250 000 *g* for 18 h at 20 °C (35 000 r.p.m. in a Beckman SW41 rotor).
9. Harvest the DNA (two visible orange bands corresponding to supercoiled and nicked-circular species) from above or below (see step 4 above) and extract the ethidium bromide by mixing three or four times with equal volumes of *n*-butanol.[a] Remove and discard the upper butanol phase after each extraction. During the butanol extraction do not shake the DNA vigorously.
10. Dialyse the DNA solution against 1 litre of sterile TE overnight at 4°C. Transfer the DNA into a sterile container and store at 4 °C.[a,b] Determine the concentration of the DNA from its OD_{260}. Remember that the final DNA will be used for transfections and therefore should be kept sterile. Do not freeze virus DNA, since this will reduce its infectivity.

[a] Care must be taken to prevent shearing when using pipettes, etc.
[b] The DNA should not be precipitated as this may cause shearing.

4. Baculovirus expression vectors

4.1 Manipulating the baculovirus genome

The large size of the baculovirus genome (AcMNPV, 128 kbp) means that it cannot be manipulated directly in order to insert foreign sequences, using restriction enzymes and DNA ligases. Until recently, unique sites for restriction enzyme cleavage were unknown in the AcMNPV genome. Consequently, it was necessary to adopt an indirect approach to manipulating baculovirus DNA.

Regions of the virus genome spanning the polyhedrin gene were inserted into bacterial plasmids. The polyhedrin gene (11) was modified using restriction enzymes and exonucleases to delete the coding sequences; the promoter and transcription termination regions were left intact. A restriction enzyme site was introduced between the regulatory sequences to facilitate the insertion of foreign DNA sequences and thus derive the transfer vector (12, 13).

The transfer vector containing the inserted foreign DNA is co-transfected into insect cells along with infectious viral DNA to allow the recombinant DNA to insert into the virus genome via homologous recombination. The site of insertion is dictated by the identity of the homologous flanking regions. Replacement of the polyhedrin coding sequences in the virus results in a polyhedrin-negative mutant. This may be distinguished as a clear plaque in a standard titration experiment and provides a visually identifiable phenotype for selection of recombinant viruses. Other methods of selection are also now available and are discussed in Section 7.

The baculovirus expression vector is the virus produced after co-transfection of insect cells with infectious viral DNA and the transfer vector.

4.2 Baculovirus transfer vectors

Most baculovirus expression vectors are based on the polyhedrin gene locus and make use of the polyhedrin gene promoter. More recently, the p10 gene locus and its promoter have been exploited. Vectors have also been constructed which permit the insertion of two foreign genes into the same virus genome (at the polyhedrin and p10 gene loci). Virus gene promoters active in the earlier stages of the virus replication cycle have also been utilized, to allow investigation of the efficacy of post-translational modifications prior to the very late phase of gene expression. A summary of some of the baculovirus transfer vectors and the appropriate viral DNA to be used in co-transfections is provided in *Table 2*.

4.2.1 Polyhedrin gene promoter-based transfer vectors

The differences between the many baculovirus transfer vectors utilizing the polyhedrin promoter are minimal, but may cause confusion to the inexperienced user of the system. Frequently, different research groups have constructed essentially the same vector and given it a new designation. Recommended transfer vectors include: pEV55 (14); pAcRP23 (15), pAcYM1 (16), and pVL941 (17). Each of these vectors retains the complete polyhedrin gene promoter. Another useful transfer vector is pAcCL29 (18). This vector was derived from pAcYM1 and has reduced virus sequences flanking the polyhedrin gene promoter, together with the bacteriophage M13 intergenic

Table 2. List of suggested transfer vectors and appropriate viral DNA to be used in co-transfections

Transfer vector	Parental viral DNA	Selection method for recombinant virus	Reference
1. *Polyhedrin gene locus*			
(a) Single expression vectors	AcMNPV	polyhedrin-negative plaques	13
		dot-blot hybridization	36
	AcRP23.lacZ	β-galactosidase-negative (clear plaques)	15
	AcRP23.lacZ digested with *Bsu*36I	β-galactosidase-negative (clear plaques)	37
pAcYM1 (polyhedrin)			16
pEV55 (")			14
pVL941 (")			17
pAcCL29 (")			18
pAcMP1 (basic)			28
(*positive selection*)	AcMNPV	β-galactosidase-positive (blue plaques)	
pJVNheI (polyhedrin)			19
pAcDZI (")			20
(b) Multiple expression vectors	AcRP6.SC digested with *Bsu*36I		
pAcUW2B			37
(p10 and normal polyhedrin gene)		polyhedrin-positive virus	23
pAcUW3 (polyhedrin and p10)		dot-blot hybridization on randomly-selected plaques	27
2. *p10 gene locus*	AcMNPV	dot-blot hybridization	36
	AcUW1.lacZ (digested with *Bsu*36I)	β-galactosidase-negative (clear plaques)	23, 37
pAcUW1 (p10)			23
(*positive selection*)	AcMNPV	β-galactosidase-positive (blue plaques)	
pAcS3 (p10)			24

region. The latter feature enables the synthesis of single-stranded DNA in *Escherichia coli* after superinfection with the mutant bacteriophage M13 KO7. It is particularly useful if alterations are to be made to a foreign gene which require sequencing to confirm the genetic structure. Two other transfer vectors include pJVNhel (19) and pAcDZ1 (20). In these vectors a copy of the *lacZ* gene coding region has been incorporated, under the control of a second promoter, upstream from the polyhedrin gene. When the transfer vector is co-transfected with viral DNA, the recombinant progeny yield a readily detectable blue phenotype in plaque assays involving staining with 5-bromo-4-chloro-3-indolyl β-D-galactopyranoside (X-gal).

4.2.2 p10 gene promoter-based transfer vectors

The p10 gene may be deleted from the virus genome without apparent effect on virus replication and the formation of infectious virus particles and polyhedra (21, 22). The role of the p10 protein is poorly understood, but it appears to function in the formation of polyhedra. There are few transfer vectors available for inserting foreign coding sequences at the p10 gene locus. This is largely because, until recently, there was no convenient way of identifying recombinant viruses with insertions at this site; the presence or absence of p10 protein synthesis cannot be identified using light microscopy.

Transfer vectors available include pAcUW1 (23), which is analogous to pAcYM1, and pAcS3 (24), which has a copy of the *lacZ* gene inserted within the transfer vectors to facilitate positive selection of recombinant viruses in a similar way to pJVNhel and pAcDZ1. Although pAcUW1 may now seem redundant, the advent of other recombinant virus selection methods (see Section 7) make it a useful vector.

4.2.3 Multiple gene promoter transfer vectors

The co-expression of multiple foreign genes in insect cells may be achieved by co-infecting cells with several baculoviruses, each containing a single coding region under the control of the polyhedrin or p10 gene promoters. Although this has been accomplished (for example influenza virus polymerase genes, ref. 25), it is inconvenient because of the need to prepare accurately titrated stocks of each virus and the difficulty of infecting every cell. To circumvent these problems, baculovirus transfer vectors have been constructed which can contain two foreign gene coding sequences. In preliminary studies, a copy of the polyhedrin gene promoter was inserted upstream of the native polyhedrin gene to derive a virus (AcVC2) in which both the polyhedrin protein and the lymphocytic choriomeningitis virus nucleoprotein were synthesized after infection of insect cells (26). In a further development, the polyhedrin gene coding sequences were deleted to enable insertion of a second foreign coding region into the transfer vector (pAcVC3; V. Emery and D. H. L. Bishop, unpublished data).

The multiple gene promoter transfer vector pAcUW3 (27) uses a copy of the p10 gene promoter, inserted upstream of the polyhedrin gene promoter, to permit the insertion of two foreign gene coding sequences. It was employed to co-synthesize the influenza virus haemagglutinin and neuraminidase glycoproteins in insect cells, and has the advantage over pAcVC3 of having less duplicated sequence, thus avoiding the instability problems associated with the latter vector.

A transfer vector is also available which has the p10 gene promoter inserted upstream of, and in the opposite orientation from, the complete polyhedrin gene (pAcUW2B; ref. 23). This is useful for constructing polyhedrin-positive viruses which are required for the production of recombinant proteins in insect larvae. It also has the advantage that recombinant viruses are selected for the polyhedrin-positive phenotype from a background of polyhedrin-negative plaques.

Other multiple expression vectors currently under development include some which will allow the insertion of two foreign genes at the p10 gene locus (R. D. Possee, unpublished data).

4.2.4 Other baculovirus transfer vectors

The level of protein production achieved using the promoters of baculovirus genes active at earlier stages of the replication cycle to drive foreign gene expression does not match that of the polyhedrin and p10 gene promoters. However, use of these early gene promoters does allow synthesis of foreign proteins in that phase of the virus growth cycle when complex glycoproteins and enzymes are normally produced. This may be an advantage if extensive post-translational modifications to the gene product are required. Unfortunately, most of the early genes encode essential virus structural or regulatory proteins which are unlikely to be dispensable for virus replication. In consequence, these gene promoters must be duplicated within the virus genome, at non-essential gene loci such as those of polyhedrin or p10.

Two AcMNPV late genes have been exploited in expression systems. The basic, or arginine-rich, protein gene promoter was duplicated at the polyhedrin gene locus to effect expression of *lacZ* coding sequences in the late phase (28). This transfer vector was designated pAcMP1. The virus capsid protein-encoding gene (p39; ref. 29) was used in a similar manner to drive expression of the chloramphenicol acetyl transferase gene (CAT) (30). In the latter study, a hybrid promoter was also constructed which contained elements of the p39 gene promoter, coupled with the polyhedrin gene promoter. This vector induced expression of the CAT gene in the late phase and continued to the very late phase. Similar studies have been performed with the basic protein and polyhedrin gene promoters (pAcMP2; Hill-Perkins and Possee, unpublished data).

4.2.5 Baculovirus vectors for secretion or efficient recovery of foreign proteins

Expression vectors are currently being developed to facilitate secretion of recombinant proteins from virus-infected cells. The issue is complicated by the fact that it is uncertain whether simply adding a signal peptide to the amino terminus of a foreign protein will cause its secretion. The intrinsic nature of the gene product itself must be considered; for instance, it is unlikely that a protein which is normally found in the nucleus of the cell can be forced into the secretion pathway.

The vector pAcJM125 (J. Millard, personal communication) is based on pAcYM1, and incorporates the secretory signal coding region of vesicular stomatitis virus glycoprotein (VSV G; ref. 31) after the polyhedrin gene promoter. This should permit secretion of the foreign protein into the culture medium, since VSV G is inserted into the plasma membrane of virus-infected cells when produced by a recombinant baculovirus (32). Another transfer vector, pAcAM1, also uses the polyhedrin promoter but incorporates the AcMNPV gp67 signal peptide coding region before the cloning site (A. T. Clarke, personal communication; ref. 33). This construction has been proven to direct secretion of foreign proteins. In these constructions it is important to consider the eventual sequence around the signal peptide cleavage site. Usually, the junction will have to be mutagenized to produce a favourable environment for cleavage (33).

Other transfer vectors are available (pAcG1/pAcG2T/pAcG3X; A. Davies and I. Jones, personal communication) which express the foreign gene fused to glutathione-S-transferase (GST). In addition, pAcG2T has a thrombin cleavage site after the GST and before the cloning site; pAcG3X has a factor XA cleavage site located in the same position. This facilitates purification of the fusion proteins on glutathione–Sepharose followed by cleavage of the GST by the appropriate endoproteinase. Another benefit of the system is the ability to identify the protein immunogenically when fused to GST, using antisera raised against GST produced in bacteria.

4.3 Preparing recombinant transfer vector

A suitable procedure for the preparation of recombinant transfer vector ready for co-transfection with viral DNA is described in *Protocol 4*.

Protocol 4. Preparing a recombinant transfer vector

Materials

- transfer vector (for example pAcYM1)
- calf intestinal phosphatase (Boehringer)
- foreign DNA insert with compatible ends (i.e. *Bam*HI or *Bgl*II)
- competent *E. coli* cells (prepared for transformation by calcium chloride treatment; ref. 34)
- selective agar plates (containing 100 μg/ml ampicillin)

- T4 DNA ligase (1 U/μl)
- 10 × ligation buffer: 660 mM Tris–HCl (pH 7.5), 50 mM $MgCl_2$, 50 mM dithiothreitol, 10 mM ATP

Method

1. Digest 5 μg of the transfer vector with the appropriate restriction endonuclease (e.g. *Bam*HI) and then treat it with 5 U of calf intestinal phosphatase at 37 °C for 15 min. Add 20% (w/v) sodium dodecyl sulphate (SDS) to 1% (w/v) and 200 mM EDTA to 25 mM, then extract twice with phenol:chloroform (1:1, v/v), ethanol precipitate, and resuspend in 50 μl of TE.
2. Ligate the required fragment to the transfer vector (e.g. 10 ng of transfer vector plus 50 ng of purified fragment in a 10–20 μl volume incorporating the ligation buffer) overnight at 16 °C.
3. Transform competent *E. coli* cells with the recombinant transfer vector (34). Transformation efficiencies are typically 10^6–10^7 colonies/μg vector DNA.
4. Spread the transformed cells on to selective agar plates and incubate at 37 °C for 16–18 h.
5. Isolate plasmid DNA from selected colonies and examine it by restriction enzyme analysis to ensure that the recombinant plasmid contains the inserted DNA fragment in the correct copy number and orientation.[a] This plasmid DNA may now be used in a co-transfection with viral DNA. Although DNA prepared by minipreparation (34) may be used, CsCl-purified DNA (34) is more reliable.

[a] It is a good idea to sequence across the junction between the virus gene promoter and the inserted foreign coding region to confirm the predicted structure.

5. The recombination process

In the first studies to develop baculovirus expression vectors, loss of the polyhedrin phenotype was used as a marker to identify recombinant viruses in plaque assays. With experience, such virus mutants can be readily isolated. However, for the inexperienced user of the system this selection procedure was frequently a source of considerable frustration; many workers found it impossible to locate the necessary plaques. The frequency of recombination between plasmid transfer vectors and infectious viral DNA has been estimated to be around 0.1–1.0% (4, 35), making identification of the recombinant virus tedious.

This problem was alleviated somewhat by the use of dot-blot hybridization analysis on dilutions of the co-transfection medium used to infect cells in 96-well microtitre plates (see *Protocol 7*; ref. 36). The method is still time consuming, however, and ultimately relies on identification of polyhedrin-negative plaques.

Another approach to the problem was the development of transfer vectors containing the *E. coli lacZ* gene in addition to the foreign gene of interest (ref. 19, see also Section 4.2.1). Insertion of this gene into the baculovirus genome confers a blue phenoype on staining plaques with X-gal. Unfortunately, it also involves the addition of extra (unwanted) DNA into the baculovirus genome and does not address the problem of the low frequency of recombination.

The most satisfactory answer to date for the efficient selection of recombinant viruses is the use of linearized baculovirus DNA (37). A unique restriction enzyme site (*Bsu*36I/*Mst*II/*Sau*I) was engineered into a polyhedrin-negative AcMNPV genome (AcRP6.SC) at the polyhedrin gene locus. Linearization of the viral DNA and subsequent transfection of insect cells demonstrated that the yield of virus was considerably reduced; this is consistent with the hypothesis that linear viral DNA is non-infectious. Furthermore, when the linear viral DNA was co-transfected with a transfer vector based on the polyhedrin gene locus, the progeny virus consisted of about 25% recombinant virus. This was concluded to be a consequence of a double recombination event repairing the break in the viral DNA and restoring infectivity. The technique has been further refined by using AcRP23.lacZ (15) a virus with the *lacZ* coding region inserted under the control of the polyhedrin gene promoter; *lacZ* fortuitously contains a unique *Bsu*36I site. Co-transfection of linearized AcRP23.lacZ DNA with the appropriate transfer vector produces viruses lacking *lacZ* which produce clear plaques against a blue background. A recombinant virus, AcUW1.lacZ is also available (23) to enable foreign genes to be readily inserted at the p10 gene locus after linearization of the viral DNA in this region. We recommend the use of these virus genomes when producing recombinant baculoviruses. The procedure for the preparation of linearized viral DNA is given in *Protocol 5*.

Protocol 5. Preparation of linearized viral DNA

Materials

- infectious DNA from AcRP6.SC, AcRP23.lacZ for polyhedrin gene locus-based vectors, or AcUW1.lacZ for p10 gene locus-based vectors (*Protocol 2*)
- restriction enzyme: *Bsu*36I (New England Biolabs)
- 10 × *Bsu*36I buffer: 100 mM Tris–HCl (pH 7.5), 1 M NaCl, 100 mM $MgCl_2$, 100 mM 2-mercaptoethanol, 1 mg/ml bovine serum albumin

Method

1. Mix 1 μg of viral DNA with 5 μl of 10 × *Bsu*36I buffer and add distilled water to a total volume of 50 μl. Add 5 U of *Bsu*36I and incubate at 37 °C for 6 h.
2. Terminate the reaction by heating at 70 °C for 15 min. Store at 4 °C.
3. Since undigested virus DNA will produce a high background of non-recombinants, check the digest by comparison with undigested DNA by electrophoresis on a 0.3% (w/v) agarose minigel (34).

6. Co-transfection of cells with viral DNA and recombinant transfer vectors

Co-transfection of insect cells with viral DNA and the appropriate recombinant transfer vector was originally mediated by the calcium phosphate method (12). This has now been largely superseded by the more efficient lipofection procedure (38), which requires less viral DNA. Both methods are described in *Protocol 6*.

Protocol 6. Co-transfection of insect cells

Materials

- recombinant transfer vector DNA (see *Protocol 3*)
- linearized viral DNA (see *Protocol 5*)
- supplemented TC100 medium (*Protocol 1*)
- logarithmic phase *S. frugiperda* cells
- 100 mM glucose; filter-sterilized
- 2 M $CaCl_2$; filter-sterilized
- 2 × Hepes-buffered saline (HBS): 40 mM–Hepes NaOH (pH 7.05), 2 mM Na_2HPO_4, 10 mM KCl, 280 mM NaCl; filter-sterilized
- Lipofectin (Gibco–BRL)

Methods

A. *Calcium phosphate precipitation*

1. Seed 35 mm diameter Petri dishes with 1 × 10^6 *S. frugiperda* cells/dish in 2 ml of TC100 (5% FCS) and incubate at 28 °C for 2 h to allow the cells to attach to the plastic.

Protocol 6. *Continued*

2. Set up the co-transfection mixture in a 10 ml sterile, capped plastic tube as follows:

• viral DNA	1 μg
• recombinant transfer vector	5 μg
• 2 × HBS	500 μl
• 100 mM glucose	100 μl
• sterile water	to 935 μl

3. Add 65 μl of 2 M $CaCl_2$ dropwise to the DNA mixture, while vortexing. Leave at room temperature for 30 min to allow the calcium phosphate–DNA co-precipitate to develop.

4. Remove the medium from each dish of cells. Gently resuspend all the calcium phosphate–DNA complexes in the mixture from step 3 and drip this on to the surface of the cells. Leave the cells at room temperature for 30 min.

5. Add 2 ml of TC100 (5% FCS) to each dish and incubate at 28°C for 24 h.

6. Remove the medium from the dishes, wash each with 2 ml of TC100 (5% FCS) and replace with 2 ml of the same medium. Continue the incubation until 48 h p.i.

7. Remove the cell culture medium and titrate the virus in a plaque assay (*Protocol 2*).

B. *Lipofectin-mediated transfection*

1. Carry out step 1 as described in *Protocol 6A*.

2. Mix 100 ng of linearized viral DNA with 500 ng of transfer vector in a polystyrene (*not* polypropylene) container, and make up the volume to 50 μl with sterile water.

3. In a separate polystyrene container, dilute Lipofectin 1:10 with sterile water.

4. Add 50 μl of diluted Lipofectin to the DNA mixture from step 2 and leave for 15 min at room temperature.

5. Remove the medium from the *S. frugiperda* cells in the dishes, wash the cells twice with serum-free TC100 and then add 1 ml of the same medium to the cells. Drip the DNA–Lipofectin mixture on to the cells and leave at 28°C for 8 h or overnight.

6. Add 1 ml of TC100 (10% FCS) to each dish and incubate for 48–60 h at 28°C.

7. Harvest the medium and transfer to a sterile container. Screen for the presence of virus by a plaque assay (*Protocol 2*).

7. Identification of recombinant viruses

7.1 Plaque titration and staining with neutral red or X-gal

The plaque assay performed on virus-infected cell culture medium from a co-transfection is very similar to that used to titrate virus stocks. When linearized DNA has been used, 10 plates should be prepared at 10^{-1} to 10^{-4} dilutions. The plaques can be identified by neutral red staining (see *Protocol 2*), for AcRP6.SC, or both X-gal and neutral red staining for AcRP23.lacZ or AcUW1.lacZ (p10 gene locus). For the latter two viruses, the recombinant plaques will be clear against a low background of blue parental viruses. In the case of AcRP6.SC, both parental and recombinant plaques have the polyhedrin-negative phenotype. However, because at least 25% of the plaques will be recombinant, it is simple to select 5–10 plaques at random and then screen for the presence of the foreign gene by dot-blot hybridization or PCR analysis.

Replace the liquid overlay from each dish with 1 ml of TC100 (5% FCS) containing 0.025% (w/v) X-gal. Incubate the dishes for 3–4 h at 28°C then add 1 ml of neutral red stain and incubate for another 2–3 h. Remove the liquid and store the dishes in the dark till the plaques have cleared. The isolation of recombinant plaques is described in Section 6.2.

7.2 Direct screening of recombinant viruses for the inserted sequence

A more direct method used in the identification of recombinant virus is to harvest the co-transfection mixture and screen it directly for the foreign gene by nucleic acid hybridization. Dilutions of the progeny from a co-transfection are used to infect cells seeded in a 96-well dish followed by dot-blot hybridization analysis of the DNA (see *Protocol 7*). This method is less likely to result in false positives since the presence of a sequence is detected rather than an associated phenotype. However, the method is becoming somewhat obsolete with the introduction of efficient ways of identifying recombinant plaques after using linearized virus DNA.

Protocol 7. Screening for recombinants using limiting dilution and dot-blot screening

Materials

- hybridization membrane (for example Schleicher and Schuell nitrocellulose or Amersham Hybond)
- dot-blot filtration apparatus (for example Gibco–BRL)
- two 96-well tissue culture dishes and lids

Protocol 7. *Continued*

- progeny virus from a co-transfection (*Protocol 6*)
- *S. frugiperda* cells at 1.5×10^5 cells/ml, in TC100 + 5% (v/v) FCS (*Protocol 1*)
- ^{32}P-radiolabelled DNA probe (specific activity 10^7–10^8 c.p.m./μg)
- reagents for hybridization and filter washing, etc. (34)

Method

1. Place 100 μl aliquots of 1.5×10^4 *S. frugiperda* cells into each well of a 96-well plate. Let the cells attach to the surface for 1 h at 28°C.
2. Serially dilute (10^{-1} to 10^{-4}) the virus from the co-transfection mixture in 1 ml of TC100 growth medium.
3. To the first row of 10 wells add 50 μl of neat co-transfection medium.
   To the 2nd and 3rd rows of 10 wells add 50 μl of the 10^{-1} dilution.
   To the 4th and 5th rows of 10 wells add 50 μl of the 10^{-2} dilution.
   To the 6th and 7th rows of 10 wells add 50 μl of the 10^{-3} dilution.
   To the 8th row of 10 wells add 50 μl of the 10^{-4} dilution.
   Add 50 μl of medium to the remaining unused wells as controls.
4. Incubate in a moist sandwich box at 28°C for 4–5 days.
5. Transfer the culture medium from each well to another 96-well dish, seal and store at 4°C. These are the master cultures. Resuspend the remaining cells in each well in 200 μl of 200 mM NaOH and allow to stand for 10 min.
6. Assemble a dot-blot apparatus and wash the wells with 400 μl of 2 M NaCl. Add 200 μl of the relevant cell lysate to each well. Also add blot denatured transfer vector and AcMNPV DNA as positive and negative controls. Remove the membrane and air dry prior to baking for 2 h at 80°C *in vacuo*.
7. Hybridize with the radiolabelled DNA probe.
8. Use autoradiography to identify wells containing recombinant virus.
9. Carry out further plaque purification on the medium harvested prior to the dot-blot analysis from the corresponding wells of the master plate.

7.3 Isolation of recombinant plaques

Plaques can be identified with the naked eye and examined in detail using a good quality light microscope. Ring the desired plaques using a fine marker pen on the base of the Petri dish. Pick out the plaques using a sterile Pasteur pipette to remove a plug of agarose containing the plaque and place this in 0.5 ml of TC100 growth medium (5% FCS) in a sterile vial. Vortex and store at 4°C. It is suggested that at least six plaques are picked and screened from the initial co-transfection (*Protocol 7*). It is also suggested that, in order to separate contaminating wild-type (WT) virus from recombinant virus, the plaques should undergo several rounds of plaque purification to eliminate false positives. In order to reduce the risk of contamination, pick well separated plaques from dishes infected with a high dilution of virus.

8. Identification of polyhedrin-positive viruses by passage *in vivo*

This approach is appropriate only when using a transfer vector such as pAcUW2B (see Section 4.2.3) (23) to restore the polyhedrin-positive phenotype to a polyhedrin-negative parental virus. The most infectious form of the baculovirus is the polyhedron. Polyhedra are resistant to the effects of ionic detergents, which may be used to inactivate the background of parental virus particles produced after a co-transfection experiment. The polyhedrin-positive phenotype may then be recovered by passage through insect larvae (see *Protocol 8*), and harvested from the larval haemolymph as infectious virus for further plaque purification in cell culture. This method is only useful if an insectary is available; otherwise the polyhedrin-positive phenotype may be selected by the usual method of plaque purification in cell culture (*Protocol 2*).

Protocol 8. Recovery of polyhedrin-positive virus from insects

Materials

- 24-well microtitre plate with cover
- progeny virus from a co-transfection (*Protocol 6*)
- late third instar larvae (*Trichoplusia ni*)
- individual cups of artificial diet (adapted from ref. 39)

Method

1. Allow the co-transfection to proceed for 3–4 days or until the cells visibly contain polyhedra.
2. Agitate the cells into the medium. Transfer the cells to a microcentrifuge tube and centrifuge for 1 min at low speed (2500–3000 *g*). Wash the cells once in 0.5 ml of PBS (*Protocol 2*) then resuspend in 0.1% (w/v) SDS for 5 min at room temperature to destroy the integrity of the cell walls and any remaining non-occluded virus.
3. Pellet the polyhedra at 10 000 *g* for 30 min at 4°C, wash in 0.5 ml of PBS and then resuspend in 20 μl of PBS. Use this to infect the larvae.
4. Place the larvae in individual wells of the microtitre plate with a small plug of artificial diet which has been inoculated with 1 μl of the SDS-treated cells.

Protocol 8. *Continued*

5. Allow the larvae to feed for 24 h then transfer them to individual containers of diet and incubate at 24°C.
6. The larvae have now been infected with the polyhedrin-positive virus present in the co-transfection. They will usually start to look very pale in colour and succumb to virus infection at around 5 days p.i. Bleed the larvae just prior to death. Surface-sterilize the larvae by immersion in cold 80% (v/v) ethanol and dry on sterile filter paper. This process also anaesthesizes the larvae prior to haemolymph extraction.
7. Snip off one of the prolegs and collect the haemolymph using a glass capillary or a pipette tip.
8. Assay the haemolymph for recombinant virus by plaque assay (*Protocol 2*).

9. Characterization of recombinant virus DNA

The presence of the foreign gene in the recombinant baculovirus genome can be confirmed by restriction enzyme analysis followed by Southern hybridization and sequencing if necessary. The yield from a DNA extraction (*Protocol 9*) should be about 5 μg of DNA. There should be very little contaminating chromosomal DNA. A simple restriction enzyme analysis followed by separation on an ethidium bromide-stained agarose gel may allow identification of any change in the pattern due to insertion of the foreign sequences (*Protocol 10*). If the fragments are then blotted on to a nitrocellulose membrane the location of foreign genes may be confirmed using an appropriate radiolabelled probe.

Protocol 9. Isolation of DNA from infected insect cells

Materials

- titrated virus stocks (*Protocol 1*)
- lysis buffer: 50 mM Tris–HCl (pH 8.0), 5% (v/v) 2-mercaptoethanol, 0.4% (w/v) SDS, 10 mM EDTA
- 10 mg/ml proteinase K in TE: made fresh and pre-incubated at 37°C for 30 min prior to use
- 10 mg/ml RNase A in TE: boil for 10 min before use and store at −20°C
- phenol:chloroform (1:1) equilibrated in 50 mM Tris–HCl (pH 8.0)

Method

1. Prepare 35 mm Petri dishes containing 1×10^6 *S. frugiperda* cells as described previously (*Protocol 1*).

2. Remove the medium and inoculate the cells with virus using a high m.o.i. (10 p.f.u./cell; also use WT virus-infected and mock-infected controls). Leave for 1 h at room temperature to allow the virus to adsorb.
3. Remove the inoculum and overlay the cells with 1.5 ml of TC100 (5% FCS). Incubate the cells at 28°C for 18 h in a moist tissue-lined sandwich box.
4. Scrape the cells into the medium. Transfer them to a microcentrifuge tube and centrifuge at low speed (2500–3000 *g*) for 1 min then wash the pellet once with 0.5 ml of PBS. The pellets may be stored at −20°C or used immediately.
5. Resuspend the cell pellet in 250 μl of TE and add 250 μl of lysis buffer.
6. Mix gently and add 12.5 μl of proteinase K and 2.5 μl of RNase A. Incubate the solutions at 30°C for 30 min and gently extract the lysate with an equal volume of phenol:chloroform and rocking for 2–3 min. Centrifuge for 2–5 min in a microcentrifuge and extract the aqueous phase with phenol:chloroform.
7. Ethanol precipitate DNA from the upper aqueous phase by adding 50 μl of 3 M sodium acetate and 1 ml of ethanol. Pellet the DNA by centrifuging for 5 min in a microcentrifuge.
8. Wash the DNA pellet twice with 75% (v/v) ethanol and air-dry.
9. Dissolve the DNA in 100 μl of TE. The purified DNA is of high molecular weight, so treat it with care. Soak the pellet overnight and resuspend it gently the next day after incubating at 37°C for 30 min.

Protocol 10. Analysis of DNA from virus-infected cells

Materials

- DNA extracted from virus-infected insect cells (*Protocol 9*)
- 10 × TBE: 108 g/l Tris, 55 g/l boric acid, 9.4 g/l EDTA
- 5 × DNA loading dye: 40% (w/v) sucrose, 2.5% (w/v) SDS, 0.25% (w/v) bromophenol blue in 5 × TBE
- size markers (e.g. λ DNA digested with *Hin*dIII)
- 0.7% (w/v) agarose gel in 1 × TBE containing 0.5 μg/ml ethidium bromide
- ^{32}P-radiolabelled DNA probe (see ref. 34)

Method

1. Digest 5–10 μl of DNA with an appropriate restriction enzyme in a volume of 20–40 μl at 37°C for 2 h.

Protocol 10. *Continued*

2. Add 5 × DNA loading dye and electrophorese the samples, with appropriate DNA size markers, in the agarose gel.
3. When the bromophenol blue marker dye is at the bottom of the gel, photograph the gel on a UV transilluminator.
4. Transfer the DNA to a nitrocellulose membrane and identify recombinant DNA sequences using the radiolabelled probe.

10. Analysis of protein synthesis in virus-infected cells

When the foreign gene is inserted under the control of a strong promoter such as that of the polyhedrin or p10 gene, it is usually possible to visualize the protein on a polyacrylamide gel stained with Coomassie blue (*Protocol 11*). If the protein is not synthesized at a high level, it may be necessary to incorporate a radiolabelled amino acid to detect the protein.

In order to obtain reproducible and comparable results, the cells must be infected synchronously by using a high m.o.i. Remember to include controls of WT AcMNPV infected- and mock-infected cells. Analyse samples prepared at 24 and 48 h p.i. initially and then conduct a more extensive time course (12–60 h p.i.).

If the recombinant gene product cannot be seen after staining, repeat the gel electrophoresis and transfer proteins to nitrocellulose membranes to allow detection (of the foreign protein) by Western blotting. If the protein is secreted into the medium, it may be necessary to concentrate it by freeze drying or by immunoprecipitation. The protein can also be identified in the supernatant by HPLC analysis. Remember that at later times in infection (i.e. after 48 h) the cells are liable to leak, so that detection of protein in the supernatant will not necessarily indicate secretion. Visualizing the protein by pulse-labelling the cells with a radiolabelled amino acid is routinely performed using the amino acid [^{35}S]methionine or more rarely [^{35}S]cysteine and [^{3}H]leucine (*Protocol 12*). The amino acid chosen will depend on the amino acid content of the protein under study.

Protocol 11. Analysis of proteins in virus-infected cell extracts

Materials

- titrated virus (*Protocol 2*)
- 12% single concentration (ref. 15) or 10–30% gradient polyacrylamide gel (ref. 40)
- 5 × dissociation mixture: 10% (w/v) SDS, 25% (v/v) 2-mercaptoethanol, 50 mM Tris–HCl (pH 6.8), 25% (w/v) glycerol, 0.25% (w/v) bromophenol blue

Method

1. Set up one Petri dish for each time point and controls for AcMNPV-infected and mock-infected cells, each containing 1×10^6 *S. frugiperda* cells/dish. When the cells have attached to the dish, remove the medium and inoculate with the virus at a m.o.i. of 5–10 p.f.u./cell. Incubate the dishes at room temperature for 1 h.
2. Remove the inoculum from each dish and replace it with 1.5 ml of TC100 (5% FCS). Incubate the cells at 28°C for the appropriate time (for example 24 h).
3. Scrape the cells from each dish into the medium and transfer to a separate microcentrifuge tube.
4. Pellet the cells by centrifugation at low speed (2500–3000 *g*) for 2 min and wash with 0.5 ml of PBS.
5. Repellet the cells and resuspend them in 80 μl of TE. Store the sample at −20°C or use immediately.
6. Prior to SDS–PAGE, mix 20 μl of the cell sample with 5 μl of dissociation mixture and boil for 5 min.
7. Electrophorese the gel at 50 V overnight or until the bromophenol blue has reached the bottom of the gel.
8. Stain the gel with Coomassie blue to visualize the separated proteins (ref. 40).

Protocol 12. Radiolabelling of virus-infected cell proteins

Materials

- starvation medium: TC100 medium lacking both the amino acid used for labelling and bactotryptose broth, but supplemented with 2% (v/v) dialysed FCS (filter-sterilized)
- radiolabelled amino acid: for each dish use 15–30 μCi of ^{35}S-labelled amino acids, 1–5 μCi of ^{14}C-labelled amino acids, or 30–50 μCi of ^{3}H-labelled amino acids

Method

1. Infect the cells as described in *Protocol 11*.
2. Carefully remove the medium from the cells 15–30 min prior to sampling. Wash the monolayer very gently with starvation medium, add 0.5 ml of starvation medium, and incubate at 28°C for 30 min.
3. At the appropriate time, add a further 0.5 ml of starvation medium containing the radiolabelled amino acid and incubate the cells for 1 h at 28°C.

Protocol 12. *Continued*

4. Scrape the cells into the medium and transfer to a microcentrifuge tube.[a]
5. Electrophorese the cell samples on a gel as for an unlabelled protein sample (*Protocol 11*).
6. Fix and destain, dry the gel, and expose it to X-ray film.

[a] If the supernatant is to be assayed, transfer the supernatant into a separate tube prior to harvesting the cells. Add 1 ml of PBS to the dish and harvest the cells. This will eliminate leakage of proteins from the cells into the medium during centrifugation.

11. Post-translational modification of proteins

Prokaryotic expression systems have the advantages of ease of manipulation and low cost of scale up, but do not have the necessary machinery to carry out post-translational modifications required by many proteins. Synthesis of proteins in the insect cell culture system using *S. frugiperda* cells has demonstrated the capacity to perform post-translational modifications such as glycosylation, phosphorylation, internal and signal peptide cleavage, and also the formation of complexes between recombinant proteins (16, 41–44).

(a) *Glycosylation*. The glycosylation events in insect cells are dissimilar from those carried out in mammalian cells (45, 46), and in some cases the proteins appear smaller. This may be because insect cells, unlike mammalian cells, use mainly high-mannose oligosaccharides in the processing of proteins (47).
(b) *Signal peptide cleavage*. Recognition and cleavage of mammalian signal sequences followed by insertion of the protein into the plasma membrane have been reported (for example influenza virus haemagglutinin (HA) (45, 47, 48)). Where appropriate, proteins are also secreted (33).
(c) *Internal cleavage*. Internal cleavage of proteins into subnits has also been demonstrated (for example cleavage of fowl plague influenza virus HA into HA1 and HA2 (48), and the product of the Sindbis virus 26S mRNA into capsid and two envelope glycoproteins (49)).
(d) *Formation of protein complexes*. In a number of cases the formation of tertiary structures between recombinant proteins has been observed in infected cells (for example formation of bluetongue virus tubules (50), complexes between the SV40 T antigen and mouse p53 (51), and construction of non-infectious poliovirus particles containing VP0/VP1/VP3 (52)).

These examples serve to indicate that the baculovirus/insect cell system can function to produce correctly modified foreign proteins with demonstrable biological activity. These proteins can produce immunogenic reactions similar to those of the original proteins (53–56).

Many of these features are demonstrated in a single example where the baculovirus system was used to produce a monoclonal IgG antibody against the lipoprotein I of *Pseudomonas aeruginosa*. Both chains were cloned into a multiple expression vector and inserted into the AcMNPV genome. Virus-infected cells secreted glycosylated antibody which bound to *P. aeruginosa* lipoprotein I (57).

12. Scale up

The successful expression of a foreign gene using the baculovirus system is often followed by attempts to scale up production. For those simply interested in producing enough protein to raise monospecific antibody in mice or rabbits, a few 75 cm^2 culture flasks may be sufficient to provide the necessary raw material. However, if the recombinant protein is to be used as a vaccine or in structure–function analyses, such as X-ray crystallography, it will almost certainly be necessary to use suspension cultures to meet the demand.

Suspension cultures of insect cells undoubtedly have the potential to yield considerably higher numbers of cells than monolayer cultures. Insect cells may be grown to high density in simple spinner cultures using flat-bottomed round flasks with a magnetic stirring bar (see *Protocol 3*). Cell densities of the order of $2–3 \times 10^6$ cells/ml may be attained in cultures up to 500 ml. Beyond this volume, the supply of oxygen to the cells becomes rate limiting as the ratio of the surface area of the culture to the volume decreases. It is, therefore, necessary to supplement the culture with an additional supply of oxygen or air.

Unfortunately, insect cells are sensitive to shear forces and are damaged by gas sparging (58). This means that if air is passed through a normal suspension culture, many of the cells will be lysed when they come into contact with small air bubbles. A further complication is that gas sparging induces foaming in the culture. Despite these problems, some success has been achieved with insect cells in large-scale culture in bioreactors and air-lift fermenters (59–63).

For the average researcher, sophisticated fermentation equipment will not be available. We have found that quite reasonable results may be obtained using simple, 'home-made' 500 ml spinner flasks (see Section 3). Cells are introduced at a density of 10^5 cells/ml, grown to about 10^6 cells/ml and then the culture is inoculated with recombinant virus at a m.o.i. of 3–10 p.f.u./cell. The time at which to harvest depends on the kinetics of protein production and the stability of the recombinant product.

13. Production of foreign proteins from insect larvae

Expression of protein *in vivo* rather than in cell culture can increase the quantity of protein recovered and may lead to a product which is biologically more active. *Bombyx mori* has been used to produce α-interferon (64), and the smaller *Trichoplusia ni* has been used in the production of Punta Toro virus N and NS proteins (65) and also hepatitis B virus surface and core antigens (66). The *B. mori* larvae are large and easy to handle, and the technology for rearing silkworms is well advanced. Purification of proteins from larval carcasses is relatively simple and high yields are obtained. Targeting the protein out of the cell and into the haemolymph may also aid in purification. Viruses which retain the polyhedrin gene are essential for this approach, but this is ceasing to be a difficulty with the emergence of more sophisticated vectors.

14. Advantages and disadvantages of the baculovirus system

14.1 Advantages

(a) The virus has two very late gene products (polyhedrin and p10) which are not required for the production of infectious virus particles in cell culture. Their coding regions may be replaced with foreign sequences.

(b) The polyhedrin and p10 gene promoters are very strong, resulting in the synthesis of large quantities of proteins in infected cells. Other baculovirus gene promoters, although less active, are also useful.

(c) Expression of the very late virus gene promoters occurs in cells after maturation of budded, infectious virus particles. Consequently, if a cytotoxic protein is synthesized it will not adversely affect virus replication.

(d) Baculovirus genomes are remarkably variable in size (88–160 kbp) and can accommodate large amounts of foreign DNA without affecting normal replication and DNA packaging.

(e) Replication of baculovirus in eukaryotic cells ensures faithful processing of foreign eukaryotic gene products, although glycosylation may present some problems (see Section 14.2).

(f) Baculoviruses replicate only in invertebrates and there is no risk to the investigator.

(g) Insect cells are amenable to large-scale volume production in fermenter systems.

14.2 Disadvantages

(a) Baculovirus infection of insect cells results in the death of the host and thus requires that fresh cultures of cells are infected for each round of protein production. This may be a problem for production on a commercial scale, although it is acceptable for research purposes.

(b) The process of glycosylation in insect cells is different compared with vertebrate cells. Insect cells produce glycoproteins with relatively simple, unbranched sugar side-chains with a high mannose content. The most obvious consequence of this processing is that the glycoproteins produced in insect cells have a greater mobility in denaturing polyacrylamide gels due to a reduction in molecular weight. The importance of this feature for the use of the recombinant protein is unclear at the present time.

15. Future work

One of the most important developments in the baculovirus expression vector system is the availability of methods for the simple isolation of recombinant viruses. The use of linearized virus DNA to promote the recovery of recombinant viruses is described in Section 5. Further improvements will undoubtedly be made in the near future. Two methods have been published recently which offer alternative ways of producing recombinants. In the first (68), yeast ARS and CEN sequences were inserted into the complete AcMNPV genome, thus ensuring stable replication in *Saccharomyces cerevisiae*. The virus DNA can still replicate in insect cells. Homologous recombination is then performed in yeast, rather than the insect cells, and is reported to completely remove the background of parental viruses. In the second method (69), the Cre-*lox* system of bacteriophage P1 was used to develop a highly efficient *in vitro* system for producing recombinant viruses. Up to 50% of the viral progeny were reported to be recombinants. These methods have yet to be tested in laboratories other than where they were originally developed, but appear to offer excellent alternatives to the techniques currently available for producing recombinant AcMNPV.

In the longer term, it would be advantageous to produce foreign proteins on a continuous basis in insect cells, rather than in the batch process required by virus infection of cells. Jarvis *et al.* (67) developed a system which was based on the use of a baculovirus early promoter to produce a stably transformed insect cell line. A foreign gene (tissue plasminogen activator) was expressed under control of a promoter from an immediate early gene (IE-1) which is transcriptionally active in these cells in the absence of other viral functions. The product is made continuously in the absence of viral infection, although the protein is present at low levels. Extensions to this idea, such as up-regulation by addition of transactivating factors or elimination of negative controls, may become possible when the factors controlling hyperexpression of late genes have been elucidated.

References

1. Burgess, S. (1977). *J. Gen. Virol.,* **37,** 501.
2. Gröner, A. (1986). In *The biology of baculoviruses,* Vol. 1 (ed. R. R. Granados and B. A. Federici), pp. 177. CRC Press, Boca Raton, Florida.

3. Volkman, L. E. and Keddie, B. A. (1990). *Seminars in Virology*. Vol. 1, 249.
4. Blissard, G. W. and Rohrmann, G. F. (1990). *Annu. Rev. Entomol.,* **35,** 125.
5. Guarino, L. A. and Summers, M. D. (1986). *J. Virol.,* **57,** 563.
6. Vlak, J. M. and Rohrmann, G. F. (1985). In *Viral insecticides for biological control,* (ed. K. E. Sherman and K. Maramorosch), p. 489. Academic Press, New York and London.
7. Van der Wilk, F., Van Lent, J. W. M., and Vlak, J. M. (1987). *J. Gen. Virol.,* **68,** 2615.
8. Vaughn, J. L., Goodwin, R. H., Thomkins, G. J., and McCawley, P. (1977). *In Vitro,* **13,** 213.
9. Summers, M. D. and Smith, G. E. (1987). *A manual of methods for baculovirus vectors and insect cell culture procedures.* Texas Agric. Exp. Sta. Bull., No. 1555.
10. Brown, M. and Faulkner, P. (1977). *J. Gen. Virol.,* **36,** 361.
11. Hooft van Iddekinge, B. J. L., Smith, G. E., and Summers, M. D. (1983). *Virology,* **131,** 561.
12. Smith, G. E., Fraser, M. J., and Summers, M. D. (1983). *J. Virol.,* **46,** 584.
13. Smith, G. E., Summers, M. D., and Fraser, M. J. (1983). *Mol. Cell. Biol.,* **3,** 2156.
14. Miller, D. W., Safer, P., and Miller, L. K. (1986). In *Genetic engineering,* Vol. 8 (ed. J. K. Setlow and A. Hollander), pp. 277. Plenum Publishing Corp., New York.
15. Possee, R. D. and Howard, S. C. (1987). *Nucl. Acids Res.,* **15,** 10233.
16. Matsuura, Y., Possee, R. D., Overton, H. A., and Bishop, D. H. L. (1987). *J. Gen. Virol.,* **68,** 1233.
17. Luckow, V. A. and Summers, M. D. (1989). *Virology,* **170,** 31.
18. Livingstone, C. and Jones, I. (1989). *Nucl. Acids Res.,* **17,** 2366.
19. Vialard, J., Lalumiere, M., Vernet, T., Briedis, D., Alkhatib, G., Henning, D., Levin, D., and Richardson, C. (1990). *J. Virol.,* **64,** 37.
20. Zuidema, D., Schouten, A., Usmany, M., Maule, A. J., Belsham, G. J., Roosien, J., Klinge-Roode, E. C., van Lent, J. W. M., and Vlak, J. M. (1990). *J. Gen. Virol.,* **71,** 2201.
21. Vlak, J. M., Klinkenberg, F. A., Zaal, K. J. M., Usmany, M., Klinge-Roode, E. C., Geervliet, J. B. F., Roosien, J., and van Lent, J. W. M. (1988). *J. Gen. Virol.,* **69,** 765.
22. Williams, G. V., Rohel, D. Z., Kuzio, J., and Faulkner, P. (1989). *J. Gen. Virol.,* **70,** 187.
23. Weyer, U., Knight, S., and Possee, R. D. (1990). *J. Gen. Virol.,* **71,** 1525.
24. Vlak, J. M., Schouten, A., Usmany, M., Belsham, G. J., Klinge-Roode, E. C., Maule, A. J., van Lent, J. W. M., and Zuidema, D. (1990). *Virology,* **179,** 312.
25. St Angelo, C., Smith, G. E., Summers, M. D., and Krug, R. M. (1987). *J. Virol.,* **61,** 361.
26. Emery, V. C. and Bishop, D. H. L. (1987). *Protein Eng.,* **1,** 359.
27. Weyer, U. and Possee, R. D. (1991). *J. Gen. Virol.,* **72,** 2967.
28. Hill-Perkins, M. S. and Possee, R. D. (1990). *J. Gen. Virol.,* **71,** 971.
29. Thiem, S. M. and Miller, L. K. (1989). *J. Virol.,* **63,** 2008.
30. Thiem, S. M. and Miller, L. K. (1990). *Gene,* **91,** 87.
31. Rose, J. K. and Gallione, C. J. (1981). *J. Virol.,* **39,** 519.

32. Bailey, M. J., McLeod, D. A., Kang, C.-L., and Bishop, D. H. L. (1989). *Virology,* **169,** 323.
33. Stewart, L. M. D., Hirst, M., Lopez-Ferber, M., Merryweather, A. T., Cayley, P. J., and Possee, R. D. (1991). *Nature,* **352,** 85.
34. Brown, T. A. (ed.) (1991). *Essential molecular biology: a practical approach.* IRL Press, Oxford.
35. Fraser, M. J. (1989). In Vitro *Cell. Dev. Biol.,* **25,** 225.
36. Fung, M.-C., Chiu, Y. M., Weber, T., Chang, T.-W., and Chang, N. T. (1988). *J. Virol. Meth.,* **19,** 33.
37. Kitts, P. A., Ayres, M. D., and Possee, R. D. (1990). *Nucl. Acids Res.,* **18,** 5667.
38. Felgner, P. L., Gadek, T. R., Holm, M., Roman, R., Chan, H. W., Wenz, M., Northrop, J. P., Ringold, G. M., and Daruelsen, M. (1987). *Proc. Natl Acad. Sci. USA,* **84,** 7413.
39. Hoffman, J. D., Lawson, F. R., and Yamamoto, R. (1966). In *Insect colonisation and mass production,* (ed. C. N. Smith), p. 479. Academic Press, New York.
40. Hames, B. D. and Rickwood, D. (ed.) (1990). *Gel electrophoresis of proteins: a practical approach,* (2nd edn). IRL Press, Oxford.
41. Quilliam, L. A., Der, C. J., Clark, R., O'Rourke, C., Zhang, K., McCormack, F. M., and Bocock, G. M. (1990). *Mol. Cell. Biol.,* **10,** 2901.
42. Landford, R. E. (1988). *Virology,* **167,** 72.
43. Peggy, H. and Robbins, P. W. (1984). *J. Biol. Chem.,* **259,** 2375.
44. Overton, H. A., Fuji, Y., Price, R., and Jones, I. (1989). *Virology,* **170,** 107.
45. Possee, R. D. (1986). *Virus Res.,* **5,** 43.
46. Jarvis, D. L. and Summers, M. D. (1989). *Mol. Cell. Biol.,* **9,** 214–223.
47. Kuroda, K., Geyer, H., Geyer, R., Doerfler, W., and Klenk, H. D. (1990). *Virology,* **174,** 418.
48. Kuroda, K., Hauser, C., Rott, R., Klenk, H. D., and Doerfler, W. (1986). *EMBO J.,* **5,** 1359.
49. Oker-Blom, C. and Summers, M. D. (1989). *J. Virol.,* **63,** 1256.
50. Urakawa, T. and Roy, P. (1988). *J. Virol.,* **62,** 3919.
51. O'Reilly, D. R. and Miller, L. K. (1988). *J. Virol.,* **62,** 3109.
52. Urakawa, T., Ferguson, M., Minor, P. D., Cooper, J., Sullivan, M., Almond, J. W., and Bishop, D. H. L. (1989). *J. Gen. Virol.,* **70,** 1453.
53. Frech, B., Zimber-Strobl, U., Suentzenich, K. O., Pavlish, O., Lenoir, G. M., Bornkamm, G. W., and Mueller-Lantzsch, N. (1990). *J. Virol.,* **64,** 2759.
54. Murphy, V. F., Rowen, W. C., Page, M. J., and Holder, A. A. (1990). *Parasitology,* **100,** 177.
55. Prehaud, C., Takehara, K., Flamand, A., and Bishop, D. H. L. (1989). *Virology,* **173,** 390.
56. Schmaljohn, C. S., Parker, M. P., Ennis, W. H., Dalrymple, J. M., Collett, M. S., Suzion, J., and Schmaljohn, A. L. (1989). *Virology,* **170,** 184.
57. Putlitz, J., Kubasek, W. L., Duchene, M., Marget, M., von Specht, B. U., and Domdey, H. (1990). *Bio/Technology,* **8,** 651.
58. Tramper, J., Williams, J. B., Joustra, D., and Vlak, J. M. (1986). *Enzyme Microb. Tech.,* **8,** 33.
59. Miltenberger, H. G. and David, P. (1980). *Develop. Biol. Stand.,* **46,** 183.
60. Maiorella, B., Inlow, D., Shauger, A., and Harano, D. (1988). *Bio/Technology,* **6,** 1406.

61. Inlow, D., Shauger, A., and Maiorella, B. (1989). *J. Tissue Culture Meth.,* **12,** 13–16.
62. Tramper, J., van den End, E. J., de Gooijer, C. D., Kompler, R., van Lier, F. L. J., Usmany, M., and Vlak, J. M. (1990). *Ann. NY Acad. Sci.,* **589,** 423.
63. Weiss, S. A., Gorfien, S., Fike, R., DiSorbio, D., and Jayme, P. (1990). *Proceedings of 9th Australian Biotechnology Conference*, p. 230. University of Queensland Press.
64. Maeda, S., Kawai, T., Obinata, M., Fujiwara, H., Horiuchi, T., Sacki, Y., Sato, Y., and Furusawa, M. (1985). *Nature,* **315,** 592.
65. Overton, H. A., Ihara, T., and Bishop, D. H. L. (1987). *Virology,* **157,** 338.
66. Takehara, K., Ireland, D., and Bishop, D. H. L. (1988). *J. Gen. Virol.,* **69,** 2763.
67. Jarvis, D. L., Fleming, J., Kovacs, G. R., Summers, M. D., and Guarino, L. (1990). *Bio/Technology,* **8,** 950.
68. Patel, G., Nasmyth, K., and Jones, N. (1992). *Nucl. Acids Res.,* **20,** 97–104.
69. Peakman, T. C., Harris, R. A., and Gewert, D. R. (1992). *Nucl. Acids Res.,* **20,** 495–500.

9

Expression of genes by vaccinia virus vectors

GEOFFREY L. SMITH

1. Introduction

Jenner introduced the use of cowpox virus as a vaccine against smallpox in 1798, and, as he predicted, this practice led to the eradication of smallpox (in 1977), although in the modern era it was vaccinia, not cowpox, virus which was used as the smallpox vaccine. Today, variola (smallpox) virus is confined to two high-security laboratories in Moscow and Atlanta, and the World Health Organization plans to destroy these last stocks in 1993 after the genomes of representative strains have been sequenced. Cowpox and vaccinia viruses have clearly played important roles in the history of human medicine, since they represent the first human vaccine and the only vaccine whose use has resulted in the eradication of a human disease. Surprisingly, interest in poxviruses in general, and vaccinia virus in particular, increased substantially after the disappearance of smallpox. This interest stemmed, in a large part, from the development of vaccinia virus as an expression vector and from the potential use of vaccinia virus recombinants as live vaccines against pathogens other than variola virus. In this chapter, concepts and protocols central to the use of vaccinia virus as an expression vector are described. Readers should refer to another chapter in this series (1), which provides an earlier account of this topic and contains several protocols which remain essentially unaltered, and to another recent review on the construction and applications of vaccinia virus recombinants (2).

2. Vaccinia virus biology

Excellent reviews have been written recently on poxvirus molecular biology (3) and pathogenesis (4) and on the enzymology of vaccinia virus transcription (5) and DNA replication (6), and these should be referred to by interested readers. Only brief outlines of these processes are included in this chapter, with emphasis on those concepts necessary for the use of vaccinia virus as an expression vector.

2.1 Virus structure

Poxviruses are the largest animal viruses and their complexity is illustrated by the presence of more than 100 proteins within the virus particle. Vaccinia virus is the prototype of the *Orthopoxvirus* genus, and is approximately 250 by 350 nm in size. The biconcave core contains the DNA genome and many virus-coded enzymes, and is flanked by lateral bodies. Two forms of infectious virus exist. One form, termed intracellular naked virus (INV), is found in the cytoplasm of virus-infected cells and represents the great majority of infectious progeny: 99% of the total progeny of the WR strain of vaccinia virus is cell-associated. The second form, termed extracellular enveloped virus (EEV), is released from infected cells and possesses an additional lipid envelope which is acquired from Golgi membrane. Several virus-coded glycoproteins are present only in this envelope (7) and give EEV its distinct immunological and biological properties. In particular, EEV production is required for plaque formation and for the efficient spread of virus *in vitro* and *in vivo* (7).

2.2 Virus genome

The vaccinia virus genome is a large double-stranded DNA molecule of 180–200 kbp, the precise size depending on virus strain. It has a base composition of 67% A+T and is non-infectious when de-proteinized. The termini of the two DNA strands are covalently joined into one polynucleotide chain by a single-stranded DNA hairpin. The terminal regions exhibit considerable heterogeneity that may result from deletions, duplications, or transpositions. In contrast, the central two-thirds of the genome show high conservation among different orthopoxviruses and contain the genes essential for virus replication. The complete 192 kbp sequence of the Copenhagen strain of vaccinia virus (8) and much of that of the laboratory WR strain have been reported. The genome contains about 200 genes whose protein-coding sequences are packed tightly, displaying only short intergenic regions or, in some cases, limited overlaps. Deletions of specific genes or of larger regions of DNA near the termini (9) have shown that approximately one-third of the gene complement is non-essential for virus replication *in vitro*.

2.3 Enzymes

Poxviruses have evolved to replicate in the cytoplasm and consequently encode a multitude of replicative and transcriptional enzymes which permit replication to proceed independently of the host cell nucleus. Many of these enzymes are packed within the virus particle and are essential for initiation of infection, while others are present only within the infected cell. The most important virion enzyme is the multi-subunit DNA-dependent RNA polymerase which recognizes promoters only from poxviruses. Thus, the protein-coding regions of foreign genes are linked to vaccinia virus promoters to enable their expression. For DNA replication, the virus encodes a DNA polymerase, a topoisomerase, and a DNA ligase, in addition to several enzymes involved in the synthesis of nucleoside triphosphate precursors. Other exotic enzymes, such as the steroid hormone biosynthetic enzyme 3β-hydroxysteroid dehydrogenase (3β-HSD), are also encoded by vaccinia virus.

2.4 Gene expression

The virus genes are expressed in strict temporal fashion. Immediately after infection, the virus-associated DNA-dependent RNA polymerase transcribes only the early class of genes. Some of these genes encode enzymes needed for DNA synthesis, and others encode virulence factors or proteins that mediate further uncoating of the particle to permit DNA replication. At the onset of DNA replication, transcription of most early genes stops and an intermediate class of genes is activated. Intermediate transcription requires prior synthesis of early proteins, but is not strictly dependent upon DNA replication, since it may occur from plasmid templates that are transfected into virus-infected cells in which DNA replication is blocked. At least three intermediate proteins are transcription factors which are required for late transcription (10). Late genes are distinct from intermediate genes in that they are absolutely dependent upon prior DNA synthesis. They encode late structural proteins which constitute the new virus particles, and also some enzymes and early transcription factors which are packaged into maturing particles. A fourth class of gene is termed constitutive because it is expressed throughout infection owing to the presence of early and late promoters upstream from the open reading frame (ORF).

2.5 Morphogenesis

Poxvirus replication occurs in distinct cytoplasmic factories. Various stages of morphogenesis may be seen in the electron microscope and these are shown in *Figure 1*. The first structures formed are lipid crescents which contain virus protein. These enlarge into complete ovals containing the condensing nucleoprotein core, and then mature into electron-dense infectious INV particles. A small fraction of INV becomes enveloped by a double layer of Golgi-derived membrane, and the particle migrates to the cell surface, where the outer layer fuses with the plasma membrane and releases EEV from the cell.

3. Construction of vaccinia virus recombinants

The large and non-infectious nature of the vaccinia virus genome has made it impracticable to ligate foreign DNA directly into the genome and recover infectious virus. Moreover, foreign protein-coding sequences need to be positioned downstream from a vaccinia virus promoter to ensure efficient

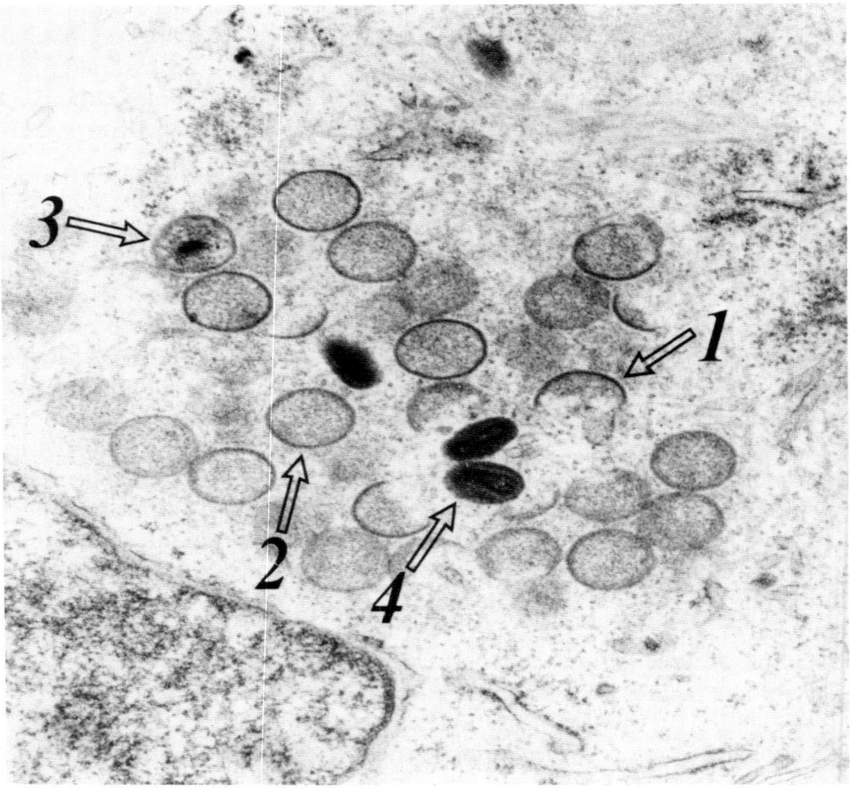

Figure 1. Electron micrograph of a vaccinia virus cytoplasmic factory. Numbers 1–4 show stages of morphogenesis starting from lipid crescent (1) to INV (4). (Reproduced from ref. 29 with permission from Oxford University Press.)

expression. In view of these practical difficulties, recombinant vaccinia viruses are constructed by the two-step procedure illustrated in *Figure 2*. A plasmid vector is constructed in which the foreign gene is linked to a vaccinia virus promoter, and then the gene is transferred into the virus genome by homologous recombination in virus-infected cells transfected with the recombinant plasmid.

3.1 Plasmid vectors

Numerous plasmid insertion vectors have been described for transfer of foreign genes into vaccinia virus. The choice of vector determines the onset, level, and duration of gene expression, the site in the genome at which the foreign gene is inserted, and the method by which recombinants are selected. Careful consideration must be given to these parameters. The simplest

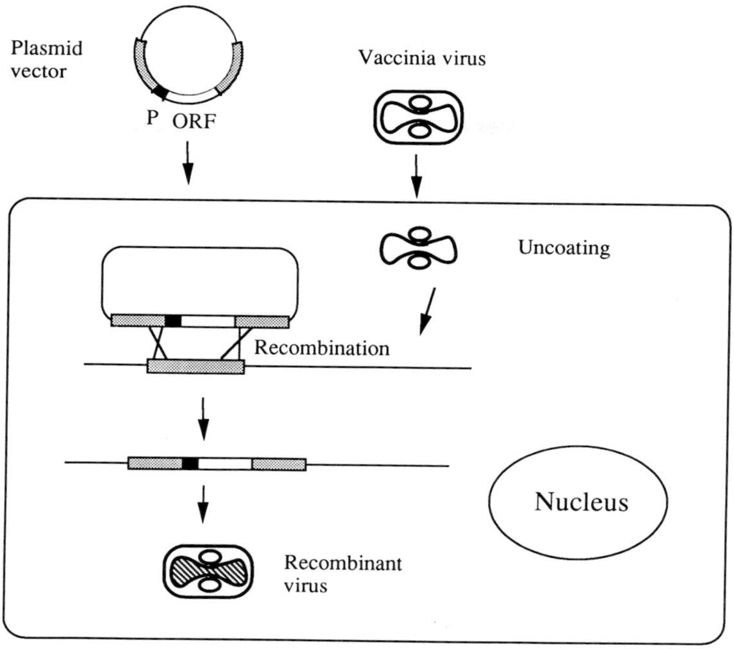

Figure 2. Schematic representation of the formation of recombinants. A plasmid vector containing the ORF to be expressed linked to a vaccinia virus promoter (P) is transfected into cells infected with vaccinia virus. Recombination between the vaccinia virus DNA flanking the gene and the corresponding DNA from the virus genome inserts the foreign gene into the virus genome. This is replicated and packaged into infectious progeny virus.

plasmid vectors have structures illustrated by vector pRK19 (11) in *Figure 3*. A vaccinia virus promoter, in this case from the vaccinia virus late gene encoding the structural protein p4b, is positioned immediately upstream from several unique restriction endonuclease sites into which a foreign gene can be inserted. The promoter and cloning sites are flanked by DNA from a non-essential locus in the virus genome, in this case the thymidine kinase gene (*tk*), which directs insertion of the foreign gene to this site. The *tk* locus has been used extensively, since interruption of this gene facilitates selection of recombinants (see Section 3.4.1).

Vector pRK19 and many other vectors, such as pGS20 (12), pSC11 (13), and pMJ601 (14), do not have an ATG codon positioned downstream from the vaccinia virus promoter to which foreign sequences can be fused, and are therefore suitable for expression of complete ORFs which provide their own translational control signals. Other vectors have an ATG codon positioned upstream from restriction sites which are arranged in all three reading frames and are followed by translational termination codons in all reading frames to enable straightforward expression of any protein-coding sequence (15, 16).

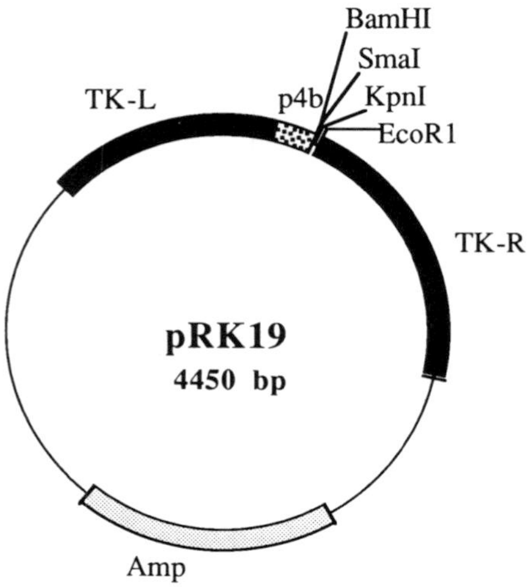

Figure 3. Structure of pRK19 (11), a typical plasmid vector for construction of a recombinant expressing a foreign gene. The p4b vaccinia virus promoter is positioned upstream of restriction endonuclease sites, and the region is flanked by DNA from the left (*tk*-L) and right (*tk*-R) of *tk*.

3.2 Choice of promoters

The promoter determines the level of foreign gene expression and the time during infection at which the gene is expressed. The strongest natural vaccinia virus promoters are from genes encoding major late proteins that form the new virus particles, such as the 11 kDa (17) and p4b genes (18). Vectors enabling even higher levels of expression are described in Section 5. A constitutively expressed promoter from a gene within the inverted terminal repetition of vaccinia virus strain WR which encodes a polypeptide of 7.5 kDa has been used widely (12). (For a list of other poxvirus promoters that have been used see ref. 2.) In constructing a transfer vector, the promoter of the foreign gene should be removed so that the protein-coding sequence can be placed as close as possible to the transcriptional start site of the vaccinia virus promoter. Additional ATG codons upstream from the ORF should also be removed.

In some situations it is important to express the foreign gene either early or late during infection. If the foreign gene contains the early transcriptional termination sequence TTTTTNT within its coding region, the majority of early transcripts will terminate approximately 50 nucleotides downstream from this site and full-length protein may not be produced. In such a situation it is desirable either to express the gene from a late promoter so that premature termination does not occur (19) or to remove termination sequences by mutagenesis (20). Conversely, there are some situations in which early expression is important. For instance, late during vaccinia virus infection the presentation of some peptide epitopes to class I restricted cytotoxic T lymphocytes (CTLs) is blocked, whereas the same epitopes may be presented early during infection (21, 22). If recombinants are constructed in order to determine whether the expressed antigen is recognized on the surface of infected cells by CTLs, it is sensible, and may be essential, to express the antigen from an early promoter.

3.3 Insertion sites

The vaccinia virus genome contains numerous sites that are non-essential for virus replication *in vitro*, into which foreign genes may be inserted. These are listed in *Table 1*. While the endogenous genes present at these sites are non-essential *in vitro*, in many cases the encoded protein plays an important role in virulence of the virus *in vivo*. Virus attenuation resulting from deletion of a specific gene, or group of genes, may diminish the immune response following virus infection, and this may be important if the recombinant is used to raise an antibody or T-cell response against the foreign antigen or is tested as a candidate vaccine. For most laboratory applications where the recombinant is to be used to infect cells *in vitro*, however, the site of insertion is less important than the kinetics and level of gene expression.

The most convenient and widely used site is *tk*, since insertion here allows tk^- recombinants to be selected on tk^- cells in the presence of 5-bromodeoxyuridine (BUdR) (23). Several other genes involved in nucleic acid metabolism, such as those encoding thymidylate kinase, ribonucleotide reductase, and DNA ligase, are also non-essential for virus replication, but, like *tk* (24), they affect virus virulence *in vivo*. Other non-essential virulence factors include the vaccinia virus growth factor (25), a 13.8 kDa secretory virokine (26), a secretory protein which interferes with complement function (27), and the biosynthetic steroid enzyme 3β-HSD (28). Since both the INV and EEV forms of virus particle are infectious, the genes required for the formation and release of EEV are also non-essential for production of infectious virus. Some, however, are required for plaque formation: the 14 kDa fusion protein, a 37 kDa acylated protein, and a lectin-like glycoprotein (29–31). Many other individual genes and blocks of genes adjacent to the genomic termini (9) are non-essential for virus replication *in vitro*. If the insertion site is not important for other reasons, however, foreign genes are better inserted into the central region of the genome, since it is more stable than the variable termini.

3.4 Methods for selecting recombinants

Recombinants are formed by homologous recombination within cells infected with vaccinia virus and transfected with a plasmid vector, as described in *Protocol 1*. This method produces recombinants at a frequency of approximately

Table 1. Sites for inserting foreign DNA into the vaccinia virus genome

Gene name[a]	Function	Reference
C23L-K1L		(9, 16)
C13L[b]	7.5 kDa protein	(57)
C11R[g]	Vaccinia virus growth factor[c]	(25)
C3L[g]	Complement control protein	(27)
N1L[g]	13.8 kDa secretory virokine	(26)
M1L		(58)
M2L		(58)
K1L	Human host range protein	(58)
K2L	Serine protease inhibitor	(58, 59)
K3L	eIF-2α homologue, interferon resistance	(60)
K4L		(30, 58)
K7R		(58)
F1L		(58)
F4L	Ribonucleotide reductase (small subunit)	(58)
F9L		(58)
F13L	EEV envelope antigen	(30)
E1L		(58)
I4L[g]	Ribonucleotide reductase (large subunit)	(61)
J2R[g]	Thymidine kinase[d]	(23)
D8L	35 kDa membrane protein	(62)
A26L/A27L[e]	A-type inclusion body protein	(50)
A27L	14 kDa fusion protein, needed for virus egress	(29)
A34R (*Sal*L4R)	Lectin-like EEV envelope glycoprotein	(31)
A40R (*Sal*L2R)	Lectin-like glycoprotein	(Duncan and Smith, unpublished data)
A42R	Profilin	(47)
A43 (*Sal*F5R)	Glycoprotein	(63)
A44L (*Sal*F7L)[g]	3-β-hydroxy steroid dehydrogenase	(28)
A48R (*Sal*F11R)[g]	Thymidylate kinase	(64)
A50R (*Sal*F13R)[g]	DNA ligase	(42, 65)
A56R	Haemagglutinin	(66, 67)
B5R	Complement control protein Host range	(68)
B12R	Protein kinase-related	(Banham and Smith, unpublished data)
B13R[e]	Serine protease inhibitor	(69)
B22R[f]	Serine protease inhibitor	(69)
B15R-B29R		(9)

[a] Gene names relate to strains Copenhagen (8) or WR (shown in brackets) (70). The nomenclature indicates the *Hin*dIII restriction fragment containing the gene (fragments are labelled alphabetically in order of decreasing size) and the direction of transcription (R or L). *Sal*I nomenclature is used for the right side of the WR *Hin*dIII A fragment because the sequence to the left is incomplete.
[b] In some virus strains (for example, WR) this gene is located within the inverted terminal repeat.
[c] In some virus strains this is a 35 kDa secreted protein and maps within the inverted terminal repeat.
[d] The majority of recombinants have been generated by insertion into this gene.
[e] In the Copenhagen strain this gene is interrupted.
[f] In the Copenhagen strain this gene is present at the left end of the genome and is named C12L.
[g] The gene product is known to increase virus virulence.

0.1%; selective methods are, therefore, needed to distinguish recombinant from parental virus. Many different approaches have been used, and only those in widespread use are considered here; for a comprehensive list the reader should see ref. 2.

Protocol 1. Transfection procedure for producing recombinants

Materials

- CV-1 cells: obtainable from the American Type Culture Collection (ATCC) as CCL-70
- MEM: minimal essential medium (Gibco) containing 100 U/ml penicillin and 100 μg/ml streptomycin
- wild type (WT) vaccinia virus: for example the WR strain obtainable from the ATCC as VR-119
- plasmid DNA containing the foreign gene cloned downstream from a vaccinia virus promoter
- carrier DNA: 1 mg/ml sonicated salmon sperm or calf thymus DNA

Method

1. Grow CV-1 cell monolayers in MEM containing 8% (v/v) fetal calf serum (FCS) in a 25 cm^2 flask to approximately 80% confluence.
2. Aspirate the medium and infect the cells with WT vaccinia virus at a multiplicity of infection (m.o.i.) of 0.05 plaque-forming units (p.f.u.)/cell. Gently rock the flask at 37°C for 2 h.
3. Precipitate 1 μg of plasmid DNA and 19 μg of carrier DNA in a total volume of 1 ml with calcium phosphate (see Chapter 8, *Protocol 6*).
4. Remove the virus inoculum and add the calcium phosphate-precipitated DNA, then incubate the flask for 30 min at 37°C.
5. Add 9 ml of MEM containing 8% FCS and incubate the flask for 3–4 h at 37°C.
6. Replace the medium with fresh MEM containing 8% FCS.
7. Scrape the cells into the medium at 48 h post infection (p.i.), pellet them at 2000 *g* for 5 min and resuspend the cell pellet in 1 ml of MEM containing 8% FCS. Aliquot and store the cells at −70°C.

3.4.1 Selection of tk^- virus

The most widely used method of generating recombinants is to insert the foreign DNA into the *tk* locus and select tk^- recombinants, as described in *Protocol 2* (12, 23). Despite its popularity, this method has some limitations:

- it requires a tk^+ parental virus
- it allows insertion only into *tk*
- recombinant plaques must be selected on tk^- cells in the presence of the mutagenic compound BUdR
- some tk^- plaques are spontaneous mutants which do not contain the foreign gene

Spontaneous tk^- mutants may be distinguished from recombinants by a variety of means, including DNA hybridization (*Protocol 3*), immunological screening for expression of the foreign antigen, or polymerase chain reaction (PCR; *Protocol 4*). The latter approach is fast and uses oligonucleotide primers that bind to either side of the insertion site within *tk*.

Protocol 2. Selection of tk^- virus

Materials

- an aliquot of infected cells produced in *Protocol 1*
- 5 mg/ml BUdR; filter-sterilize, aliquot, and store at −20°C
- 2% (w/v) low gelling temperature agarose: autoclave
- human tk^-143 cells
- 2 × MEM (Gibco)
- 1% (w/v) neutral red stain
- 20 mg/ml 5-bromo-4-chloro-3-indolyl β-D-galactopyranoside (X-gal) in dimethylformamide

Method

1. Thaw the aliquot of infected cells and freeze–thaw twice more. Sonicate briefly in a sonic bath to complete cell lysis and disaggregate the virus.
2. Prepare ten-fold dilutions of virus to a final dilution of 10^{-4} in MEM containing 8% FCS.
3. Remove the medium from confluent monolayers of tk^-143 cells in 50 mm Petri dishes and add 0.5 ml of diluted virus. Plate out the 10^{-2}, 10^{-3}, and 10^{-4} dilutions in duplicate. Gently rock the dishes for 2 h at 37°C.
4. Remove the virus inocula and overlay the cells with 4 ml of MEM containing 1% low gelling temperature agarose, 2.5% FCS, and 25 μg/ml BUdR. Leave the dishes at room temperature in the dark until the agarose has set and then incubate them for 2 days at 37°C.

5. Stain the virus plaques by overlaying the dishes with 4 ml of MEM containing 0.01% neutral red and 1% low gelling temperature agarose, and incubate for 2–4 h at 37°C. (If the recombinant is derived from vector pSC11 or other plasmids which insert *Escherichia coli lacZ* into *tk*, include 200 μg/ml X-gal in the overlay. This compound is converted by β-galactosidase into a deep blue colour, allowing recombinants to be distinguished from spontaneous tk^- mutants.) Use Pasteur pipettes to pick plaques from the highest dilution at which they are visible. Store the plaques at −70°C in a small volume of MEM containing 8% FCS.

Protocol 3. Screening tk^- plaques by DNA hybridization

Materials

- tk^- plaques isolated in *Protocol 2*
- phosphate-buffered saline (PBS): 170 mM NaCl, 3.4 mM KCl, 10 mM Na_2HPO_4, 1.8 mM KH_2PO_4 (pH 7.2)
- nitrocellulose membrane (Schleicher and Schuell)
- 20 × SSC: 3 M NaCl, 0.3 M trisodium citrate
- 50 × Denhardt's solution: 1% (w/v) bovine serum albumin (fraction V), 1% (w/v) Ficoll, 1% (w/v) polyvinylpyrrolidone
- 10% (w/v) sodium dodecyl sulphate (SDS)
- a ^{32}P-labelled DNA probe specific for the foreign gene

Method

1. Use half of the yield from tk^- plaques to infect tk^-143 monolayers in 24-well plates. Incubate the plates for 2 h at 37°C. Add 1 ml of MEM containing 2.5% FCS and 25 μg/ml BUdR. Incubate the plates for 2 days at 37°C.
2. Check the cells by microscopy for cytopathic effect (c.p.e.). If more than 50% of the cells show c.p.e., scrape the cells into the medium using the plunger from a 1 ml syringe.
3. Transfer the cells to a 1.5 ml tube and centrifuge at 12 000 *g* for 1 min.
4. Discard the supernatant and resuspend the cells in 200 μl of PBS. Freeze–thaw three times, sonicate briefly, and then load 50 μl on to a nitrocellulose membrane. Place the membrane for 3 min on a filter paper soaked with 0.5 M NaOH, transfer for 3 min to a second filter paper soaked with 1 M Tris–HCl (pH 7.5), and then for 3 min to a third filter paper soaked with 2 × SSC. Air-dry the membrane and bake it for 2 h at 80°C in a vacuum oven.

Protocol 3. *Continued*

5. Hybridize the membrane with the radioactive probe.
   (a) Incubate the filter in 6 × SSC containing 5 × Denhardt's solution for 4 h at 65°C.
   (b) Denature the probe by boiling for 2 min, add it to the solution and incubate for 12–18 at 65°C.
   (c) Wash the membrane twice for 15 min each in 2 × SSC, 0.1% SDS at 65°C, and twice for 15 min each in 0.2 × SSC at 65°C.
   (d) Air-dry the membrane and expose it to X-ray film.
6. Select positive isolates by reference to the developed film.
7. Plaque-purify the recombinants further. A total of three cycles of plaque purification usually gives homogeneous isolates. Prepare and titrate virus stocks. Store them at −70°C.

Protocol 4. Screening *tk*⁻ plaques by PCR

Materials

- *tk*⁻ plaques isolated in *Protocol 2*
- 20 μM solutions of oligonucleotide primers 1 and 2 (flanking either side of the insertion site within *tk*)
- 5 U/ml *Taq* polymerase
- 10 × PCR buffer: 100 mM Tris–HCl (pH 8.3), 500 mM KCl, 15 mM $MgCl_2$, 0.01% (w/v) gelatin
- dNTP solution: 2 mM each of dATP, dCTP, dGTP, and dTTP

Method

1. Carry out steps 1–3 of *Protocol 3*. Resuspend the infected cells in 400 μl of water. Freeze–thaw the cells twice and sonicate them.
2. Heat the disrupted cells at 95°C for 5 min and then place them on ice.
3. Set up the PCR in a final volume of 50 μl by mixing the following:

• infected cell extract	3 μl
• oligonucleotide primer 1	2.5 μl
• oligonucleotide primer 2	2.5 μl
• dNTP solution	5 μl
• 10 × PCR buffer	5 μl
• *Taq* polymerase	0.25 μl
• water	32 μl

4. Incubate the mixture at 92°C for 1 min, 60°C for 1 min, and 72°C for 1.5 min. Repeat this cycle 24 times. The hybridization temperature depends upon the primers used.

5. Electrophorese 10–20 μl of the product on a 1% (w/v) agarose gel including size markers. The size of the PCR product indicates whether the plaque is a spontaneous *tk*$^-$ mutant or a recombinant with exogenous DNA inserted into *tk*.

6. Carry out step 7 of *Protocol 3*.

3.4.2 Expression of β-galactosidase

To expedite the distinction between *tk*$^-$ recombinants and spontaneous *tk*$^-$ mutants selected in the presence of BUdR, some vectors contain *lacZ* linked to a vaccinia virus promoter which may be co-inserted into *tk* (13). In the presence of the chromogenic substrate X-gal, plaques formed by recombinants can be distinguished visually from spontaneous mutants by their blue colour (see *Protocol 2*). Indeed, expression of β-galactosidase can be used alone to detect recombinants without the need for other selection methods (32). Moreover, subsequent replacement of *lacZ* by another foreign gene enables the selection of recombinants which no longer form blue plaques (32).

3.4.3 Selection of *tk*$^+$ virus

An alternative method for selecting recombinants is to insert herpes simplex virus type 1 (HSV-1) *tk* linked to a vaccinia virus promoter into any non-essential site of a *tk*$^-$ vaccinia virus genome (23). Recombinants are *tk*$^+$ and will form plaques in *tk*$^-$ cells in the presence of aminopterin (methotrexate), as described in *Protocol 5*. A second foreign gene may be inserted simultaneously with HSV-1 *tk* and selected in the same way (33). Unlike the selection of *tk*$^-$ virus using BUdR, this system produces only true recombinants and does not require the use of mutagenic compounds. It remains dependent, however, upon use of a *tk*$^-$ cell line.

Protocol 5. Selection of *tk*$^+$ virus

Materials

- 1 mM aminopterin (methotrexate)
- 40 × TAGG: 0.6 mM thymidine, 2 mM adenosine, 2 mM guanosine, 0.4 mM glycine
- filter-sterilize, aliquot, and store these solutions at −20°C
- a *tk*$^-$ vaccinia virus

Protocol 5. *Continued*

Method

1. Complete *Protocol 1* using the tk^- virus and an appropriate plasmid (such as pVPHTK2: see ref. 23) in steps 1–3 of *Protocol 2*.
2. Remove the virus inoculum and overlay the cells with MEM containing 1% low gelling temperature agarose, 2.5% FCS, 0.1 mM non-essential amino acids, 1 × TAGG and 1 μM aminopterin. Leave the plates at room temperature in the dark until the agarose has set and then incubate them for 2 days at 37°C.
3. Carry out step 5 of *Protocol 2*.
4. Carry out step 7 of *Protocol 3*.

3.4.4 Neomycin resistance

Another dominant marker gene used to select recombinants is the neomycin resistance gene. The antibiotic G418 inhibits production of WT virus but allows plaque formation by virus expressing this gene (34). This selection system has been used infrequently (16).

3.4.5 Expression of the *E. coli* guanine phosphoribosyltransferase gene (*Ecogpt*)

This selectable marker gene is now widely used to make recombinants and for experimental manipulation of the vaccinia virus genome. Mycophenolic acid (MPA) inhibits inosine monophosphate dehydrogenase and results in blockage of purine synthesis and inhibition of vaccinia virus replication in most cell lines. The blockage may be overcome by expressing *Ecogpt* from a constitutive vaccinia virus promoter and providing the substrates xanthine and hypoxanthine (15, 35), as described in *Protocol 6*. Some workers have found it necessary to include aminopterin, to block *de novo* synthesis of purines, and thymidine (35). The system is versatile and overcomes the deficiencies of selecting tk^- viruses with BUdR. It is not restricted to insertion into a particular site in the virus genome, does not require the use of specific cell lines or mutagenic compounds, and allows plaque formation by recombinants but not by spontaneous mutants.

An additional advantage of *Ecogpt* is that it is possible to select against its expression, as described in *Protocol 7*, so that $Ecogpt^-$ recombinants may be derived from an $Ecogpt^+$ parent (36). Reverse selection requires the use of a hypoxanthine phosphoribosyltransferase negative ($hrpt^-$) cell line, since the nucleoside analogue 6-thioguanine is toxic for mammalian cells expressing *hprt*. Combined with positive selection for *Ecogpt*, reverse *Ecogpt* selection enables a mutation to be introduced into a specific site without leaving the selectable marker in the vaccinia virus genome. Consequently, multiple mutations may be built sequentially into the same recombinant.

Protocol 6. Selection of recombinants expressing *Ecogpt*

Materials

- CV-1, human *tk*$^-$143 or BSC-1 cells: BSC-1 cells are available from the ATCC as CCL-26
- 10 mg/ml MPA: filter-sterilize, aliquot, and store at −20°C
- 10 mg/ml xanthine in 0.1 M NaOH: filter-sterilize, aliquot, and store at −20°C
- 10 mg/ml hypoxanthine: filter-sterilize, aliquot, and store at −20°C

Method

1. Construct a plasmid containing *Ecogpt* linked to a vaccinia virus promoter (derived, for instance, from pGpt07/14: see ref. 35) inserted into the vaccinia virus gene of interest. Carry out *Protocol 1* to obtain recombinants containing *Ecogpt.*
2. Prepare monolayers of CV-1, BSC-1, or *tk*$^-$143 cells in 50 mm Petri dishes. Replace the growth medium with MEM containing 2.5% FCS, 25 μg/ml MPA, 250 μg/ml xanthine, and 15 μg/ml hypoxanthine. Incubate the dishes for 12–24 h at 37°C.
3. Remove the medium and infect the cells with appropriate dilutions of virus (10^{-2}, 10^{-3}, and 10^{-4} dilutions if the virus is obtained directly from *Protocol 1*). Gently rock the dishes for 2 h at 37°C.
4. Remove the virus inocula and overlay the cells with MEM containing 1% low gelling temperature agarose, 2.5% FCS, 25 μg/ml MPA, 250 μg/ml xanthine, and 15 μg/ml hypoxanthine. Leave the dishes at room temperature in the dark until the agarose has set and then incubate them for 2–3 days at 37°C.
5. Carry out step 5 of *Protocol 2.*
6. Carry out step 7 of *Protocol 3.*

This method is taken from ref. 15. Other investigators (35) have reported that the selection system does not work on CV-1 cells in the absence of aminopterin (0.2 μg/ml) and thymidine (4 μg/ml).

Protocol 7. Selection against *Ecogpt* expression

Materials

- 1 mg/ml 6-thioguanine: filter-sterilize, aliquot, and store at −20°C
- an *hprt*$^-$ cell line such as HeLa D98R or murine STO: available from the Sir William Dunn School of Pathology or the ATCC (CRL 1503), respectively

Protocol 7. *Continued*

Method

1. Construct a plasmid containing sequence for replacing *Ecogpt* present in a recombinant obtained in *Protocol 6*. Carry out *Protocol 1* to obtain recombinants lacking *Ecogpt*.
2. Grow monolayers of *hprt*$^-$ cells in 50 mm Petri dishes. Remove the growth medium and infect the cells with the virus yield from step 1, using a range of virus dilutions to obtain an appropriate number of *Ecogpt*$^-$ plaques. Gently rock the dishes for 2 h at 37°C.
3. Remove the virus inocula and overlay the cells with MEM containing 1% low gelling temperature agarose, 2.5% FCS, and 1 μg/ml 6-thioguanine. Allow the agarose to set at room temperature and incubate them for 3 days at 37°C. Plaques formed on HeLa D98R cells are smaller than on those formed on BSC-1 cells, and it is therefore necessary to incubate infected D98R cells for a minimum of 3 days.
4. Carry out step 5 of *Protocol 2*.
5. Carry out step 7 of *Protocol 3*.

3.4.6 Use of conditional lethal mutant viruses

Identification of vaccinia virus genes that are required for growth on certain cell types has provided another dominant system for the selection of recombinants. The parental virus, which lacks the host range gene and must be grown on an alternative permissive cell type, is recombined with a plasmid that contains the host range gene and a second gene for co-insertion. Recombinants are selected on the cell line non-permissive for the parental virus. Vectors have been described which use a vaccinia virus human host range gene for selection and which allow expression of foreign protein-coding sequences linked to another vaccinia virus promoter (16).

Other conditional lethal viruses, such as temperature-sensitive or drug-dependent mutants, have been used to select for recombinants (37, 38). If cells are infected with such a virus under non-permissive conditions and then transfected with WT virus DNA together with a recombinant plasmid, viable virus is produced only in those cells which take up both WT virus DNA and the recombinant plasmid. The WT genome is transcribed by *trans*-acting functions provided by abortive infection with the conditional lethal mutant and recombines very efficiently with the co-transfected plasmid, so that up to 30% of the progeny virus is recombinant (38).

3.4.7 Transient dominant selection

The mechanism by which recombinants are formed from parental virus and transfected plasmid was originally thought to occur by double homologous recombination on either side of the insertion site. In this way, the WT allele present in the virus genome is simply replaced by the mutant allele from the plasmid. It has become clear, however, that recombinant genomes can also be formed by a single homologous recombination event on one side of the foreign gene, so that the entire plasmid is inserted into the virus genome (39, 40). As shown in *Figure 4*, such virus genomes contain direct repeats of vaccinia virus DNA and are consequently unstable. They may resolve by recombination between the direct repeats to remove one copy of the repeat and the intervening DNA, yielding either parental or recombinant virus.

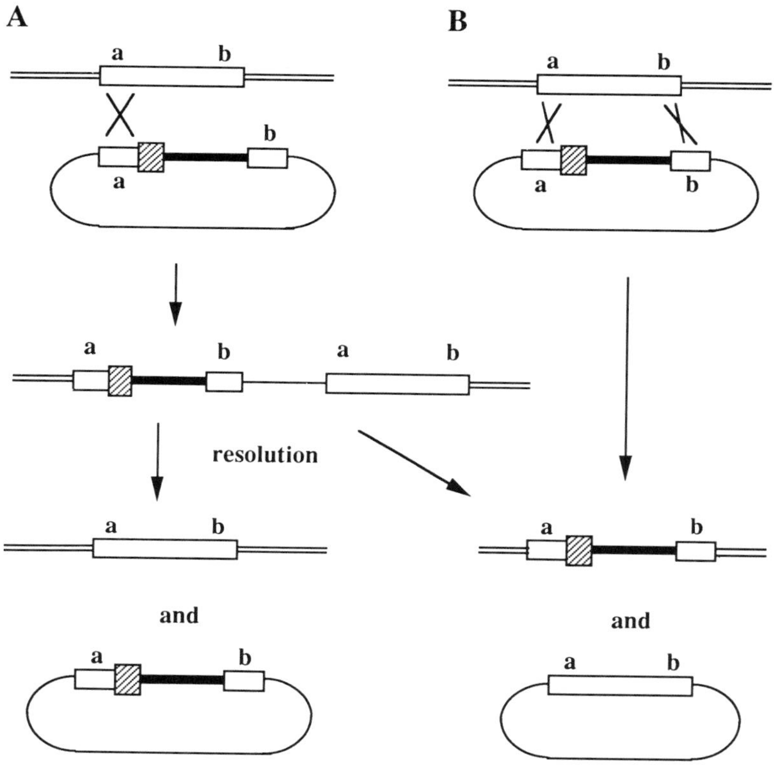

Figure 4. Formation of recombinants by single (A) or double (B) homologous recombination events. In (A) a single recombination event is shown to the left of the foreign gene between homologous virus sequences designated 'a'. This produces an intermediate virus that contains direct repeats of sequences 'a' and 'b'. Owing to the instability of the direct repeats, the virus undergoes recombination between the 'a' or 'b' repeats. In the former case, the products of the resolution event are the WT virus genome and the plasmid that was used for transfection. In the latter case the products are a recombinant genome and a plasmid containing the wild type allele of the insertion site. In (B) a double homologous recombination event produces the recombinant directly. The open boxes represent flanking vaccinia virus DNA, the hatched box a vaccinia virus promoter, the solid line a foreign gene, the double line the virus genome, and the single line plasmid sequences.

Although the intermediate virus is unstable, it can be retained if the sequence between the direct repeats contains a dominant selectable maker for which selective pressure is maintained. This is the basis of a technique termed transient dominant selection, which enables mutations to be introduced into specific sites without leaving the genetic marker in the virus genome (41).

A plasmid is constructed that contains the desired mutated allele of the virus gene and distal to this a dominant selectable marker gene (for example *Ecogpt*) linked to a vaccinia virus promoter (*Figure 5*). Selection of recombinants in the presence of MPA (*Protocol 6*) ensures that all viruses have the entire plasmid integrated into the virus genome If selection is removed, unstable structures resolve and plaque isolates representing either parental or

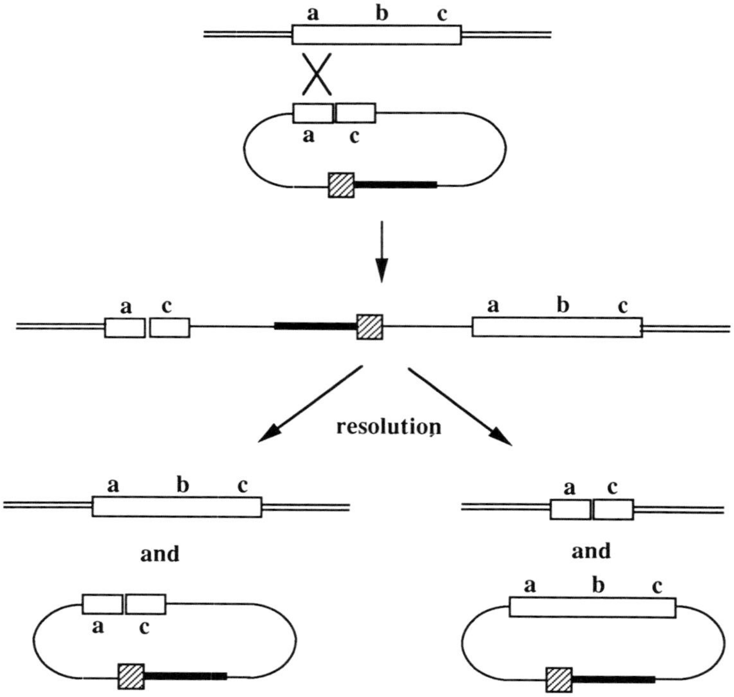

Figure 5. Formation of recombinants by transient dominant selection. The plasmid contains a vaccinia virus gene (open box) with an internal region ('b') deleted and a dominant selectable marker (solid line) such as *Ecogpt* linked to a vaccinia virus promoter (hatched box). Recombination with the homologous sequence 'a' of the virus genome produces an intermediate recombinant with direct repeats of sequences 'a' and 'c'. If selection for the marker gene is removed the intermediate virus resolves by recombination between repeated sequences 'a' or 'c'. If resolution occurs by recombination through sequence 'a' the WT virus genome and the plasmid that was used for transfection are formed. Alternatively, if recombination occurs between 'c' sequences the products are a recombinant with 'b' deleted and a plasmid bearing the complete vaccinia virus gene.

mutant virus can be distinguished by PCR using oligonucleotide primers that span the mutated site. If only WT virus plaques are obtained it is likely that the mutation is lethal for virus replication or plaque formation. The resolution event can be speeded by plaquing on *hprt*$^{-}$ cells and selecting against the presence of *Ecogpt* using 6-thioguanine (*Protocol 7*).

During resolution of the intermediate virus, plasmids are formed in addition to progeny virus genomes (42) (*Figures 4* and *5*). These represent either the original plasmid used to construct the recombinant, or a plasmid containing the WT allele rescued from the recombinant. Since all plasmids are replicated in vaccinia virus-infected cells, irrespective of whether they contain vaccinia virus sequences, plasmids formed during resolution events will be amplified and may then recombine with virus genomes causing virus heterogeneity (42). A screen for plasmid sequences should therefore be included during selection of recombinants to ensure that resolution has already occurred and that plasmid sequences do not persist. In addition, the plasmid should be digested prior to transfection with two restriction enzymes which produce incompatible termini in order to avoid formation of plasmids in virus-infected cells, since linear plasmids with compatible termini could be circularized by the virus DNA ligase. With this precaution, viable virus can only be produced by double homologous recombination.

4. Inducible gene expression from vaccinia virus vectors

Conventional vaccinia virus vectors allow the expression of a target gene at a time during infection and at a level determined by the selected vaccinia virus promoter. In the last few years, new vectors have been developed which allow genes to be regulated in an inducible manner in vaccinia virus-infected cells (29, 31, 43–46). As illustrated in *Figure 6*, these vectors use the *E. coli* lacI repressor protein and the operator sequence to which it binds. A chimeric gene is constructed which contains a late vaccinia virus promoter, one or two copies of the *lac* operator, and the open reading frame to be regulated. This gene is inserted into *tk* together with a constitutive vaccinia virus promoter driving the *lacI* ORF. The lacI protein binds to the operator and prevents transcription of the downstream ORF. This blockage may be released by the addition of isopropyl β-D-thiogalactopyranoside (IPTG), which causes the lacI protein to dissociate from the operator. The stringency of repression and degree of inducibility with IPTG depend upon the nature of the operator sequence, the proximity of the operator to the RNA start site of the vaccinia virus promoter, the number of operator copies, and the concentration of IPTG. One vector (pPR34) allows 97% inhibition of gene expression and 90% induction with 5 mM IPTG, and another (pPR35) has two copies of the operator and inhibits gene expression by 99.9% but permits only 50% induction (44).

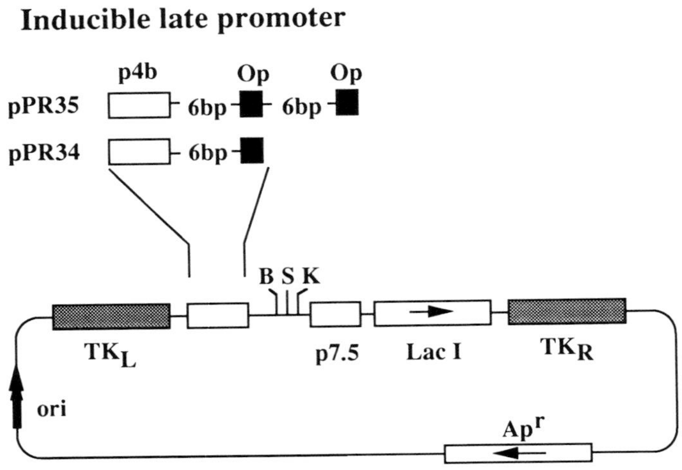

Figure 6. Plasmid vectors for IPTG-dependent gene expression from vaccinia virus. Plasmids pPR34 and pPR35 contain *lacI* downstream from the vaccinia virus 7.5 kDa promoter, the p4b promoter upstream of the *lac* operator, and flanking DNA from the left and right of *tk*. The plasmids differ only in the presence of a second copy of the operator sequence in pPR35. (This figure has been reproduced from ref. 44 with permission from Academic Press.)

These vectors potentially allow regulation of genes that would otherwise be toxic to vaccinia virus replication, and in addition have been most valuable for the study of vaccinia virus gene function. Two approaches have been taken. In one method, which is described in *Protocol 8* and illustrated in *Figure 7* (29, 31), a copy of the ORF to be studied is produced by PCR and cloned downstream from the IPTG-inducible promoter. This is inserted into vaccinia virus *tk* with *lacI* to form a *tk*$^-$ intermediate virus which contains an inducible copy of the ORF plus the endogenous copy. The endogenous copy is then disrupted by insertion of *Ecogpt* and a recombinant is selected in the presence of MPA and IPTG. This virus will have only a single inducible copy of the ORF and will be dependent on IPTG for replication if the ORF encodes an essential protein. To study the function of the protein during virus replication, a stock of virus is grown in the presence of IPTG and then used to infect cells in the absence of IPTG. The stage at which virus replication is aborted is then determined biochemically or by electron microscopy. This method may be used to study the function of any late gene, irrespective of whether a conditional lethal mutant exists, and has the advantage that the function of a protein with the WT amino acid sequence is studied rather than proteins bearing conditional lethal amino acid substitutions.

An alternative approach to studying vaccinia virus gene function using IPTG-inducible vectors uses the transient dominant selection method. First, a

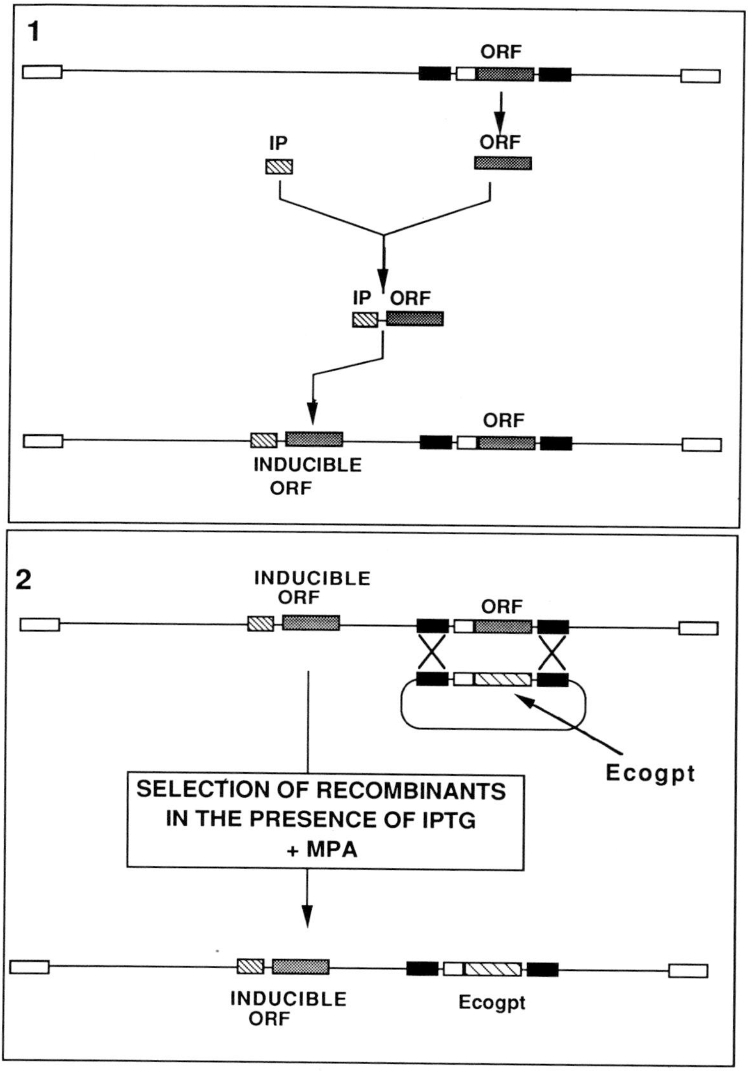

Figure 7. General scheme for the formation of IPTG-dependent vaccinia virus. (1) A PCR version of an ORF (stippled box) is cloned downstream of an IPTG-inducible promoter (hatched box) and then inserted into *tk* to form an intermediate virus. (2) Cells infected with the intermediate virus are transfected with a plasmid in which the ORF has been replaced by *Ecogpt* linked to a vaccinia virus promoter. Recombination between the vaccinia virus sequences flanking the ORF (filled box) forms a recombinant which is selected in the presence of MPA and IPTG.

virus is constructed that expresses the lacI protein from within *tk*. Next, the *lac* operator is inserted between the transcriptional initiation site of the gene to be regulated and the downstream ORF within a second plasmid vector. The *Ecogpt* gene linked to a constitutive vaccinia virus promoter is then cloned into this plasmid distal to the vaccinia virus gene. The resulting plasmid is transfected into cells infected with the virus expressing *lacI*, and recombinants are selected in the presence of MPA (*Protocol 6*). Progeny viruses retain the plasmid sequences and, after removal of MPA selection, resolve in the presence of IPTG to form either parental or recombinant virus, which may be distinguished by PCR across the site at which the operator has been inserted. If the gene being tested is required for replication or plaque formation, these events will be dependent upon IPTG.

Recombinant viruses which express IPTG-inducible copies of the 11 kDa core protein (gene *F18R*), the 14 kDa fusion protein (gene *A27L*), or a lectin-like glycoprotein (gene *SalL4R*) have been described. In the absence of IPTG the former virus produces no infectious progeny owing to a blockage in nucleoid condensation (47), while the latter two viruses produce normal amounts of INV but are dependent upon IPTG for plaque development and have altered EEV formation (29, 31).

To date, only vaccinia virus late genes have been regulated by the *lac* operator and lacI protein, as attempts to regulate early genes have not provided a sufficient degree of repression (J. F. Rodriguez, unpublished data). This presumably reflects the need to package lacI bound to the *lac* operator within all infectious virus particles, so that no expression occurs in freshly-infected cells without the addition of IPTG. Additionally, IPTG has to reach the virus genome within the core to induce efficiently. These problems may prevent efficient IPTG-inducible regulation of early genes, but adequate regulation of intermediate genes should be possible.

Protocol 8. Construction of IPTG-dependent vaccinia virus

Materials

- 0.1 M IPTG: filter-sterilize, aliquot, and store at −20°C

Method

1. Select the late gene to be regulated in an IPTG-inducible manner. Produce a copy of the protein-coding region by PCR and clone it downstream of the IPTG-inducible promoter in plasmid vector pPR34 or pPR35 (44). These vectors allow insertion of the foreign gene into *tk* together with *lacI* linked to the vaccinia virus 7.5 kDa promoter.
2. Infect CV-1 cells with WT vaccinia virus and transfect with the plasmid produced, as described in *Protocol 1*.
3. Isolate tk^- recombinants as described in *Protocol 2*.

4. Identify recombinants by PCR screen (*Protocol 4*).
5. Grow a stock of virus, titrate it, and infect CV-1 cells at an m.o.i. of 0.05 p.f.u./cell. Transfect the infected cells with a plasmid containing the target gene interrupted by *Ecogpt* linked to a vaccinia virus promoter (for example for interruption of the *SalL4R* gene use plasmid pSAD15: see ref. 31).
6. Select recombinants in the presence of MPA, xanthine, and hypoxanthine (*Protocol 6*) and 5 mM IPTG. Plaque purify the recombinants three times and grow a stock of virus in the presence of 5 mM IPTG.

5. High-level gene expression

5.1 Powerful poxvirus promoters

Several approaches have been used to increase the level of expression from vaccinia virus expression vectors. Natural vaccinia virus promoters from genes encoding the most abundant late structural proteins, such as those encoding for 11 kDa and p4b, give levels of expression approximately three- to five-fold higher than the widely-used 7.5 kDa promoter. Synthetic promoters based upon the results of saturation mutagenesis of vaccinia virus promoters (48, 49) produce higher levels of expression than natural strong promoters (14). Promoters from other poxviruses which produce a cytoplasmic inclusion body composed predominantly of a single abundant virus protein, such as those from cowpox virus (50) and the *Choristoneura biennis* entomopoxvirus (51), are both highly active in vaccinia virus.

5.2 T7 RNA polymerase

The most successful method for high-level expression of foreign genes by vaccinia virus has utilized bacteriophage T7 RNA polymerase (52–56). The system is illustrated in *Figure 8*. Two recombinants are constructed, one which expresses the bacteriophage T7 RNA polymers from a vaccinia virus promoter, and a second which contains the gene for high-level expression under the control of T7 transcriptional initiation and termination sequences. Co-infection of cells with the two viruses enables expression of T7 RNA polymerase and high-level transcription of the target gene from the T7 RNA polymerase promoter. The transcripts are very abundant, representing 30% of total cellular RNA at 24 h p.i., but most are uncapped and therefore poorly translated in vaccinia virus-infected cells. This difficulty was overcome by insertion of the 5′ non-coding sequences of the picornavirus encephalomyocarditis virus, allowing cap-independent entry of ribosomes on to mRNA between the 5′ end of the T7 transcript and the ORF to be expressed. With this modification and the use of hypertonic medium to favour translation of virus mRNAs, chloramphenicol acetyltransferase was expressed at a level representing 10% of total cellular protein (54).

Although expression levels are highest if two recombinants are used, namely one that expresses T7 RNA polymerase and one containing the target gene, satisfactory levels of expression are obtained by infecting cells with the virus expressing T7 RNA polymerase and then transfecting them with a plasmid containing the target gene flanked by T7 RNA polymerase transcriptional control signals. This has the advantage of speed and is especially useful for screening multiple gene variants containing specific mutations. After selection of a suitable clone, the same plasmid can be used to construct a recombinant to achieve optimal levels of expression.

Further refinements in the expression system have been made by combining T7 RNA polymerase-based expression with IPTG-inducibility (56). Also, both the T7 RNA polymerase gene and the T7 promoter and termination sequences can be accommodated in a single virus, if expression of the polymerase is made IPTG-dependent (71). Previously, it had not been possible to have both components in the same virus, presumably because the very high level of transcription by T7 RNA polymerase prevented virus replication.

6. Safety considerations

Vaccinia virus is an established human vaccine but is also a class II human pathogen, and should be handled under category 2 laboratory containment.

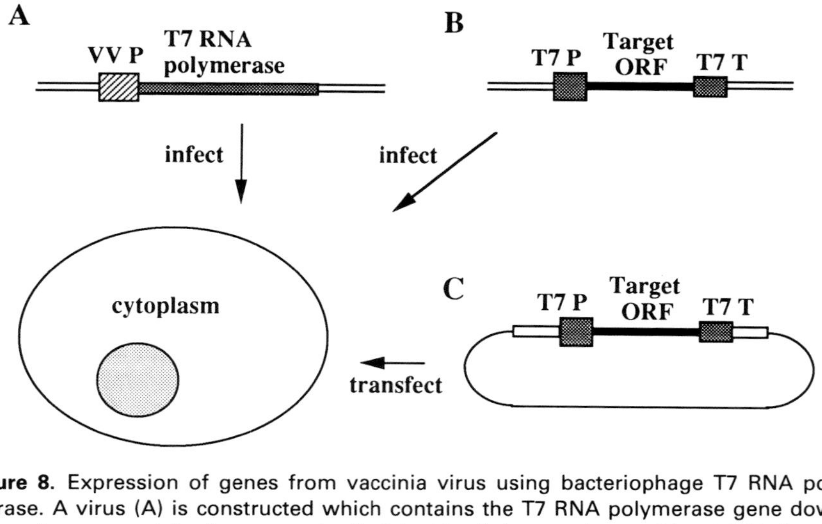

Figure 8. Expression of genes from vaccinia virus using bacteriophage T7 RNA polymerase. A virus (A) is constructed which contains the T7 RNA polymerase gene downstream from a vaccinia virus promoter (hatched box). A second recombinant (B) is formed which contains the target ORF (solid line) controlled by T7 promoter (T7 P) and terminator (T7 T) transcriptional sequences. Co-infection of cells with the two viruses allows high-level expression of the target ORF. For rapid screening of the target ORF, or variants of it, the plasmid (C) may be transfected into cells infected with virus (A).

Vaccination is no longer recommended in the UK unless the manipulation of the virus is considered likely to enhance its virulence, very large amounts of virus are being used, or the work involves animals. In the latter situation, a case-by-case evaluation of the need for vaccination should be undertaken. Vaccination is strictly contraindicated in potential vaccinees with eczema or any form of immunological deficiency.

Strains of vaccinia virus vary considerably in their virulence. For routine laboratory use, a virus with a deliberately engineered attenuating phenotype should be selected, such as one lacking *tk* (see also *Table 1*).

Acknowledgements

The author is a Lister-Institute Jenner Research Fellow. He thanks Antonio Alcamí for critical reading of the manuscript.

References

1. Mackett, M., Smith, G. L., and Moss, B. (1985). In *DNA cloning: a practical approach* (ed. D. M. Glover), p. 191. IRL Press, Oxford.
2. Smith, G. L. and Mackett, M. (1992). In *Recombinant poxviruses* (ed. M. M. Binns and G. L. Smith), pp. 81–122. CRC Press, Boca Raton, Florida.
3. Moss, B. (1990). In *Virology* (ed. B. N. Fields and D. M. Knipe), p. 2079. Raven Press, New York.
4. Buller, R. M. L. and Palumbo, G. J. (1991). *Microbiol. Rev.,* **55,** 80.
5. Moss, B. (1990). *Annu. Rev. Biochem.,* **59,** 661.
6. Traktman, P. (1990). *Curr. Top. Microbiol. Immunol.,* **163,** 93.
7. Payne, L. G. (1980). *J. Gen. Virol.,* **50,** 89.
8. Goebel, S. J., Johnson, G. P., Perkus, M. E., Davis, S. W., Winslow, J. P., and Paoletti, E. (1990). *Virology,* **179,** 247.
9. Perkus, M. E., Goebel, S. J., Davis, S. W., Johnson, G. P., Norton, E. K., and Paoletti, E. (1991). *Virology,* **180,** 406.
10. Keck, J. G., Baldick, C. J., and Moss, B. (1990). *Cell,* **61,** 801.
11. Kent, R. K. (1988). Isolation and analysis of the vaccinia virus p4b gene promoter. PhD thesis, University of Cambridge, UK.
12. Mackett, M., Smith, G. L., and Moss, B. (1984). *J. Virol.,* **49,** 857.
13. Chakrabarti, S., Brechling, K., and Moss, B. (1985). *Mol. Cell. Biol.,* **5,** 3403.
14. Davison, A. J. and Moss, B. (1990). *Nucl. Acids Res.,* **18,** 4285.
15. Falkner, F. G. and Moss, B. (1988). *J. Virol.,* **62,** 1849.
16. Perkus, M. E., Limbach, K., and Paoletti, E. (1989). *J. Virol.,* **63,** 3829.
17. Bertholet, C., Drillien, R., and Wittek, R. (1985). *Proc. Natl Acad. Sci. USA,* **82,** 2096.
18. Rosel, J. L. and Moss, B. (1985). *J. Virol.,* **56,** 830.
19. Browne, H., Churcher, M., Stanley, M., Smith, G. L., and Minson, A. C. (1988). *J. Gen. Virol.,* **69,** 1263.
20. Earl, P. L., Hugin, A. W., and Moss, B. (1990). *J. Virol.,* **64,** 2448.
21. Coupar, B. E. H., Andrew, M. E., Both, G. W., and Boyle, D. B. (1986). *Eur. J. Immunol.,* **16,** 1479.

22. Townsend, A., Bastin, J., Gould, K., Brownlee, G., Andrew, A., Boyle, D. B., Chan, Y., and Smith, G. (1988). *J. Exp. Med.,* **168,** 1211.
23. Mackett, M., Smith, G. L., and Moss, B. (1982). *Proc. Natl Acad Sci. USA,* **79,** 7415.
24. Buller, R. M. L., Smith, G. L., Cremer, K., Notkins, A. L., and Moss, B. (1985). *Nature,* **317,** 813.
25. Buller, R. M. L., Chakrabarti, S., Cooper, J. A., Twardzik, D. R., and Moss, B. (1988). *J. Virol.,* **62,** 866.
26. Kotwal, G. J., Hugin, A. W., and Moss, B. (1989). *Virology,* **171,** 579.
27. Kotwal, G. J., Isaacs, S. N., McKenzie, R., Frank, M. M., and Moss, B. (1990). *Science,* **250,** 827.
28. Moore, J. B. and Smith, G. L. (1992). *EMBO J.,* **11,** 1973.
29. Rodriguez, J. F. and Smith, G. L. (1990). *Nucl. Acids Res.,* **18,** 5347.
30. Blasco, R. and Moss, B. (1991). *J. Virol.,* **65,** 5910.
31. Duncan, S. A. and Smith, G. L. (1992). *J. Virol.,* **66,** 1610.
32. Panicali, D., Grzelecki, A., and Huang, C. (1986). *Gene,* **47,** 193.
33. Ramshaw, I. A., Andrew, M. E., Philips, S. M., Boyle, D. B., and Coupar, B. E. H. (1987). *Nature,* **329,** 545.
34. Franke, C. A., Rice, C. M., Strauss, J. H., and Hruby, D. E. (1985). *Mol. Cell. Biol.,* **5,** 1918.
35. Boyle, D. B. and Coupar, B. E. H. (1988). *Gene,* **65,** 123.
36. Isaacs, S. N., Kotwal, G. J., and Moss, B. (1990). *Virology,* **178,** 626.
37. Wiktor, T. J., Macfarlan, R. I., Reagan, K. J., Dietzschold, B., Curtis, P. J., Wunner, W. H., Kieny, M.-P., Lathe, R., Lecocq, J.-P., Mackett, M., Moss, B., and Koprowski, H. (1984). *Proc. Natl Acad. Sci. USA,* **81,** 7194.
38. Fathi, Z., Sridhar, P., Pacha, R. F., and Condit, R. C. (1986). *Virology,* **155,** 97.
39. Ball, L. A. (1987). *J. Virol.,* **61,** 1788.
40. Spyropoulos, D. D., Roberts, B. E., Panicali, D., and Cohen, L. K. (1988). *J. Virol.,* **62,** 1046.
41. Falkner, F. G. and Moss, B. (1990). *J. Virol.,* **64,** 3108.
42. Kerr, S. M. and Smith, G. L. (1991). *Virology,* **180,** 625.
43. Fuerst, T. R., Fernandez, M. P., and Moss, B. (1989). *Proc. Natl Acad. Sci. USA,* **86,** 2549.
44. Rodriguez, J. F. and Smith, G. L. (1990). *Virology,* **177,** 239.
45. Zhang, Y. and Moss, B. (1991). *Proc. Natl Acad. Sci. USA,* **88,** 1511.
46. Zhang, Y. and Moss, B. (1991). *J. Virol.,* **65,** 6101.
47. Blasco, R., Cole, N. B., and Moss, B. (1991). *J. Virol.,* **65,** 4598.
48. Davison, A. J. and Moss, B. (1989). *J. Mol. Biol.,* **210,** 771.
49. Davison, A. J. and Moss, B. (1989). *J. Mol. Biol.,* **210,** 749.
50. Patel, D. D., Ray, C. A., Drucker, R. P., and Pickup, D. J. (1988). *Proc. Natl Acad. Sci. USA,* **85,** 9431.
51. Pearson, A., Richardson, C., and Yuen, L. (1991). *Virology,* **180,** 561.
52. Fuerst, T. R., Niles, E. G., Studier, F. W., and Moss, B. (1986). *Proc. Natl Acad. Sci. USA,* **83,** 8122.
53. Fuerst, T. R., Earl, P. L., and Moss, B. (1987). *Mol. Cell. Biol.,* **7,** 2538.
54. Elroy-Stein, O., Fuerst, T. F., and Moss, B. (1989). *Proc. Natl Acad. Sci. USA,* **86,** 6126.
55. Fuerst, T. R. and Moss, B. (1989). *J. Mol. Biol.,* **206,** 333.

56. Moss, B., Elroy-Stein, O., Mizukami, T., Alexander, W. A., and Fuerst, T. (1990). *Nature,* **348,** 91.
57. Patel, A. H., Gaffney, D. F., Subak-Sharpe, J. H., and Stow, N. D. (1990). *J. Gen. Virol.,* **71,** 2013.
58. Perkus, M. E., Panicali, D., Mercer, S., and Paoletti, E. (1986). *Virology,* **152,** 285.
59. Law, K. M. and Smith, G. L. (1992). *J. Gen. Virol.,* **73,** 549.
60. Beattie, E., Tartaglia, J., and Paoletti, E. (1991). *Virology,* **183,** 419.
61. Child, S. J., Palumbo, G. J., Buller, R. M. L., and Hruby, D. E. (1990). *Virology,* **174,** 625.
62. Niles, E. G. and Seto, J. (1988). *J. Virol.,* **62,** 3772.
63. Duncan, S. A. and Smith, G. L. (1992). *J. Gen. Virol.,* **73,** 1235.
64. Hughes, S. J., Johnston, L. H., de Carlos, A., and Smith, G. L. (1991). *J. Biol. Chem.,* **266,** 20103.
65. Colinas, R. J., Goebel, S. J., Davis, S. W., Johnson, G., Norton, E. K., and Paoletti, E. (1990). *Virology,* **179,** 267.
66. Shida, H., Tochikura, T., Sato, T., Konno, T., Hirayoshi, K., Seki, M., Ito, Y., Hatanaka, M., Hinuma, Y., Sugimoto, M., Takahashi-Nishimaki, F., Maruyama, T., Miki, K., Suzuki, K., Morita, M., Sashiyama, H., and Hayami, M. (1987). *EMBO J.,* **6,** 3379.
67. Shida, H., Hinuma, Y., Hatanaka, M., Morita, M., Kidokoro, M., Suzuki, K., Maruyama, T., Takahashi-Nishimaki, F., Sugimoto, M., Kitamura, R., Miyazawa, T., and Hayami, M. (1988). *J. Virol.,* **62,** 4474.
68. Takahashi-Nishimaki, F., Funahashi, S.-I., Miki, K., Hashizume, S., and Sugimoto, M. (1991). *Virology,* **181,** 158.
69. Zhou, J., McLean, L., Sun, X.-Y., Stanley, M., Almond, N., Crawford, L., and Smith, G. L. (1990). *J. Gen. Virol.,* **71,** 2185.
70. Smith, G. L., Chan, Y. S., and Howard, S. T. (1991). *J. Gen. Virol.,* **72,** 1349.
71. Alexander, W. A., Moss, B., and Fuerst, T. R. (1992). *J. Virol.,* **66,** 2934.

10

Herpes simplex virus vectors

FRAZER J. RIXON and JOHN MCLAUCHLAN

1. Introduction

As with other virus vector systems, one reason for using herpesviruses is to allow expression of proteins outside their normal environment so that their properties can be examined, either alone or in combination with other defined proteins. However, there is additional interest attached to herpes simplex viruses as vectors which relates to their ability to establish long-term latent infections in neuronal cells. This neurotropism implies that herpes simplex virus-based vectors may be used as tools to deliver genes of interest into neurones.

2. Properties of HSV

2.1 Host range and safety

Herpes simplex virus (HSV) is limited, as a natural infection, to man, where serological studies suggest an incidence of greater than 90%, but under laboratory conditions it has a wide host range in experimental animals and in tissue culture cells. Two serotypes of HSV are known: HSV-1 and HSV-2, which are typically associated with facial and genital lesions, respectively. Initial infections in man are normally inapparent but may result in localized lesions which are also typical of reactivating latent virus. While not usually life-threatening, except in rare cases where a viral encephalitis occurs, HSV infections are frequently uncomfortable and unsightly. More seriously, where the eye is involved, repeated recurrences may lead to corneal scarring and permanent damage to the sight. Therefore, it is essential that appropriate precautions are taken when handling HSV. Guidelines for the handling of such pathogens are available from the relevant regulatory bodies. In the UK, HSV is listed as a hazard group 2 pathogen and both the virus and infected animals are handled under category 2 conditions (1). Furthermore, there are separate regulations concerning the construction and handling of recombinant viruses. Before embarking on such work, approval from the relevant regulatory bodies *must* be obtained.

2.2 Virus structure and genome content

The HSV virion is a large and complicated structure which has a distinctive morphology, comprising an icosahedral capsid (containing the DNA), a glycoprotein-containing lipid envelope and, separating these two, an amorphous proteinaceous layer designated the tegument. The viral genome is a double-stranded DNA molecule of 152 260 bp, the sequence of which has been determined (2). The genome contains a total of 73 genes (three are present in two copies), of which at least 38 are known to be non-essential for virus growth in tissue culture (3), although most, if not all, are likely to be important in the life cycle of the virus in humans. The sizeable number of non-essential genes provides many potential target sites for insertion of foreign genes to produce recombinants that are viable in tissue culture. In practice, however, only a few of the possible candidates have been utilized (see Section 4.1).

2.3 Virus life cycle

HSV has two phases in its natural life cycle: the lytic or productive phase and the latent phase (4, 5). Lytic infections, which occur in tissue culture and in animals, have been characterized extensively. Virus transcription is regulated in three broad temporal classes: immediate early (IE), early (E), and late (L). IE transcription takes place in the absence of *de novo* protein synthesis, and several IE proteins have been shown to function as activators of early and late transcription. Early mRNAs are generally considered as those which do not require virus DNA replication for their expression, whereas replication is required for late mRNA synthesis. However, these are not clear-cut definitions, and sub-classes of genes (for example delayed early) have been described. As a general (though not invariable) rule, early functions are involved in virus replication while late functions are predominantly structural. Capsid assembly and DNA packaging take place in the cell nucleus, and the virion acquires tegument and envelope as it passes out of the nucleus and through the cytoplasm. Virus release is by exocytosis and is not accompanied by cell lysis.

Latent infections are characteristic features of herpesviruses and various cell types can harbour latent virus genomes. The site of HSV latency is primarily in the neurones of sensory ganglia, although there is evidence for latent virus in the central nervous system (CNS) and in certain non-neuronal tissues. Following infection, the virus travels up the nerve to the ganglia where a limited lytic infection may occur before latency is established. The ganglia which harbour latent virus are those which innervate the site of inoculation. During latency, lytic gene transcription is shut down and only one viral promoter remains active, generating a number of RNAs (6) which have been designated the latency-associated transcripts (LATs). The function of LATs is uncertain, but the existence of the LAT promoter is important for the potential use of HSV-1 vectors in neurones. HSV can remain latent within sensory ganglia for the lifetime of the host animal. Spontaneous reactivations are not a feature of mouse latency models, although they do occur in other animal systems (rabbit, guinea-pig). The initial and final stages of latent infections resemble those of lytic infections, and the latent phase can be thought of as a prolonged and sometimes indefinite suspension of the infectious process.

3. Growth of HSV-1

3.1 Cell culture

In our laboratory, virus is grown in baby hamster kidney (BHK) cells which can be maintained as monolayers on glass or plastic (*Protocols 1* and *2*). For large-scale preparations, BHK cells are grown in 80 fl. oz (~2.25 l) plastic roller bottles which are rotated at 0.3–0.8 r.p.m. (as illustrated in ref. 7). This method provides sufficient contact between cells and medium to permit normal growth. A confluent monolayer of BHK cells grown under these conditions contains 1–2 $\times$ 10^8 cells. Large areas of space need not be dedicated to growing cells, since bench-top incubators and roller bottle machines are available commercially.

Protocol 1. Passage of BHK cells

Materials

- sterile plastic 80 fl. oz (~2.25 l) roller bottles (Becton-Dickinson)
- BHK-21 C13 cells as confluent monolayers in roller bottles: BHK cells are available from the American Type Culture Collection as CCL-10
- newborn calf serum (NCS)
- tryptose phosphate broth (TPB; Life Technologies)
- ETC10: Glasgow minimal essential medium (GMEM; Life Technologies) supplemented with 10% (v/v) NCS, 5% (v/v) TPB, 100 U/ml penicillin, 100 μg/ml streptomycin
- trypsin–EDTA (Life Technologies)
- haemocytometer (improved Neubauer ruling)
- supply of CO_2

Method

1. Decant the medium from each roller bottle and wash the cell sheet briefly with 20 ml of trypsin–EDTA by manually rotating the bottle. Discard the trypsin–EDTA.
2. Wash the cell sheet again with another 20 ml of trypsin–EDTA and replace the roller bottle in the roller bottle machine at 37°C for approximately 5 min. The cells should detach readily.

Protocol 1. *Continued*

3. Add 20 ml of ETC10 and resuspend the cells by pipetting with a narrow-bore 5 ml pipette. Ensure that the cells are evenly dispersed and transfer the suspension to a sterile 20 ml container.
4. Dilute 0.1 ml of the cell suspension 10-fold with ETC10 and determine the cell density using a haemocytometer. Dilute the cell suspension to 4×10^6 cells/ml with fresh ETC10.
5. At this stage, cells can be stored in a sterile container at 4°C for up to a week without significant loss of viability.
6. To passage cells, add 2.5 ml of cell suspension (10^7 cells) to a fresh roller bottle containing 100 ml of ETC10.
7. Gas the roller bottle by adding 100 ml of CO_2 to give a final concentration of 5% (v/v) CO_2. Place the sealed bottles in the roller bottle apparatus and rotate at 37°C.

One confluent roller bottle provides sufficient cells to seed ten fresh roller bottles. Under these conditions, cells will be confluent after growth at 37°C for 3–4 days. BHK cells are not immortal and can typically be passaged up to 15 times before becoming senescent. Consequently, frozen stocks, which can be recovered at regular intervals, must be stored as shown in *Protocol 2*. Stored cells should be of low passage number and high viability.

Protocol 2. Storage and recovery of cells

Materials

- storage medium: 40% (v/v) GMEM, 40% (v/v) NCS, 20% (v/v) glycerol
- 1.8 ml sterile cryotubes (Life Technologies)
- 175 cm^2 tissue culture flasks (Life Technologies)

Method

1. Remove cells from a roller bottle as described in steps 1 to 4 of *Protocol 1*.
2. Transfer 25 ml of cell suspension (10^8 cells) to a sterile container and pellet the cells by centrifuging at 700 *g* for 5 min.
3. Discard the supernatant and resuspend the cells in 25 ml of storage medium to give a final concentration of 4×10^6 cells/ml. Disperse the cell pellet by pipetting with a 5 ml narrow-bore pipette.
4. Dispense 1 ml aliquots of the cell suspension into cryotubes and place the tubes upright in an expanded polystyrene container. Allow the tubes to cool slowly by placing the container overnight in a −70°C freezer, then transfer them to the gaseous phase over liquid nitrogen (−190°C) for long-term storage.

5. To recover cells, thaw 1 ml of suspension rapidly at 37°C and transfer to a tissue culture flask containing 50 ml of ETC10. Add CO_2 to 5% and incubate the flask overnight at 37°C. Replace the medium with fresh ETC10 and continue incubating at 37°C until the cells are confluent.
6. Remove the cells from the flask and use them to seed a single roller bottle as described in *Protocol 1*.

3.2 Growth, titration, and plaque-purification of virus

Growing HSV-1 is relatively straightforward and has been described in a previous volume in this series (7). A number of laboratory-adapted virus strains which grow to high titre are in current use. We use the Glasgow isolate of HSV-1, strain 17, for which the complete DNA sequence is known. However, differences between this and other laboratory strains are likely to be minor, and the descriptions given here should prove generally applicable. We describe our methods for growing virus in *Protocols 3–5*.

Protocol 3. Growth of virus

Materials

- 200 ml centrifuge bottles (Falcon or Becton-Dickinson)
- blood agar plates: poured using 8% (v/v) defibrinated horse blood (Becton-Dickinson) in 4% (w/v) blood agar base

Method

1. Grow ten 80 fl. oz (~2.25 l) roller bottles of BHK cells (*Protocol 1*) to approximately 70–80% confluence as assessed by microscopy.
2. Remove the medium and add a virus inoculum at a multiplicity of infection (m.o.i.) of 0.002 plaque forming units (p.f.u.)/cell (i.e. 2×10^5 p.f.u./bottle) in 30 ml of ETC10.[a]
3. Incubate at 37°C for 3 days, by which time all the cells should have rounded and will have detached from the bottle or can be removed by gently swirling the medium. When growing a temperature-sensitive (ts) mutant of the virus, incubate the cells at the recommended permissive temperature (PT; for example 31°C) for 4–5 days.
4. After shaking off the cells, decant the mixture of cells and medium into sterile 200 ml centrifuge bottles. Separate cells from the medium by centrifugation at 1400 *g* for 10 min at 4°C.
5. Decant the medium into sterile 200 ml containers. The medium contains virus which has been released from the cells and is a good source of relatively pure virions. Further handling of this material is described

Protocol 3. *Continued*

from step 9 onwards. The cell pellet also contains large amounts of infectious virus, which is prepared as follows.

6. Resuspend the cell pellets in 5 ml of ETC10, transfer to a sterile 20 ml container and sonicate in a sonicator bath to disrupt the cells. The sample should be kept cool by precooling the water in the sonicator bath and storing the sample on ice. Pellet the debris at 1400 *g* for 10 min at 4°C and retain the supernatant.
7. Repeat step 6 and pool the supernatants, this forms the cell-associated virus (CAV) stock.[b]
8. Dispense the CAV into cryotubes and store at −70°C.[c]
9. Pellet virus in the culture medium from step 4 by centrifuging in sterile centrifuge bottles at 23 000 *g* for 2 h at 4°C.
10. Resuspend the virus pellet in 2 ml of ETC10. Detach the pellet from the wall of the bottle by pipetting and transfer the suspension to a sterile 5 ml container. This material forms the cell-released virus (CRV) stock. It is usually cleaner and has a higher titre than the CAV. Sonicate the pellet following the precautions mentioned in step 6. Follow *Protocols 6* and *7* if it is to be used to prepare virus DNA. Otherwise, dispense 0.5 ml aliquots into sterile cryotubes and store at −70°C.

[a] HSV can generate populations of defective particles containing incomplete genomes if consistently passaged at high m.o.i.

[b] The CAV is useful as a source of virus for subsequent infection of cells, but it is heavily contaminated with cellular debris. We do not recommend it as a source of viral DNA for cloning purposes.

[c] Titrate the virus stocks before freezing (see *Protocol 4*). Also, check them for sterility (8) by sterilizing a platinum wire loop in a Bunsen flame and streaking the virus preparation on to blood agar plates to test for yeast and bacterial contamination. Incubate the plates at 37°C for 7 days.

Protocol 4. Virus titration[a]

Materials

- BHK cells applied at 2×10^6 cells/50 mm plate in 4 ml of ETC10 and grown overnight at 37°C
- phosphate-buffered saline containing divalent cations (PBS^+): 140 mM NaCl, 2.7 mM KCl, 8 mM Na_2HPO_4, 1.4 mM KH_2PO_4, 0.7 mM $CaCl_2$, 0.5 mM $MgCl_2$ (pH 7.2)
- PBS^+/NCS: PBS^+ supplemented with 5% (v/v) NCS
- ETC10 containing 1.5% (w/v) carboxymethyl cellulose (CMC) or 0.5% (w/v) agar

- Giemsa stain (Gurr)
- 2.5% (v/v) glutaraldehyde (Cidex, Surgikos Ltd)
- screened pooled human serum (Life Technologies)

Method

1. Serially dilute virus stocks by first diluting 10 μl of the virus stock in 1 ml of PBS$^+$/NCS (10^{-2} dilution) and thereafter by consecutive 10-fold dilution steps (200 μl added to 1800 μl of PBS$^+$/NCS). Dilutions in the range 10^{-5}–10^{-9} are appropriate for most virus stocks. Use a fresh pipette or tip for each dilution to prevent carry-over of virus which will result in anomalously high titres.
2. Remove the medium from the plates and apply 100 μl of diluted virus to the centre of each of two duplicate plates. Ensure that the plates are on a level surface to allow the inoculum to spread uniformly over the surface. Incubate for 1 h at 37°C.
3. Remove the inoculum and overlay the plates with medium.[b] Incubate the cells at 37°C for 3 days. If a ts mutant is being titrated, incubate duplicate pairs of plates at the PT (typically 31–34°C) and the non-permissive temperature (NPT; usually in the range 38–41°C).
4. Fix and stain the cells. For plates overlayed with medium containing CMC or human serum, add Giemsa stain directly to the plates. Leave for at least 1 h at room temperature then wash under a uniform flow of tap water. If the CMC proves difficult to remove completely, leave water on the plates for 15 min after the initial wash before rinsing off the residue. For agar overlays, first fix the cells for 1 h with 2.5% glutaraldehyde. Remove the agar by washing under a tap and stain the fixed cells as before.
5. Drain the plates and count the plaques using a stereomicroscope.

[a] Where accurate titres are critical, virus should be re-titrated at the time of the experiment.
[b] A variety of overlay media can be used (7), the main requirement being that they must prevent virus spread other than through direct cell-to-cell contact. For standard titrations and picking plaques, we prefer to overlay with CMC or agar medium. In the latter case, particular care should be taken to remove all the inoculum, since a thin layer of liquid between the cell sheet and agar renders the monolayer very susceptible to damage during manipulation. If required, NCS can be replaced in the medium by 2% (v/v) pooled human serum, which contains neutralizing antibodies against HSV.

Protocol 5. Selection and growth of individual plaques

Materials

- plugged plastic pipette tips (Continental Laboratory Products)

Protocol 5. *Continued*

Method

1. Infect and overlay cells with the appropriate medium, typically CMC or agar medium, as described in steps 1–3 of *Protocol 4*. Allow the plaques to grow to a reasonable size to facilitate selection.
2. Select plaques from plates containing a small number of well-separated plaques by viewing under a stereomicroscope. Use a fresh sterile Pasteur pipette or plastic tip for each plaque. Withdraw cells from the area of the plaque and transfer them to a 1 ml cryotube containing 200 μl of PBS^+/NCS. If plastic tips are employed, use plugged tips in order to prevent virus being drawn into the barrel of the pipette.
3. To release virus from the cells, freeze the cryotubes in dry ice–ethanol and thaw at 37 °C. Repeat this step twice more and store the plaques at −70 °C.
4. Replate the plaques at 10-fold dilutions in the range 10^{-1}–10^{-3}, as described in *Protocol 4*, and repeat steps 1–3. Repeat the process until the stock consists of homogeneous virus.
5. Inoculate confluent 50 mm plates of BHK cells with a 50 μl aliquot of the selected final plaque isolate. Overlay the cells 1 h later with ETC10 and incubate at 37 °C. Harvest the medium and cells when the cells show complete cytopathic effect (c.p.e.), transfer to a 20 ml container, and disrupt the cells in a sonicator bath. Dispense 0.5 ml aliquots into sterile cryotubes and store at −70 °C. These serve as the seed stock for growing large-scale preparations of virus (*Protocol 3*).

3.3 Preparation of virus DNA

Virion DNA can be prepared by caesium chloride density gradient centrifugation of DNA isolated directly from CRV (9), or by extraction of gradient-banded virions without further purification (see *Figure 1*; ref. 10). We prefer the latter option (*Protocol 6*).

Protocol 6. Purification of virions

Materials

- GMEM−PR: GMEM lacking phenol red (Life Technologies)
- 5% and 15% (w/v) Ficoll 400 in GMEM−PR

Method

1. Cool the Ficoll solutions to 4 °C and pour two 35 ml linear gradients in cellulose nitrate ultracentrifuge tubes. The gradients can be stored overnight at 4 °C. Virions from up to ten roller bottles can be purified on each gradient.

2. Layer 1 ml of CRV (see step 10 of *Protocol 3*) on to each gradient.
3. Centrifuge at 19 000 *g* (12 000 r.p.m. in a Sorvall AH629 rotor) for 2 h at 4°C. Using overhead illumination two bands are visible in each tube (*Figure 1*): a clearly defined intense band (V, virions) about halfway down the gradient and a diffuse band (L) above this.
4. Collect the virion band by side puncture through an 18 gauge syringe needle and transfer it to a fresh ultracentrifuge tube.
5. Combine the virion bands from the two gradients and dilute with fresh GMEM–PR to a final volume of 35 ml.
6. Pellet the virions by centrifuging at 65 000 *g* (22 000 r.p.m. in a Sorvall AH629 rotor) for 2 h at 4°C. If the virus is to be used as a stock, resuspend the pellet carefully in sterile GMEM–PR. Aliquot and store the virus suspension at −70°C and titrate it as described in *Protocol 4*.

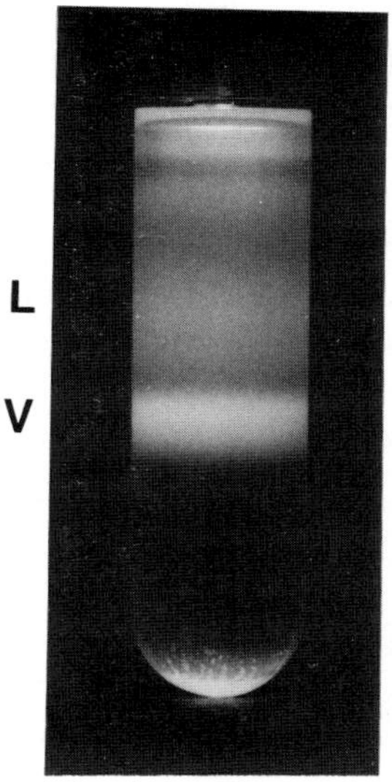

Figure 1. Purification of HSV-1 virions. A suspension of CRV was applied to a 5–15% (w/v) Ficoll gradient and banded as described in *Protocol 6*. The positions of the virion band (V) and the diffuse L-particle band (L) are indicated.

Protocol 7. Preparation of virion DNA

Materials

- virion pellet from *Protocol 6*
- TE: 10 mM Tris–HCl (pH 7.5), 1 mM EDTA

Protocol 7. *Continued*

- 20% (w/v) sodium dodecyl sulphate (SDS)
- 20 mg/ml proteinase K (Boehringer–Mannheim)

Method

1. Resuspend the virion pellet in 300 μl of TE. Transfer the suspension to a 1.5 ml polypropylene tube and add SDS and proteinase K to final concentrations of 0.2%[a] and 250 μg/ml, respectively. Incubate at 37°C for 1 h.
2. Add an equal volume of phenol:chloroform saturated with TE and rotate the sample slowly for 5 min. Centrifuge at 12 000 *g* for 5 min and transfer the aqueous layer to a fresh tube. Repeat the extraction twice more and then extract the aqueous phase once with chloroform.
3. Ethanol-precipitate the DNA, which should be seen to spool out. Allow the DNA to settle to the bottom of the tube and pellet it at 12 000 *g* for 5 min. Discard the supernatant and rinse the pellet with 200 μl of 70% (v/v) ethanol. Allow the pellet to air dry.
4. Dissolve the DNA overnight at 37°C in 100 μl of sterile distilled water. The yield of HSV-1 DNA should be about 15 μg for each roller bottle preparation of virions.
5. Store the DNA as 20 μl aliquots at −20°C. If the DNA is to be treated with restriction enzymes for use in the direct ligation cloning strategy (see *Protocol 11*), this should be done before aliquoting the DNA.

[a] Once SDS has been added, the DNA is no longer protected within virions and becomes very susceptible to shearing. Cut off the ends of plastic tips with a clean scalpel before pipetting the DNA solution gently. Do not vortex the DNA solution or shake it vigorously.

3.4 Reconstitution of virus from DNA

The calcium phosphate precipitation technique is used routinely to transfect cells. Recently, transfection using synthetic lipid mixtures (lipofection) has become used widely. This is a simpler procedure which is less harmful to cells and transfects DNA with greater efficiency in certain cell types. Both approaches are described in *Protocols 8* and *9*.

Protocol 8. Transfection with calcium phosphate

Materials

- cells plated 12–24 h in advance (2×10^6 cells/50 mm dish)
- Hepes-buffered saline (HBS): see Chapter 3, *Protocol 3*
- HSV-1 DNA (from *Protocol 7*)

- carrier DNA: 2 mg/ml calf thymus or salmon sperm DNA
- ETC5: as ETC10 but containing 5% (v/v) NCS
- DMSO solution: 20% (v/v) dimethyl sulphoxide (DMSO) in HBS

Method

1. Mix 1 ml of HBS, approximately 5 μl of carrier DNA, and 0.5 μg of HSV-1 DNA. Add 70 μl of 2 M $CaCl_2$ dropwise. Mix carefully but do not vortex.
2. Replace the medium on duplicate plates with 500 μl of the mixture. Incubate at 37°C for 40 min.
3. Without removing the inoculum, add 4 ml of ETC5 to each plate and incubate for 200 min at 37°C.
4. Remove the medium and wash the cells with GMEM. Add 1 ml of DMSO solution and incubate at room temperature for 4 min. Overlay the plates at 30 sec intervals to facilitate precise timing. The DMSO solution is toxic to cells and should be added carefully by applying to the side of a tilted plate then laying the plate flat.
5. Remove the DMSO solution and wash the cells carefully with GMEM, adding the GMEM as described in step 4 for the DMSO solution. Overlay the cells with ETC5 and incubate at 37°C. Plaques should appear within 2–3 days.

Protocol 9. Lipofection

Materials

- cells plated 12–24 h in advance (2×10^6 cells/50 mm dish)
- Hepes buffer: 20 mM Hepes–NaOH (pH 7.4), 150 mM NaCl; filter-sterilized
- lipofection reagent[a]
- OptiMEM 1 medium (Life Technologies)
- HSV-1 DNA (*Protocol 7*)

Method

1. Add 30 μl of lipofection reagent to 70 μl of Hepes buffer in a 15 ml polystyrene tube and mix vigorously.
2. Dilute 0.5 μg of HSV-1 DNA to 100 μl with Hepes buffer in a 15 ml polystyrene tube. Mix gently.
3. Gently mix the solutions prepared in steps 1 and 2 and incubate at room temperature for 10 min.

Protocol 9. *Continued*

4. Add 2 ml of OptiMEM 1 medium, mix gently, add the mixture to the drained plates of cells, and incubate at 37°C for 5 h.
5. Add 2 ml of ETC10 to each plate and incubate at 37°C. Plaques should appear within 2–3 days.

[a] Lipofection reagent is available commercially from a number of suppliers: DOTAP (Boehringer–Mannheim), Lipofectin (Life Technologies). Alternatively, it can be prepared easily by following published procedures (11, 12).

4. Insertion of DNA into herpesvirus genomes

The construction of herpesvirus recombinants is a relatively straightforward task. Herpesvirus genomes are able to accommodate additional sequences of at least 10 kbp without affecting genome stability or significantly reducing virus viability. Three basic methods have been described for introducing sequences into herpesvirus genomes: homologous recombination, *in vitro* recombination, and direct ligation. Each approach has its own inherent advantages and disadvantages, and the method chosen for any potential application will reflect these.

The standard approach for generating recombinants is homologous recombination, which can be used to insert DNA fragments at any desired location in the genome. The chosen insertion site is usually a region which is non-essential for virus growth in tissue culture. An example is given in Section 4.1. This method does not require specialized reagents other than a suitably constructed plasmid. It is, therefore, the method of choice for most applications. Homologous recombination is simple and flexible, but recombinants are generated at low frequency (less than 1%), and this feature has led to the development of alternative methods for incorporating DNA sequences with higher efficiency.

A method developed for generating recombinants from pseudorabies virus (PRV), a herpesvirus related to HSV, uses the Cre-*lox* system from coliphage P1 (13). A plasmid which carries the *lox* recombination site within a non-essential glycoprotein gene is incubated *in vitro* with viral DNA in the presence of Cre protein. This mediates insertion of plasmid sequences into the viral genome. Upon transfection into cells, approximately 5% of the virus progeny are recombinants. To facilitate identification of recombinants, an immunological screen is used to distinguish viruses lacking the target glycoprotein. This method leads to insertion of the entire plasmid, including the vector sequences. One advantage of this is that the Cre protein can also excise the plasmid from recombinant virus DNA, allowing the inserted sequences to be recloned and analysed. In the system as described, the *lox* site is located within a glycoprotein gene, but other loci could be used. This system could also be applied to other herpesviruses, including HSV.

We have developed a method in which DNA fragments are inserted into a unique restriction enzyme site in the HSV-1 genome by direct ligation (14). The efficiency of generating virus progeny containing DNA inserts is high (in excess of 20%), and recombinants can be detected by screening for β-galactosidase expression from an inserted *lacZ* gene. Details for this system are given in Section 4.2.

van Zijl *et al.* (15) described a method for reconstructing the PRV genome from cosmids which contain overlapping fragments representing the entire genome. Inserts from the cosmids (typically four or five in number) are transfected into cells, and infectious virus is produced by recombination between regions of overlap. This method was developed principally to simplify manipulation of the virus genome, but it could in principle be modified to permit the introduction of foreign genes. Since the virus genomes generated by this approach are derived solely from the cloned DNA fragments, all progeny should be recombinant. One potential problem is the need to manipulate plasmids containing viral DNA fragments of about 40 kbp.

A somewhat different approach from those described above involves the construction of defective vectors (16–18). Foreign DNA is not inserted directly into the virus genome but is present in a plasmid which also contains an origin of virus DNA replication and a packaging signal. These sequences allow the plasmid to be replicated and packaged into virions when other necessary functions (replication enzymes and structural proteins) are supplied by superinfecting helper virus. Thus, the progeny from cells transfected with plasmid and superinfected with helper virus comprises a mixture of defective and helper virus. If passaged at high m.o.i., the defective viruses are maintained and accumulate in the population. Defective vectors have the theoretical advantages that there are likely to be fewer constraints on the nature or size of DNA sequences that can be maintained in this way. They may also have particular advantages when used as neuronal expression vectors (see Section 4.3). However, there are major disadvantages with this approach, the chief being that defective viruses require the continuous presence of infectious helper virus. Thus, they retain the potential problems implicit in the use of infectious virus (for example see Section 5), and in addition have the associated drawback that the ratio of defective to helper virus is difficult to monitor or control.

4.1 Insertion of DNA by recombination

The HSV-1 thymidine kinase (*tk*) locus is frequently chosen as the site for inserting foreign DNA sequences; *tk* is dispensable for virus growth in tissue culture (in dividing cells) and a selection system for tk^- virus (i.e. growth in the presence of 5-bromo-deoxycytidine (BCdR)) aids in the isolation of recombinants. Under certain circumstances, however, *tk* may not be an appropriate

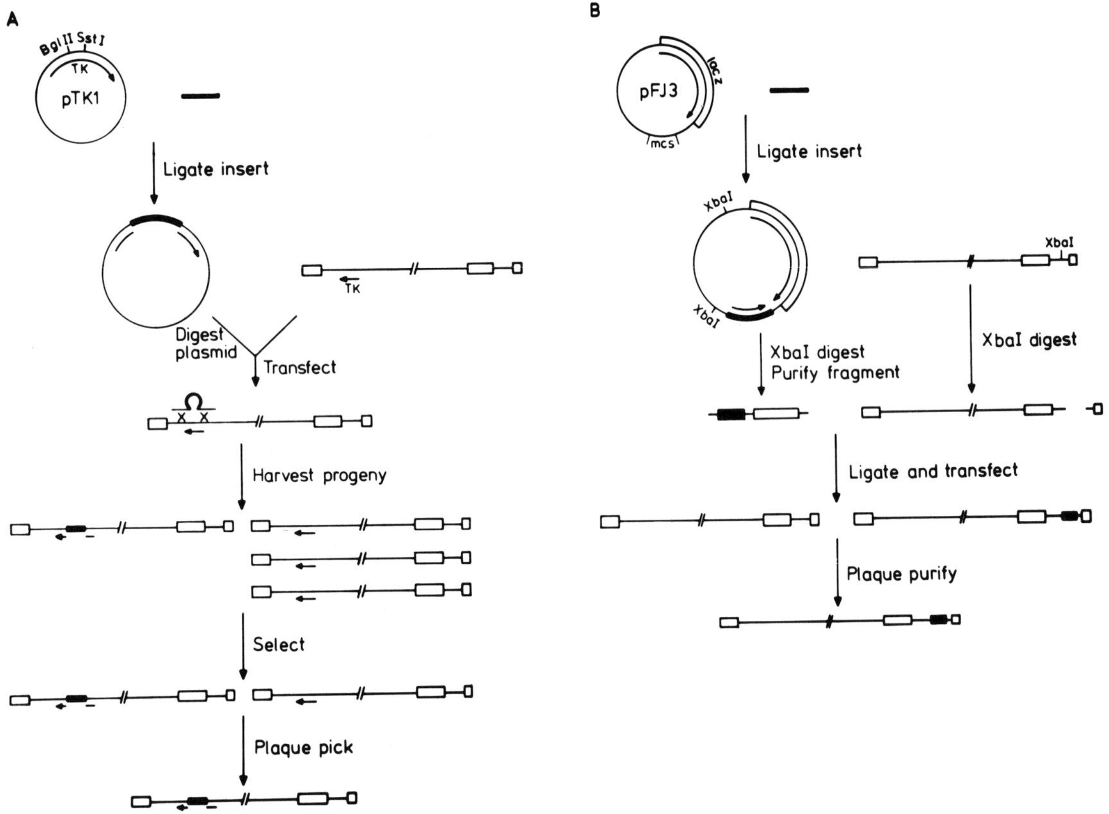
A
Bgl II
Sst I
TK
pTK1
Ligate insert
TK
Digest plasmid
Transfect
Harvest progeny
Select
Plaque pick
B
lacz
pFJ3
mcs
Ligate insert
XbaI
XbaI
XbaI
XbaI digest
Purify fragment
XbaI digest
Ligate and transfect
Plaque purify

Figure 2. Insertion of DNA into the HSV-1 genome by homologous recombination (A) and direct ligation (B). (A) Foreign DNA is inserted into unique sites (for example *Bgl*II, *Bst*EII, *Sst*I, and *Tth*111I) within the thymidine kinase (*tk*) sequences contained on plasmid pTK1. Following co-transfection with HSV-1 DNA, homologous recombination results in incorporation of the foreign DNA into the endogenous *tk* gene rendering the recombinant tk^-. Selection for growth in the presence of BCdR eliminates tk^+ virus and the recombinant viruses are purified to homogeneity by picking individual plaques and screening for the presence of foreign DNA sequences (*Protocol 10*). (B) Foreign DNA is cloned into the multiple cloning site (mcs) of pFJ3. From the resultant plasmid, the *Xba*I fragment containing the foreign DNA and *lacZ* is purified and ligated into the *Xba*I-digested 1802 DNA. Following transfection of the ligated DNA into cells, progeny virus is harvested and recombinants containing foreign DNA are identified by screening for blue plaques in the presence of X-gal (*Protocol 11*).

choice. For example, studies have shown that expression of this gene influences the ability of virus to establish and reactivate from latent infections. Therefore, alternative sites have been chosen when the recombinant is used in latency studies (19). In such cases, the procedure follows the scheme described in *Protocol 10* except that selection with BCdR is not applied.

Figure 2 shows a transfer plasmid, pTK1 (20), suitable for insertion into *tk*. This plasmid contains the 3.7 kbp HSV-1 *Bam*HI p fragment cloned into the *Bam*HI site of vector pAT153. *Bam*HI p contains the entire HSV-1 *tk* gene (alternatively designated as *UL23*) and portions of the adjacent genes, *UL22* and *UL24*. It also contains unique restriction sites within *tk* that are suitable for the insertion of DNA fragments. Sites should be chosen such that the sequences to be inserted are flanked by at least 500 bp of viral DNA, since DNA sequences flanked by shorter stretches give lower recombination efficiencies.

Protocol 10. Production of tk^- recombinants

Materials

- cells plated 12–24 h in advance (2×10^6 cells/50 mm dish)
- plasmid derived from pTK1
- HSV-1 DNA (*Protocol 7*)
- 10 mg/ml BCdR: filter-sterilized

Method

1. Transfect a mixture of 0.5 μg of digested plasmid DNA[a] and 0.5 μg of HSV-1 DNA into BHK cells using either of the methods described in *Protocols 8* and *9*.
2. Incubate the cells until complete c.p.e. is apparent and then harvest the cells and medium. The yield from each plate harvest represents clonally

Protocol 10. *Continued*

unrelated virus and should be stored separately as 1 ml aliquots in sterile cryotubes at −70°C.

3. Infect drained cell monolayers with virus at a m.o.i. of approximately 0.01 p.f.u./cell in 100 μl PBS$^+$/NCS. The yield of virus from a 50 mm plate is typically in the range of 10^7–10^8 p.f.u. The cells are infected at low m.o.i. to reduce the possibility of complementing *tk*$^+$ virus masking the effects of BCdR selection. Incubate the cells for 1 h at 37°C.
4. Overlay the cells with ETC5 supplemented with 100 μg/ml BCdR and incubate until complete c.p.e. is apparent. Harvest the cells and medium and store at −70°C.
5. Repeat steps 3–4.

After two rounds of selection, the resultant stock should be sufficiently enriched for *tk*$^-$ virus to allow individual plaques to be picked (*Protocol 5*). Since *tk*$^-$ viruses also occur spontaneously, check individual plaque isolates after selection in BCdR by Southern blot analysis (21). Virus can be grown in 24-well dishes which give sufficient DNA for Southern blot analysis. Choose probes which will hybridize to inserted sequences and to pTK1. Following *Bam*HI digestion, the presence of *Bam*HI p indicates that the isolate is not free of wild type virus.

[a] Recombination is more efficient with linearized plasmid. Therefore, the plasmid should be digested with a restriction enzyme which cuts ouside the inserted DNA more than 500 bp from the site of insertion.

4.2 Insertion of DNA by ligation

This vector system is based on an HSV-1 genome (1802) which has been modified to contain a unique restriction enzyme site for *Xba*I (ref. 14; *Protocol 11*). The unique site can in principle be placed at any suitable location in the genome, but in 1802 it is located in an intergenic region. *Figure 2* shows a plasmid, pFJ3, which has been designed for use with 1802. pFJ3 can be digested with *Xba*I to release a fragment containing *lacZ* flanked by SV40 transcriptional signals. In addition, it contains a multiple cloning site into which a second gene can be inserted. Expression of β-galactosidase allows visualization of blue plaques in the presence of an indicator compound, 5-bromo-4-chloro-3-indolyl β-D-galactopyranoside (X-gal).

Protocol 11. Insertion of DNA fragments by ligation

Materials

- HSV-1 1802 DNA (prepared as in *Protocol 7*)
- pFJ3 containing a foreign gene

- 5 × ligase buffer: 250 mM Tris–HCl (pH 7.5), 40 mM $MgCl_2$, 50 mM dithiothreitol
- 150 mg/ml X-gal in dimethylformamide

Method

1. Digest 10 μg of virus DNA with an excess of *Xba*I. Phenol:chloroform extract and ethanol-precipitate the DNA. Resuspend the DNA at 200 μg/ml in TE, aliquot in 5 μl amounts, and store at −20°C.
2. Digest the plasmid with *Xba*I and purify the appropriate fragment by agarose gel electrophoresis (21).
3. Mix 1 μg each of the digested viral DNA and the purified *Xba*I fragment in a final volume of 20 μl containing 1 × ligase buffer, 1 mM ATP and 1 U of T4 DNA ligase. Incubate at 15°C overnight.
4. Transfect 10 μl of the ligation products into BHK cells using either of the methods described in *Protocols 8* and *9*. Store the remaining 10 μl of ligation mixture at −20°C in case the transfection needs to be repeated.
5. When complete c.p.e. has developed, harvest the cells and medium. Store 1 ml of harvested material in a sterile cryotube at −70°C.
6. Purify recombinant virus by plaque selection as described in *Protocol 5*.[a]
7. Repeat the plaque selection until all plaques from an inoculum are blue. Test virus isolates for the presence of inserted sequences by Southern blot analysis (21).

[a] For virus expressing β-galactosidase, overlay the cells with CMC or agar medium (*Protocol 4*) containing 300 μg/ml X-gal.

4.3 Selection of promoters for expression of inserted genes

It is not possible to specify one particular promoter which will be ideal for directing expression under all circumstances, as the choice of promoter depends largely on the purpose for which the vector is required. Since HSV gene expression is temporally regulated during the lytic cycle (Section 2.3), different promoters can be used to achieve expression of inserted genes at different stages of infection. The behaviour of a number of promoters from each temporal class has been studied in detail (22), and among those most frequently used to direct heterologous gene expression are those for the immediate-early genes *RS1* and *US1*/*US12* and for the early genes *US6*, *UL40*, and *UL23* (*tk*). Under lytic conditions, high levels of expression can be achieved using any of these. Added flexibility is provided by the availability of ts mutants which affect essential regulatory functions and express only a limited spectrum of genes at the NPT. The most useful of these are certain mutants in the *RS1* gene which express only IE functions at the NPT. One *RS1* mutant, tsK, has been used in the vector systems described in Sections 4.1 and 4.2 to achieve expression of proteins under conditions where only HSV-1 IE proteins are produced (23, 24).

A variety of non-HSV promoters have been analysed following insertion into the HSV genome, but they appear to offer no benefits over endogenous HSV promoters when expression is required during normal lytic infection. However, we have shown that the mouse metallothionein promoter continues to respond to stimulation by heavy metals when present in the HSV genome. Therefore, it is possible that non-viral promoters with particular properties may have advantages in certain circumstances.

One property of HSV-based vectors, which is generating increasing interest, is their ability to direct gene expression in neuronal cells. During lytic infection of neuronal cells *in vitro* or in animals, gene expression appears to be normal and the lytic promoters mentioned above can be used to drive foreign gene expression. Once latency is established, the lytic promoters are inactive. To date, only two promoters have been shown to retain activity when present in a latent HSV genome. These are the endogenous viral promoter responsible for expression of the LATs (ref. 19; Section 2.3) and the promoter present in the Moloney murine leukaemia virus long-terminal repeat (LTR; ref. 25). Both have been used to direct expression of foreign genes during latency in sensory ganglia, although the LTR promoter was not demonstrably active in latently infected motor neurones. In mouse peripheral ganglia, LAT promoter-directed expression can be detected for at least one year after the initial infection. At present, the signals which direct LAT transcription are under scrutiny and their characterization is expected to extend the potential for using HSV vectors in neurobiological studies.

The human cytomegalovirus IE promoter has been shown to be active in CNS neurones following introduction into the brain of a rat using a defective vector (ref. 26; Section 4). It is not clear whether this represents activity during latency, since expression was monitored for only a two-week-period. Moreover, constraints on the expression of genes in defective vectors may differ from those in viral vectors, and thus may allow a wider range of promoters to function during latency.

5. Use of HSV as a neuronal expression vector in animals

Experiments involving live animals are subject to national and local regulations and approval from the relevant regulatory bodies must be obtained. The ability to perform such experiments also depends on the nature of the facilities available and on the experience and classification of the experimenter. These aspects are not covered here.

With experimental animals, the question of viral pathogenesis has to be considered. This depends to a large extent on the route of infection and the strains of mice and virus used. For example, HSV-1 strain 17 has an LD_{50} of around 10^5 p.f.u. following inoculation into the footpad of Balb/c mice, whereas a lethal dose can be less than 10 p.f.u. if injected directly into the brain. The virulence of the virus can be attenuated or abolished by inactivating individual genes, without diminishing the capacity of the virus to establish latency (5, 27).

Most work to date has involved the peripheral nervous system with inoculation via the footpad, nose, ear, or eye. We describe the use of the footpad model. Four-week-old female Balb/c mice are injected with 25 μl of virus subcutaneously in the right rear footpad. If only one side is injected the other side can normally be used as a negative control since lateral spread of the virus is limited. Mice are monitored closely after injection for evidence of hindquarters paralysis, which is an early sign of distress. Under standard UK licence conditions, animals showing evidence of paralysis are immediately killed painlessly to prevent further distress. The lytic phase of virus replication in the ganglion lasts up to 8 days after infection, and mice which survive this period are unlikely to die from HSV-related causes.

Protocols 12 and *13* describe methods for examining ganglia for the presence of virus. In addition, latent virus can be detected in ganglia by *in situ* hybridization using a LAT-specific probe (28). The procedures for wax embedding and sectioning tissue and for *in situ* hybridization have been described in detail elsewhere (29, 30). To examine events during the lytic stage of infection, ganglia can be explanted at 4 days after infection. For latency studies allow at least 28 days after infection.

Protocol 12. Virus detection by reactivation following explantation

Materials

- female Balb/c mice
- fetal calf serum (FCS)
- PBS^+/FCS: PBS^+ containing 10% (v/v) FCS
- 96-well microtitre plates
- EFC50: GMEM supplemented with 50% (v/v) FCS
- confluent BHK cells in microtitre plates

Method

1. Dissect out ganglia[a] from thoracic 10 to lumbar 6, wash in PBS^+/FCS, and transfer them to individual wells of a 96-well microtitre plate containing 200 μl/well of EFC50. Incubate at 37°C for 2 days.
2. Replace the medium with fresh EFC50 and continue incubation at 37°C.

Protocol 12. *Continued*

3. Transfer the culture medium to BHK monolayers in microtitre plates. Incubate at 37°C for 2 days.
4. Examine the monolayers microscopically for signs of infection. If no c.p.e. is visible, score the well as negative.
5. Repeat steps 2–4 at 2 day intervals for a total of 20 days. Ganglia which do not release virus during this period are scored as reactivation negative.

[a] Dissecting out ganglia should be done under direction until competence is gained. If the ganglia are to be cultured, sterile procedures should be followed.

Protocol 13. Detection of β-galactosidase activity

Materials

- paraformaldehyde
- phosphate buffer: mix 9.5 ml of 0.4 M NaH_2PO_4 with 40.5 ml of 0.4 M Na_2HPO_4, adjust the pH to 7.4, and dilute to 100 ml
- stain solution: 0.1% (w/v) sodium deoxycholate, 0.1% (v/v) NP40, 5 mM potassium ferricyanide, 5 mM potassium ferrocyanide, 2 mM $MgCl_2$, 1 mg/ml X-gal

Method

1. To prepare formaldehyde solution, add 1 g of paraformaldehyde to 10 ml of deionized water containing 2 drops of 1 M NaOH. Heat to 60°C with stirring until dissolved. Cool the solution, add 12.5 ml of phosphate buffer, and make up to 25 ml with deionized water.
2. At the required time after infection, sacrifice the mice and dissect out the ganglia from thoracic 10 to lumbar 6. Transfer each ganglion directly into a separate well of a 96-well microtitre plate containing 100 μl/well of formaldehyde solution.[a] Fix the ganglia at 0°C for 1 h and then wash them twice with PBS^+.
3. Incubate the ganglia at 31°C for 18 h in 100 μl/well of stain solution. The blue colour should start to develop in 2–3 h.[b]
4. Replace the stain solution with 100 μl/well of stain solution lacking X-gal. Seal the microtitre plate with Parafilm to prevent evaporation and store it at 4°C.
5. Place intact ganglia on a glass slide and examine them using a light microscope (see *Figure 3*).

[a] Make up formaldehyde solution fresh on the day of use and store it on ice until required.
[b] A generalized blue colour may develop spontaneously if the incubation conditions are not correct. This can be checked by including control ganglia from uninfected mice.

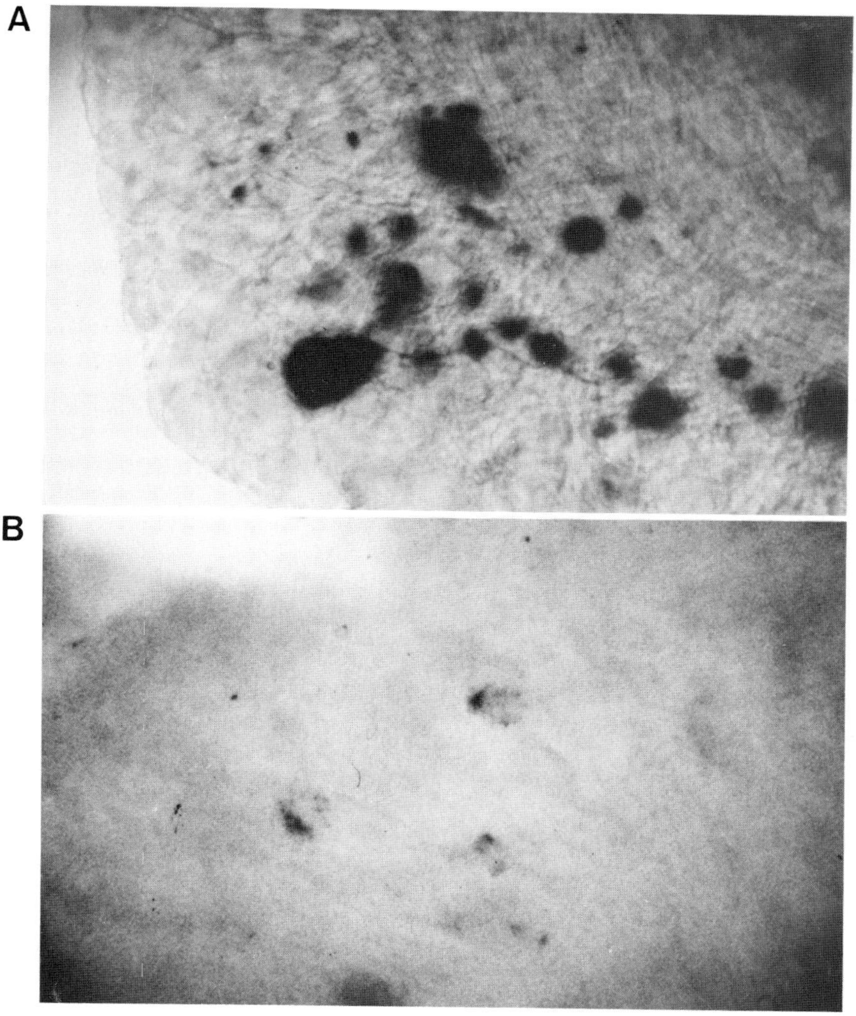

Figure 3. β-galactosidase expression in lytically (A) and latently (B) infected dorsal root ganglia. Mice were infected in the footpad with 10^7 p.f.u. of HSV-1 recombinants expressing *lacZ* under control of the HSV-2 US1/US12 IE promoter (A) or the HSV-1 LAT promoter (B). The fifth lumbar ganglia were dissected out 4 days (A) or 70 days (B) after infection. Ganglia were fixed and incubated for 18 h in the presence of X-gal (*Protocol 13*). Magnification = 1750 ×.

References

1. Advisory Committee on Dangerous Pathogens (1990). *Categorisation of pathogens according to hazard and categories of containment* (2nd edn). HMSO, London.
2. McGeoch, D. J., Dalrymple, M. A., Davison, A. J., Dolan, A., Frame, M. C., McNab, D., Perry, L. J., Scott, J. E., and Taylor, P. (1988). *J. Gen. Virol.,* **69,** 1531.
3. Roizman, B., Desroisiers, R. C., Fleckenstein, B., Lopez, C., Minson, A. C., and Studdert, M. J. (1992). *Arch. Virol.,* **123,** 425.
4. Whitley, R. J. (1990). In *Virology* (ed. B. N. Fields, D. M. Knipe, *et al.*), pp. 1843–1887. Raven Press Ltd, New York.
5. Roizman, B. and Sears, A. E. (1990). In *Virology* (ed. B. N. Fields, D. M. Knipe, *et al.*), pp. 1795–1841. Raven Press Ltd, New York.
6. Spivack, J. G. and Fraser, N. W. (1987). *J. Virol.,* **61,** 3841.
7. Killington, R. A. and Powell, K. L. (1985). In *Virology: a practical approach* (ed. B. W. J. Mahy), pp. 207–236. IRL Press, Oxford.
8. Hay, R. J. (1986). In *Animal cell culture: a practical approach* (ed. R. I. Freshney), pp. 71–112. IRL Press, Oxford.
9. Wilkie, N. M. (1973). *J. Gen. Virol.,* **21,** 453.
10. Szilagyi, J. F. and Cunningham, C. (1991). *J. Gen. Virol.,* **72,** 661.
11. Felgner, P. L., Gadek, T. R., Holm, M., Roman, R., Chan, H. W., Wenz, M., Northrop, J. P., Ringold, G. M., and Danielsen, M. (1987). *Proc. Natl Acad. Sci. USA,* **84,** 7413.
12. Rose, J. K., Buonocore, L., and Whitt, A. (1991). *Biotechniques,* **10,** 520.
13. Sauer, B., Whealy, M., Robbins, A., and Enquist, L. (1987). *Proc. Natl Acad. Sci. USA,* **84,** 9108.
14. Rixon, F. J. and McLauchlan, J. (1990). *J. Gen. Virol.,* **71,** 2931.
15. van Zijl, M., Quint, W., Briaire, J., De Rover, T., Gielkens, A., and Berns, A. (1988). *J. Virol.,* **62,** 2191.
16. Frenkel, N., Spaete, R. R., Vlazny, D. A., Deiss, L. P., and Locker, H. (1982). In *Eucaryotic viral vectors* (ed. Y. Gluzman), pp. 205–209. Cold Spring Harbor Press, Cold Spring Harbor, New York.
17. Geller, A. I. and Breakefield, X. O. (1988). *Science,* **241,** 1667.
18. Stow, N. D. and McMonagle, E. C. (1982). In *Eucaryotic viral vectors* (ed. Y. Gluzman), pp. 199–204. Cold Spring Harbor Press, Cold Spring Harbor, New York.
19. Sawtell, N. M. and Thompson, R. L. (1992). *J. Virol.,* **66,** 2157.
20. Saunders, P. G., Wilkie, N. M., and Davison, A. J. (1982). *J. Gen. Virol.,* **63,** 277.
21. Sealey, P. G. and Southern, E. M. (1990). In *Gel electrophoresis of nucleic acids: a practical approach* (2nd edn) (ed. D. Rickwood and B. D. Hames), pp. 51–100. IRL Press, Oxford.
22. Everett, R. D. (1987). *Anticancer Res.,* **7,** 589.
23. Calder, J. M., Stow, E. C., and Stow, N. D. (1992). *J. Gen. Virol.,* **73,** 531.
24. Preston, V. G., Rixon, F. J., McDougall, I. M., McGregor, M., and Al Kobaisi, M. F. (1992). *Virology,* **186,** 87.
25. Dobson, A. T., Margolis, T. P., Sedarati, F., Stevens, J. G., and Feldman, L. T. (1990). *Neuron,* **5,** 353.

26. Kaplitt, M. G., Pfaus, J. G., Kleopoulos, S. P., Hanlon, B. A., Rabkin, S. D., and Pfaff, D. W. (1991). *Mol. Cell. Neurosci.,* **2,** 320.
27. MacLean, A. R., Ul-Fareed, M., Robertson, L., Harland, J., and Brown, M. (1991). *J. Gen. Virol.,* **72,** 631.
28. Stevens, J. G., Wagner, E. K., Devi-Rao, G. B., Cook, M. L., and Feldman, L. T. (1987). *Science,* **235,** 1056.
29. Lohr, M. and Nerenberg, M. I. (1990). In *Animal virus pathogenesis: a practical approach* (ed. M. B. A. Oldstone), pp. 55–62. IRL Press, Oxford.
30. Conkie, D. (1986). In *Animal cell culture: a practical approach* (ed. R. I. Freshney), pp. 217–226. IRL Press, Oxford.

A

Suppliers of specialist items

Aalto Bioreagents, 14 Main Street, Rathfarnham, Dublin 14, Republic of Ireland.

Abbott Laboratories Ltd, Abbott House, Moorbridge Road, Maidenhead, SL6 8XZ, UK.

AIDS Directed Programme Reagent Repository, National Institute of Biological Standards and Control, Blanche Lane, South Mimms, Potters Bar, Hertfordshire, EN6 3QG, UK.

American Type Culture Collection, 12301 Parklawn Drive, Rockville, MD 20852, USA.

Amersham, Little Chalfont, Amersham, HP7 9NA, UK; 2636 South Clearbrook Drive, Arlington Heights, IL 60005, USA.

Amicon, Upper Mill, Stonehouse, Gloucester, GL10 2BJ, UK; 17 Cherry Hill Drive, Danvers, MA 01923, USA.

Applied Biosystems, Kelvin Close, Birchwood Science Park North, Warrington, WA3 7PB, UK; 850 Lincoln Center Drive, Foster City, CA 94404, USA.

BDH, Broom Road, Poole, Dorset, BH12 4NN, UK.

Beckman Instruments, Progress Road, Sands Industrial Estate, High Wycombe, Buckinghamshire, HP12 4JL, UK; 2500 Harbor Boulevard, PO Box 3100, Fullerton, CA 92634, USA.

Beckton-Dickinson, Between Towns Road, Cowley, Oxford, OX4 31Y, UK; 2 Bridgewater Lane, Lincoln Park, NJ 07035, USA.

Bio101 Inc., PO Box 2284, La Jolla, California, 92038–2284, USA; Stratech Ltd, 61 Dudley Street, Luton, Bedfordshire, LU2 0NP, UK.

Bio-Rad Laboratories Ltd, Bio-Rad House, Maylands Avenue, Hemel Hempstead, Hertfordshire, HP2 7TD, UK; 3300 Regatta Boulevard, Richmond, CA 94804, USA.

Boehringer–Mannheim, Bell Lane, Lewes, East Sussex, BN7 1LG, UK; PO Box 50414, Indianapolis, IN 46250, USA.

Calbiochem, PO Box 12087, San Diego, CA 92112–4180, USA; Novabiochem (UK) Ltd, 3 Heathcoat Building, Highfields Science Park, University Boulevard, Nottingham, NG7 2QJ, UK.

Cetus: see Perkin Elmer

Cinna/Biotecx: Biogenesis, 12 Yeomans Park, Bournemouth, BH8 0BJ, UK.

Dako Diagnostics, 22 The Arcade, The Octagon, High Wycombe, Buckinghamshire, HP11 2HT, UK.

Dupont–NEN, Wedgwood Way, Stevenage, SG1 4QN, UK; 549 Albany Street, Boston MA 02118, USA.

Dynatech Laboratories Ltd, Daux Road, Billingshurst, West Sussex, RH14 GSJ, UK; 900 Slaters Lane, Alexandria, VA 22314, USA.

Environmental Diagnostics Inc., Burlington, NC 27215, USA.

Falcon: see Becton-Dickinson.

FMC Bioproducts, 5 Maple Street, Rockland, ME 04841, USA; Flowgen Instruments Ltd, Broad Oak Enterprise Village, Broad Oak Road, Sittingbourne, Kent, ME9 8AQ, UK.

Gibco–BRL, PO Box 35, Trident House, Renfrew Road, Paisley PA3 4EF, UK; PO Box 68, Grand Island, NY, USA.

Gurr: see BDH.

Hoeffer Scientific Instruments, Unit 12, Croft Road Workshops, Croft Road, Newcastle-under-Lyme, ST5 0TH, UK; 654 Minnesota Street, PO Box 77387, San Francisco, CA 94107, USA.

Hybaid Ltd, 111–113 Waldegrave Road, Teddington, Middlesex, TW11 8LL, UK; National Labnet Co., PO Box 841, Woodbridge, NJ 07095, USA.

ICN Biomedicals, Eagle House, Peregrine Business Park, Gomm Road, High Wycombe, Buckinghamshire HP13 7DL, UK; PO Box 19536, Irvine, CA 92713–9921, USA.

ICN–Flow, 3300 Hyland Avenue, Costa Mesa, CA 92626, USA.

Invitrogen, 3985B Sorrento Valley Boulevard, San Diego, CA 92121, USA; AMS Biotechnology, Unit 6, Tannery Yard, Whitney Street, Burford, OX8 3DN, UK.

Life Technologies, see Gibco–BRL.

Millipore, The Boulevard, Blackmoor Lane, Watford, Hertfordshire, WD1 2RA, UK; 80 Ashby Road, Bedford, MA 01730, USA.

New England Biolabs, 32 Tozer Road, Beverly, MA 01915–5599, USA; CP Laboratories, PO Box 22, Bishop's Stortford, Hertfordshire, CM23 3DX, UK.

Northumbria Biologicals, Nelson Industrial Estate, Cramlington, Northumberland, NE23 9BL, UK.

Organon Teknika Corporation, Science Park, Milton Road, Cambridge, CB4 4BH, UK; Treyburn, 100 Akzo Avenue, Durham, NC 27704, USA.

Oswel DNA Service, Department of Chemistry, University of Edinburgh, West Mains Road, Edinburgh, EH9 3JJ, UK.

Narishige Scientific Instruments Laboratory, 9.28 Kasuya, 4 Chome Setagaya-ku, Tokyo, Japan.

Perkin Elmer Cetus, Maxwell Road, Beaconsfield, Buckinghamshire, HP9 1QA, UK; 761 Main Avenue, Norwalk, CT 06859, USA.

Pharmacia Ltd, Pharmacia House, Midsummer Boulevard, Central Milton Keynes, Buckinghamshire, MK9 3HP, UK; 800 Centennial Avenue, Piscataway, New Jersey, 08854, USA.

Pierce Chemical Company, PO Box 117, Rockford, IL 61105, USA; Pierce and Wariner, 44 Upper Northgate Street, Chester, CH1 4EF, UK.

Promega Corporation, Epsilon House, Enterprise Road, Chilworth Research Centre, Southampton, SO1 7N, UK; 2800 S. Fish Hatchery Road, Madison, WI 53711, USA.

Quiagen, 9259 Eton Avenue, Chatsworth, CA 91311, USA; Hybaid Ltd, 111 Waldegrave Road, Teddington, TW11 8LL, UK.

Schleicher & Schuell, 10 Optical Avenue, Keene, NH 03431, USA; Anderman and Co., 145 London Road, Kingston-upon-Thames, Surrey, KT2 6NH, UK.

Seralab Ltd, Crawley Down, Sussex, RH10 4FF, UK; Accurate Chemical Scientific Corporation, 300 Shames Drive, Westbury, NY 11590, USA.

Sigma Chemical Company, Fancy Road, Poole, Dorset, BH17 7NH, UK; PO Box 14508, St Louis, MO 63178, USA.

Sterilin, Lampton House, Lampton Road, Hounslow, Middlesex, TW3 4EE, UK.

Stratagene, 140 Cambridge Innovation Centre, Cambridge Science Park, Milton Road, Cambridge, CB4 4GF, UK; 11099 North Torrey Pines Road, La Jolla, CA 92037, USA.

Surfachem, 16 Market Street, Brighouse, HD6 1AP, UK.

Surgicos Ltd, Livingstone, Scotland, UK.

Techne (Cambridge), Duxford, Cambridge, CD2 4PZ, UK; 3700 Brunswick Pike, Princeton, NJ 08540–6192, USA.

United States Biochemical Corporation (USB), PO Box 22400, Cleveland, OH 44122, USA; Cambridge BioScience, 25 Signet Court, Stourbridge Common Business Centre, Swan's Road, Cambridge, CB5 8LA, UK.

Whatman, Whatman House, St Leonard's Road, Maidstone, ME16 0LS, UK; 5285 NE Elam Young Parkway, Suite A/400, Hillsboro, OR 97124, USA.

Worthington Biochemical, Halls Mill Road, Freehold, NJ 07728, USA.

Index

ORDER OTHER TITLES OF INTEREST TODAY

Price list for: UK, Europe, Rest of World (excluding US and Canada)

Forthcoming Titles

124. Human Genetic Disease Analysis Davies, K.E. (Ed)
.... Spiralbound hardback 0-19-963309-6 **£30.00**
.... Paperback 0-19-963308-8 **£18.50**
123. Protein Phosphorylation Hardie, G. (Ed)
.... Spiralbound hardback 0-19-963306-1 **£32.50**
.... Paperback 0-19-963305-3 **£22.50**
122. Immunocytochemistry Beesley, J. (Ed)
.... Spiralbound hardback 0-19-963270-7 **£32.50**
.... Paperback 0-19-963269-3 **£22.50**
121. Tumour Immunobiology Gallagher, G., Rees, R.C. & others (Eds)
.... Spiralbound hardback 0-19-963370-3 **£35.00**
.... Paperback 0-19-963369-X **£25.00**
120. Transcription Factors Latchman, D.S. (Ed)
.... Spiralbound hardback 0-19-963342-8 **£30.00**
.... Paperback 0-19-963341-X **£19.50**
119. Growth Factors McKay, I.A. & Leigh, I. (Eds)
.... Spiralbound hardback 0-19-963360-6 **£30.00**
.... Paperback 0-19-963359-2 **£19.50**
118. Histocompatibility Testing Dyer, P. & Middleton, D. (Eds)
.... Spiralbound hardback 0-19-963364-9 **£32.50**
.... Paperback 0-19-963363-0 **£22.50**
117. Gene Transcription Hames, D.B. & Higgins, S.J. (Eds)
.... Spiralbound hardback 0-19-963292-8 **£35.00**
.... Paperback 0-19-963291-X **£25.00**
116. Electrophysiology Wallis, D.I. (Ed)
.... Spiralbound hardback 0-19-963348-7 **£32.50**
.... Paperback 0-19-963347-9 **£22.50**
115. Biological Data Analysis Fry, J.C. (Ed)
.... Spiralbound hardback 0-19-963340-1 **£50.00**
.... Paperback 0-19-963339-8 **£27.50**
114. Experimental Neuroanatomy Bolam, J.P. (Ed)
.... Spiralbound hardback 0-19-963326-6 **£32.50**
.... Paperback 0-19-963325-8 **£22.50**
112. Lipid Analysis Hamilton, R.J. & Hamilton, S.J. (Eds)
.... Spiralbound hardback 0-19-963098-4 **£35.00**
.... Paperback 0-19-963099-2 **£25.00**
111. Haemopoiesis Testa, N.G. & Molineux, G. (Eds)
.... Spiralbound hardback 0-19-963366-5 **£32.50**
.... Paperback 0-19-963365-7 **£22.50**

Published Titles

113. Preparative Centrifugation Rickwood, D. (Ed)
.... Spiralbound hardback 0-19-963208-1 **£45.00**
.... Paperback 0-19-963211-1 **£25.00**
110. Pollination Ecology Dafni, A.
.... Spiralbound hardback 0-19-963299-5 **£32.50**
.... Paperback 0-19-963298-7 **£22.50**
109. In Situ Hybridization Wilkinson, D.G. (Ed)
.... Spiralbound hardback 0-19-963328-2 **£30.00**
.... Paperback 0-19-963327-4 **£18.50**
108. Protein Engineering Rees, A.R., Sternberg, M.J.E. & others (Eds)
.... Spiralbound hardback 0-19-963139-5 **£35.00**
.... Paperback 0-19-963138-7 **£25.00**
107. Cell-Cell Interactions Stevenson, B.R., Gallin, W.J. & others (Eds)
.... Spiralbound hardback 0-19-963319-3 **£32.50**
.... Paperback 0-19-963318-5 **£22.50**
106. Diagnostic Molecular Pathology: Volume I Herrington, C.S. & McGee, J. O'D. (Eds)
.... Spiralbound hardback 0-19-963237-5 **£30.00**
.... Paperback 0-19-963236-7 **£19.50**
105. Biomechanics-Materials Vincent, J.F.V. (Ed)
.... Spiralbound hardback 0-19-963223-5 **£35.00**
.... Paperback 0-19-963222-7 **£25.00**
104. Animal Cell Culture (2/e) Freshney, R.I. (Ed)
.... Spiralbound hardback 0-19-963212-X **£30.00**
.... Paperback 0-19-963213-8 **£19.50**
103. Molecular Plant Pathology: Volume II Gurr, S.J., McPherson, M.J. & others (Eds)
.... Spiralbound hardback 0-19-963352-5 **£32.50**
.... Paperback 0-19-963351-7 **£22.50**
101. Protein Targeting Magee, A.I. & Wileman, T. (Eds)
.... Spiralbound hardback 0-19-963206-5 **£32.50**
.... Paperback 0-19-963210-3 **£22.50**
100. Diagnostic Molecular Pathology: Volume II: Cell and Tissue Genotyping Herrington, C.S. & McGee, J.O'D. (Eds)
.... Spiralbound hardback 0-19-963239-1 **£30.00**
.... Paperback 0-19-963238-3 **£19.50**
99. Neuronal Cell Lines Wood, J.N. (Ed)
.... Spiralbound hardback 0-19-963346-0 **£32.50**
.... Paperback 0-19-963345-2 **£22.50**
98. Neural Transplantation Dunnett, S.B. & Björklund, A. (Eds)
.... Spiralbound hardback 0-19-963286-3 **£30.00**
.... Paperback 0-19-963285-5 **£19.50**
97. Human Cytogenetics: Volume II: Malignancy and Acquired Abnormalities (2/e) Rooney, D.E. & Czepulkowski, B.H. (Eds)
.... Spiralbound hardback 0-19-963290-1 **£30.00**
.... Paperback 0-19-963289-8 **£22.50**
96. Human Cytogenetics: Volume I: Constitutional Analysis (2/e) Rooney, D.E. & Czepulkowski, B.H. (Eds)
.... Spiralbound hardback 0-19-963288-X **£30.00**
.... Paperback 0-19-963287-1 **£22.50**
95. Lipid Modification of Proteins Hooper, N.M. & Turner, A.J. (Eds)
.... Spiralbound hardback 0-19-963274-X **£32.50**
.... Paperback 0-19-963273-1 **£22.50**
94. Biomechanics-Structures and Systems Biewener, A.A. (Ed)
.... Spiralbound hardback 0-19-963268-5 **£42.50**
.... Paperback 0-19-963267-7 **£25.00**
93. Lipoprotein Analysis Converse, C.A. & Skinner, E.R. (Eds)
.... Spiralbound hardback 0-19-963192-1 **£30.00**
.... Paperback 0-19-963231-6 **£19.50**
92. Receptor-Ligand Interactions Hulme, E.C. (Ed)
.... Spiralbound hardback 0-19-963090-9 **£35.00**
.... Paperback 0-19-963091-7 **£25.00**
91. Molecular Genetic Analysis of Populations Hoelzel, A.R. (Ed)
.... Spiralbound hardback 0-19-963278-2 **£32.50**
.... Paperback 0-19-963277-4 **£22.50**

90. **Enzyme Assays** Eisenthal, R. & Danson, M.J. (Eds)
.... Spiralbound hardback 0-19-963142-5 **£35.00**
.... Paperback 0-19-963143-3 **£25.00**

89. **Microcomputers in Biochemistry** Bryce, C.F.A. (Ed)
.... Spiralbound hardback 0-19-963253-7 **£30.00**
.... Paperback 0-19-963252-9 **£19.50**

88. **The Cytoskeleton** Carraway, K.L. & Carraway, C.A.C. (Eds)
.... Spiralbound hardback 0-19-963257-X **£30.00**
.... Paperback 0-19-963256-1 **£19.50**

87. **Monitoring Neuronal Activity** Stamford, J.A. (Ed)
.... Spiralbound hardback 0-19-963244-8 **£30.00**
.... Paperback 0-19-963243-X **£19.50**

86. **Crystallization of Nucleic Acids and Proteins** Ducruix, A. & Gieg‹130›, R. (Eds)
.... Spiralbound hardback 0-19-963245-6 **£35.00**
.... Paperback 0-19-963246-4 **£25.00**

85. **Molecular Plant Pathology: Volume I** Gurr, S.J., McPherson, M.J. & others (Eds)
.... Spiralbound hardback 0-19-963103-4 **£30.00**
.... Paperback 0-19-963102-6 **£19.50**

84. **Anaerobic Microbiology** Levett, P.N. (Ed)
.... Spiralbound hardback 0-19-963204-9 **£32.50**
.... Paperback 0-19-963262-6 **£22.50**

83. **Oligonucleotides and Analogues** Eckstein, F. (Ed)
.... Spiralbound hardback 0-19-963280-4 **£32.50**
.... Paperback 0-19-963279-0 **£22.50**

82. **Electron Microscopy in Biology** Harris, R. (Ed)
.... Spiralbound hardback 0-19-963219-7 **£32.50**
.... Paperback 0-19-963215-4 **£22.50**

81. **Essential Molecular Biology: Volume II** Brown, T.A. (Ed)
.... Spiralbound hardback 0-19-963112-3 **£32.50**
.... Paperback 0-19-963113-1 **£22.50**

80. **Cellular Calcium** McCormack, J.G. & Cobbold, P.H. (Eds)
.... Spiralbound hardback 0-19-963131-X **£35.00**
.... Paperback 0-19-963130-1 **£25.00**

79. **Protein Architecture** Lesk, A.M.
.... Spiralbound hardback 0-19-963054-2 **£32.50**
.... Paperback 0-19-963055-0 **£22.50**

78. **Cellular Neurobiology** Chad, J. & Wheal, H. (Eds)
.... Spiralbound hardback 0-19-963106-9 **£32.50**
.... Paperback 0-19-963107-7 **£22.50**

77. **PCR** McPherson, M.J., Quirke, P. & others (Eds)
.... Spiralbound hardback 0-19-963226-X **£30.00**
.... Paperback 0-19-963196-4 **£19.50**

76. **Mammalian Cell Biotechnology** Butler, M. (Ed)
.... Spiralbound hardback 0-19-963207-3 **£30.00**
.... Paperback 0-19-963209-X **£19.50**

75. **Cytokines** Balkwill, F.R. (Ed)
.... Spiralbound hardback 0-19-963218-9 **£35.00**
.... Paperback 0-19-963214-6 **£25.00**

74. **Molecular Neurobiology** Chad, J. & Wheal, H. (Eds)
.... Spiralbound hardback 0-19-963108-5 **£30.00**
.... Paperback 0-19-963109-3 **£19.50**

73. **Directed Mutagenesis** McPherson, M.J. (Ed)
.... Spiralbound hardback 0-19-963141-7 **£30.00**
.... Paperback 0-19-963140-9 **£19.50**

72. **Essential Molecular Biology: Volume I** Brown, T.A. (Ed)
.... Spiralbound hardback 0-19-963110-7 **£32.50**
.... Paperback 0-19-963111-5 **£22.50**

71. **Peptide Hormone Action** Siddle, K. & Hutton, J.C.
.... Spiralbound hardback 0-19-963070-4 **£32.50**
.... Paperback 0-19-963071-2 **£22.50**

70. **Peptide Hormone Secretion** Hutton, J.C. & Siddle, K. (Eds)
.... Spiralbound hardback 0-19-963068-2 **£35.00**
.... Paperback 0-19-963069-0 **£25.00**

69. **Postimplantation Mammalian Embryos** Copp, A.J. & Cockroft, D.L. (Eds)
.... Spiralbound hardback 0-19-963088-7 **£35.00**
.... Paperback 0-19-963089-5 **£25.00**

68. **Receptor-Effector Coupling** Hulme, E.C. (Ed)
.... Spiralbound hardback 0-19-963094-1 **£30.00**
.... Paperback 0-19-963095-X **£19.50**

67. **Gel Electrophoresis of Proteins (2/e)** Hames, B.D. & Rickwood, D. (Eds)
.... Spiralbound hardback 0-19-963074-7 **£35.00**
.... Paperback 0-19-963075-5 **£25.00**

66. **Clinical Immunology** Gooi, H.C. & Chapel, H. (Eds)
.... Spiralbound hardback 0-19-963086-0 **£32.50**
.... Paperback 0-19-963087-9 **£22.50**

65. **Receptor Biochemistry** Hulme, E.C. (Ed)
.... Spiralbound hardback 0-19-963092-5 **£35.00**
.... Paperback 0-19-963093-3 **£25.00**

64. **Gel Electrophoresis of Nucleic Acids (2/e)** Rickwood, D. & Hames, B.D. (Eds)
.... Spiralbound hardback 0-19-963082-8 **£32.50**
.... Paperback 0-19-963083-6 **£22.50**

63. **Animal Virus Pathogenesis** Oldstone, M.B.A. (Ed)
.... Spiralbound hardback 0-19-963100-X **£30.00**
.... Paperback 0-19-963101-8 **£18.50**

62. **Flow Cytometry** Ormerod, M.G. (Ed)
.... Paperback 0-19-963053-4 **£22.50**

61. **Radioisotopes in Biology** Slater, R.J. (Ed)
.... Spiralbound hardback 0-19-963080-1 **£32.50**
.... Paperback 0-19-963081-X **£22.50**

60. **Biosensors** Cass, A.E.G. (Ed)
.... Spiralbound hardback 0-19-963046-1 **£30.00**
.... Paperback 0-19-963047-X **£19.50**

59. **Ribosomes and Protein Synthesis** Spedding, G. (Ed)
.... Spiralbound hardback 0-19-963104-2 **£32.50**
.... Paperback 0-19-963105-0 **£22.50**

58. **Liposomes** New, R.R.C. (Ed)
.... Spiralbound hardback 0-19-963076-3 **£35.00**
.... Paperback 0-19-963077-1 **£22.50**

57. **Fermentation** McNeil, B. & Harvey, L.M. (Eds)
.... Spiralbound hardback 0-19-963044-5 **£30.00**
.... Paperback 0-19-963045-3 **£19.50**

56. **Protein Purification Applications** Harris, E.L.V. & Angal, S. (Eds)
.... Spiralbound hardback 0-19-963022-4 **£30.00**
.... Paperback 0-19-963023-2 **£18.50**

55. **Nucleic Acids Sequencing** Howe, C.J. & Ward, E.S. (Eds)
.... Spiralbound hardback 0-19-963056-9 **£30.00**
.... Paperback 0-19-963057-7 **£19.50**

54. **Protein Purification Methods** Harris, E.L.V. & Angal, S. (Eds)
.... Spiralbound hardback 0-19-963002-X **£30.00**
.... Paperback 0-19-963003-8 **£20.00**

53. **Solid Phase Peptide Synthesis** Atherton, E. & Sheppard, R.C.
.... Spiralbound hardback 0-19-963066-6 **£30.00**
.... Paperback 0-19-963067-4 **£18.50**

52. **Medical Bacteriology** Hawkey, P.M. & Lewis, D.A. (Eds)
.... Spiralbound hardback 0-19-963008-9 **£38.00**
.... Paperback 0-19-963009-7 **£25.00**

51. **Proteolytic Enzymes** Beynon, R.J. & Bond, J.S. (Eds)
.... Spiralbound hardback 0-19-963058-5 **£30.00**
.... Paperback 0-19-963059-3 **£19.50**

50. **Medical Mycology** Evans, E.G.V. & Richardson, M.D. (Eds)
.... Spiralbound hardback 0-19-963010-0 **£37.50**
.... Paperback 0-19-963011-9 **£25.00**

49. **Computers in Microbiology** Bryant, T.N. & Wimpenny, J.W.T. (Eds)
.... Paperback 0-19-963015-1 **£19.50**

48. **Protein Sequencing** Findlay, J.B.C. & Geisow, M.J. (Eds)
.... Spiralbound hardback 0-19-963012-7 **£30.00**
.... Paperback 0-19-963013-5 **£18.50**

47. **Cell Growth and Division** Baserga, R. (Ed)
.... Spiralbound hardback 0-19-963026-7 **£30.00**
.... Paperback 0-19-963027-5 **£18.50**

46. **Protein Function** Creighton, T.E. (Ed)
.... Spiralbound hardback 0-19-963006-2 **£32.50**
.... Paperback 0-19-963007-0 **£22.50**

45. **Protein Structure** Creighton, T.E. (Ed)
.... Spiralbound hardback 0-19-963000-3 **£32.50**
.... Paperback 0-19-963001-1 **£22.50**

44. **Antibodies: Volume II** Catty, D. (Ed)
.... Spiralbound hardback 0-19-963018-6 **£30.00**
.... Paperback 0-19-963019-4 **£19.50**

43. **HPLC of Macromolecules** Oliver, R.W.A. (Ed)
.... Spiralbound hardback 0-19-963020-8 **£30.00**
.... Paperback 0-19-963021-6 **£19.50**
42. **Light Microscopy in Biology** Lacey, A.J. (Ed)
.... Spiralbound hardback 0-19-963036-4 **£30.00**
.... Paperback 0-19-963037-2 **£19.50**
41. **Plant Molecular Biology** Shaw, C.H. (Ed)
.... Paperback 1-85221-056-7 **£22.50**
40. **Microcomputers in Physiology** Fraser, P.J. (Ed)
.... Spiralbound hardback 1-85221-129-6 **£30.00**
.... Paperback 1-85221-130-X **£19.50**
39. **Genome Analysis** Davies, K.E. (Ed)
.... Spiralbound hardback 1-85221-109-1 **£30.00**
.... Paperback 1-85221-110-5 **£18.50**
38. **Antibodies: Volume I** Catty, D. (Ed)
.... Paperback 0-947946-85-3 **£19.50**
37. **Yeast** Campbell, I. & Duffus, J.H. (Eds)
.... Paperback 0-947946-79-9 **£19.50**
36. **Mammalian Development** Monk, M. (Ed)
.... Hardback 1-85221-030-3 **£30.50**
.... Paperback 1-85221-029-X **£22.50**
35. **Lymphocytes** Klaus, G.G.B. (Ed)
.... Hardback 1-85221-018-4 **£30.00**
34. **Lymphokines and Interferons** Clemens, M.J., Morris, A.G. & others (Eds)
.... Paperback 1-85221-035-4 **£22.50**
33. **Mitochondria** Darley-Usmar, V.M., Rickwood, D. & others (Eds)
.... Hardback 1-85221-034-6 **£32.50**
.... Paperback 1-85221-033-8 **£22.50**
32. **Prostaglandins and Related Substances** Benedetto, C., McDonald-Gibson, R.G. & others (Eds)
.... Hardback 1-85221-032-X **£32.50**
.... Paperback 1-85221-031-1 **£22.50**
31. **DNA Cloning: Volume III** Glover, D.M. (Ed)
.... Hardback 1-85221-049-4 **£30.00**
.... Paperback 1-85221-048-6 **£19.50**
30. **Steroid Hormones** Green, B. & Leake, R.E. (Eds)
.... Paperback 0-947946-53-5 **£19.50**
29. **Neurochemistry** Turner, A.J. & Bachelard, H.S. (Eds)
.... Hardback 1-85221-028-1 **£30.00**
.... Paperback 1-85221-027-3 **£19.50**
28. **Biological Membranes** Findlay, J.B.C. & Evans, W.H. (Eds)
.... Hardback 0-947946-84-5 **£32.50**
.... Paperback 0-947946-83-7 **£22.50**
27. **Nucleic Acid and Protein Sequence Analysis** Bishop, M.J. & Rawlings, C.J. (Eds)
.... Hardback 1-85221-007-9 **£35.00**
.... Paperback 1-85221-006-0 **£25.00**
26. **Electron Microscopy in Molecular Biology** Sommerville, J. & Scheer, U. (Eds)
.... Hardback 0-947946-64-0 **£30.00**
.... Paperback 0-947946-54-3 **£19.50**
25. **Teratocarcinomas and Embryonic Stem Cells** Robertson, E.J. (Ed)
.... Hardback 1-85221-005-2 **£19.50**
.... Paperback 1-85221-004-4 **£19.50**
24. **Spectrophotometry and Spectrofluorimetry** Harris, D.A. & Bashford, C.L. (Eds)
.... Hardback 0-947946-69-1 **£30.00**
.... Paperback 0-947946-46-2 **£18.50**
23. **Plasmids** Hardy, K.G. (Ed)
.... Paperback 0-947946-81-0 **£18.50**
22. **Biochemical Toxicology** Snell, K. & Mullock, B. (Eds)
.... Paperback 0-947946-52-7 **£19.50**
19. **Drosophila** Roberts, D.B. (Ed)
.... Hardback 0-947946-66-7 **£32.50**
.... Paperback 0-947946-45-4 **£22.50**
17. **Photosynthesis: Energy Transduction** Hipkins, M.F. & Baker, N.R. (Eds)
.... Hardback 0-947946-63-2 **£30.00**
.... Paperback 0-947946-51-9 **£18.50**
16. **Human Genetic Diseases** Davies, K.E. (Ed)
.... Hardback 0-947946-76-4 **£30.00**
.... Paperback 0-947946-75-6 **£18.50**
14. **Nucleic Acid Hybridisation** Hames, B.D. & Higgins, S.J. (Eds)
.... Hardback 0-947946-61-6 **£30.00**
.... Paperback 0-947946-23-3 **£19.50**
13. **Immobilised Cells and Enzymes** Woodward, J. (Ed)
.... Hardback 0-947946-60-8 **£18.50**
12. **Plant Cell Culture** Dixon, R.A. (Ed)
.... Paperback 0-947946-22-5 **£19.50**
11a. **DNA Cloning: Volume I** Glover, D.M. (Ed)
.... Paperback 0-947946-18-7 **£18.50**
11b. **DNA Cloning: Volume II** Glover, D.M. (Ed)
.... Paperback 0-947946-19-5 **£19.50**
10. **Virology** Mahy, B.W.J. (Ed)
.... Paperback 0-904147-78-9 **£19.50**
9. **Affinity Chromatography** Dean, P.D.G., Johnson, W.S. & others (Eds)
.... Paperback 0-904147-71-1 **£19.50**
7. **Microcomputers in Biology** Ireland, C.R. & Long, S.P. (Eds)
.... Paperback 0-904147-57-6 **£18.00**
6. **Oligonucleotide Synthesis** Gait, M.J. (Ed)
.... Paperback 0-904147-74-6 **£18.50**
5. **Transcription and Translation** Hames, B.D. & Higgins, S.J. (Eds)
.... Paperback 0-904147-52-5 **£22.50**
3. **Iodinated Density Gradient Media** Rickwood, D. (Ed)
.... Paperback 0-904147-51-7 **£19.50**

Sets

Essential Molecular Biology: Volumes I and II as a set Brown, T.A. (Ed)
.... Spiralbound hardback 0-19-963114-X **£58.00**
.... Paperback 0-19-963115-8 **£40.00**
Antibodies: Volumes I and II as a set Catty, D. (Ed)
.... Paperback 0-19-963063-1 **£33.00**
Cellular and Molecular Neurobiology Chad, J. & Wheal, H. (Eds)
.... Spiralbound hardback 0-19-963255-3 **£56.00**
.... Paperback 0-19-963254-5 **£38.00**
Protein Structure and Protein Function: Two-volume set Creighton, T.E. (Ed)
.... Spiralbound hardback 0-19-963064-X **£55.00**
.... Paperback 0-19-963065-8 **£38.00**
DNA Cloning: Volumes I, II, III as a set Glover, D.M. (Ed)
.... Paperback 1-85221-069-9 **£46.00**
Molecular Plant Pathology: Volumes I and II as a set Gurr, S.J., McPherson, M.J. & others (Eds)
.... Spiralbound hardback 0-19-963354-1 **£56.00**
.... Paperback 0-19-963353-3 **£37.00**
Protein Purification Methods, and Protein Purification Applications, two-volume set Harris, E.L.V. & Angal, S. (Eds)
.... Spiralbound hardback 0-19-963048-8 **£48.00**
.... Paperback 0-19-963049-6 **£32.00**
Diagnostic Molecular Pathology: Volumes I and II as a set Herrington, C.S. & McGee, J. O'D. (Eds)
.... Spiralbound hardback 0-19-963241-3 **£54.00**
.... Paperback 0-19-963240-5 **£35.00**
Receptor Biochemistry; Receptor-Effector Coupling; Receptor-Ligand Interactions Hulme, E.C. (Ed)
.... Spiralbound hardback 0-19-963096-8 **£90.00**
.... Paperback 0-19-963097-6 **£62.50**
Signal Transduction Milligan, G. (Ed)
.... Spiralbound hardback 0-19-963296-0 **£30.00**
.... Paperback 0-19-963295-2 **£18.50**
Human Cytogenetics: Volumes I and II as a set (2/e) Rooney, D.E. & Czepulkowski, B.H. (Eds)
.... Hardback 0-19-963314-2 **£58.50**
.... Paperback 0-19-963313-4 **£40.50**
Peptide Hormone Secretion/Peptide Hormone Action Siddle, K. & Hutton, J.C. (Eds)
.... Spiralbound hardback 0-19-963072-0 **£55.00**
.... Paperback 0-19-963073-9 **£38.00**

ORDER FORM for UK, Europe and Rest of World

(Excluding USA and Canada)

Qty	ISBN	Author	Title	Amount
			P&P	
			TOTAL	

Please add postage and packing: £1.75 for UK orders under £20; £2.75 for UK orders over £20; overseas orders add 10% of total.

Name ....

Address ....

....

.... Post code ....

[ ] Please charge £ .... to my credit card

Access/VISA/Eurocard/AMEX/Diners Club (circle appropriate card)

Card No .... Expiry date ....

Signature ....

Credit card account address if different from above:

....

.... Postcode ....

[ ] I enclose a cheque for £....................

Please return this form to: OUP Distribution Services, Saxon Way West, Corby, Northants NN18 9ES

OR ORDER BY CREDIT CARD HOTLINE: Tel +44-(0)536-741519 or Fax +44-(0)536-746337

ORDER OTHER TITLES OF INTEREST TODAY

Price list for: USA and Canada

123. **Protein Phosphorylation** Hardie, G. (Ed)
.... Spiralbound hardback 0-19-963306-1 **$65.00**
.... Paperback 0-19-963305-3 **$45.00**
121. **Tumour Immunobiology** Gallagher, G., Rees, R.C. & others (Eds)
.... Spiralbound hardback 0-19-963370-3 **$72.00**
.... Paperback 0-19-963369-X **$50.00**
117. **Gene Transcription** Hames, D.B. & Higgins, S.J. (Eds)
.... Spiralbound hardback 0-19-963292-8 **$72.00**
.... Paperback 0-19-963291-X **$50.00**
116. **Electrophysiology** Wallis, D.I. (Ed)
.... Spiralbound hardback 0-19-963348-7 **$66.50**
.... Paperback 0-19-963347-9 **$45.95**
115. **Biological Data Analysis** Fry, J.C. (Ed)
.... Spiralbound hardback 0-19-963340-1 **$80.00**
.... Paperback 0-19-963339-8 **$60.00**
114. **Experimental Neuroanatomy** Bolam, J.P. (Ed)
.... Spiralbound hardback 0-19-963326-6 **$65.00**
.... Paperback 0-19-963325-8 **$40.00**
111. **Haemopoiesis** Testa, N.G. & Molineux, G. (Eds)
.... Spiralbound hardback 0-19-963366-5 **$65.00**
.... Paperback 0-19-963365-7 **$45.00**
113. **Preparative Centrifugation** Rickwood, D. (Ed)
.... Spiralbound hardback 0-19-963208-1 **$90.00**
.... Paperback 0-19-963211-1 **$50.00**
110. **Pollination Ecology** Dafni, A.
.... Spiralbound hardback 0-19-963299-5 **$65.00**
.... Paperback 0-19-963298-7 **$45.00**
109. **In Situ Hybridization** Wilkinson, D.G. (Ed)
.... Spiralbound hardback 0-19-963328-2 **$58.00**
.... Paperback 0-19-963327-4 **$36.00**
108. **Protein Engineering** Rees, A.R., Sternberg, M.J.E. & others (Eds)
.... Spiralbound hardback 0-19-963139-5 **$75.00**
.... Paperback 0-19-963138-7 **$50.00**
107. **Cell-Cell Interactions** Stevenson, B.R., Gallin, W.J. & others (Eds)
.... Spiralbound hardback 0-19-963319-3 **$60.00**
.... Paperback 0-19-963318-5 **$40.00**
106. **Diagnostic Molecular Pathology: Volume I** Herrington, C.S. & McGee, J. O'D. (Eds)
.... Spiralbound hardback 0-19-963237-5 **$58.00**
.... Paperback 0-19-963236-7 **$38.00**
105. **Biomechanics-Materials** Vincent, J.F.V. (Ed)
.... Spiralbound hardback 0-19-963223-5 **$70.00**
.... Paperback 0-19-963222-7 **$50.00**
104. **Animal Cell Culture (2/e)** Freshney, R.I. (Ed)
.... Spiralbound hardback 0-19-963212-X **$60.00**
.... Paperback 0-19-963213-8 **$40.00**
103. **Molecular Plant Pathology: Volume II** Gurr, S.J., McPherson, M.J. & others (Eds)
.... Spiralbound hardback 0-19-963352-5 **$65.00**
.... Paperback 0-19-963351-7 **$45.00**
101. **Protein Targeting** Magee, A.I. & Wileman, T. (Eds)
.... Spiralbound hardback 0-19-963206-5 **$75.00**
.... Paperback 0-19-963210-3 **$50.00**
100. **Diagnostic Molecular Pathology: Volume II: Cell and Tissue Genotyping** Herrington, C.S. & McGee, J.O'D. (Eds)
.... Spiralbound hardback 0-19-963239-1 **$60.00**
.... Paperback 0-19-963238-3 **$39.00**
99. **Neuronal Cell Lines** Wood, J.N. (Ed)
.... Spiralbound hardback 0-19-963346-0 **$68.00**
.... Paperback 0-19-963345-2 **$48.00**
98. **Neural Transplantation** Dunnett, S.B. & Björklund, A. (Eds)
.... Spiralbound hardback 0-19-963286-3 **$69.00**
.... Paperback 0-19-963285-5 **$42.00**
97. **Human Cytogenetics: Volume II: Malignancy and Acquired Abnormalities (2/e)** Rooney, D.E. & Czepulkowski, B.H. (Eds)
.... Spiralbound hardback 0-19-963290-1 **$75.00**
.... Paperback 0-19-963289-8 **$50.00**
96. **Human Cytogenetics: Volume I: Constitutional Analysis (2/e)** Rooney, D.E. & Czepulkowski, B.H. (Eds)
.... Spiralbound hardback 0-19-963288-X **$75.00**
.... Paperback 0-19-963287-1 **$50.00**
95. **Lipid Modification of Proteins** Hooper, N.M. & Turner, A.J. (Eds)
.... Spiralbound hardback 0-19-963274-X **$75.00**
.... Paperback 0-19-963273-1 **$50.00**
94. **Biomechanics-Structures and Systems** Biewener, A.A. (Ed)
.... Spiralbound hardback 0-19-963268-5 **$85.00**
.... Paperback 0-19-963267-7 **$50.00**
93. **Lipoprotein Analysis** Converse, C.A. & Skinner, E.R. (Eds)
.... Spiralbound hardback 0-19-963192-1 **$65.00**
.... Paperback 0-19-963231-6 **$42.00**
92. **Receptor-Ligand Interactions** Hulme, E.C. (Ed)
.... Spiralbound hardback 0-19-963090-9 **$75.00**
.... Paperback 0-19-963091-7 **$50.00**
91. **Molecular Genetic Analysis of Populations** Hoelzel, A.R. (Ed)
.... Spiralbound hardback 0-19-963278-2 **$65.00**
.... Paperback 0-19-963277-4 **$45.00**
90. **Enzyme Assays** Eisenthal, R. & Danson, M.J. (Eds)
.... Spiralbound hardback 0-19-963142-5 **$68.00**
.... Paperback 0-19-963143-3 **$48.00**
89. **Microcomputers in Biochemistry** Bryce, C.F.A. (Ed)
.... Spiralbound hardback 0-19-963253-7 **$60.00**
.... Paperback 0-19-963252-9 **$40.00**
88. **The Cytoskeleton** Carraway, K.L. & Carraway, C.A.C. (Eds)
.... Spiralbound hardback 0-19-963257-X **$60.00**
.... Paperback 0-19-963256-1 **$40.00**
87. **Monitoring Neuronal Activity** Stamford, J.A. (Ed)
.... Spiralbound hardback 0-19-963244-8 **$60.00**
.... Paperback 0-19-963243-X **$40.00**
86. **Crystallization of Nucleic Acids and Proteins** Ducruix, A. & Gieg‹130›, R. (Eds)
.... Spiralbound hardback 0-19-963245-6 **$60.00**
.... Paperback 0-19-963246-4 **$50.00**
85. **Molecular Plant Pathology: Volume I** Gurr, S.J., McPherson, M.J. & others (Eds)
.... Spiralbound hardback 0-19-963103-4 **$60.00**
.... Paperback 0-19-963102-6 **$40.00**
84. **Anaerobic Microbiology** Levett, P.N. (Ed)
.... Spiralbound hardback 0-19-963204-9 **$75.00**
.... Paperback 0-19-963262-6 **$45.00**

83. Oligonucleotides and Analogues Eckstein, F. (Ed)
.... Spiralbound hardback 0-19-963280-4 **$65.00**
.... Paperback 0-19-963279-0 **$45.00**
82. Electron Microscopy in Biology Harris, R. (Ed)
.... Spiralbound hardback 0-19-963219-7 **$65.00**
.... Paperback 0-19-963215-4 **$45.00**
81. Essential Molecular Biology: Volume II Brown, T.A. (Ed)
.... Spiralbound hardback 0-19-963112-3 **$65.00**
.... Paperback 0-19-963113-1 **$45.00**
80. Cellular Calcium McCormack, J.G. & Cobbold, P.H. (Eds)
.... Spiralbound hardback 0-19-963131-X **$75.00**
.... Paperback 0-19-963130-1 **$50.00**
79. Protein Architecture Lesk, A.M.
.... Spiralbound hardback 0-19-963054-2 **$65.00**
.... Paperback 0-19-963055-0 **$45.00**
78. Cellular Neurobiology Chad, J. & Wheal, H. (Eds)
.... Spiralbound hardback 0-19-963106-9 **$73.00**
.... Paperback 0-19-963107-7 **$43.00**
77. PCR McPherson, M.J., Quirke, P. & others (Eds)
.... Spiralbound hardback 0-19-963226-X **$55.00**
.... Paperback 0-19-963196-4 **$40.00**
76. Mammalian Cell Biotechnology Butler, M. (Ed)
.... Spiralbound hardback 0-19-963207-3 **$60.00**
.... Paperback 0-19-963209-X **$40.00**
75. Cytokines Balkwill, F.R. (Ed)
.... Spiralbound hardback 0-19-963218-9 **$64.00**
.... Paperback 0-19-963214-6 **$44.00**
74. Molecular Neurobiology Chad, J. & Wheal, H. (Eds)
.... Spiralbound hardback 0-19-963108-5 **$56.00**
.... Paperback 0-19-963109-3 **$36.00**
73. Directed Mutagenesis McPherson, M.J. (Ed)
.... Spiralbound hardback 0-19-963141-7 **$55.00**
.... Paperback 0-19-963140-9 **$35.00**
72. Essential Molecular Biology: Volume I Brown, T.A. (Ed)
.... Spiralbound hardback 0-19-963110-7 **$65.00**
.... Paperback 0-19-963111-5 **$45.00**
71. Peptide Hormone Action Siddle, K. & Hutton, J.C.
.... Spiralbound hardback 0-19-963070-4 **$70.00**
.... Paperback 0-19-963071-2 **$50.00**
70. Peptide Hormone Secretion Hutton, J.C. & Siddle, K. (Eds)
.... Spiralbound hardback 0-19-963068-2 **$70.00**
.... Paperback 0-19-963069-0 **$50.00**
69. Postimplantation Mammalian Embryos Copp, A.J. & Cockroft, D.L. (Eds)
.... Spiralbound hardback 0-19-963088-7 **$70.00**
.... Paperback 0-19-963089-5 **$50.00**
68. Receptor-Effector Coupling Hulme, E.C. (Ed)
.... Spiralbound hardback 0-19-963094-1 **$70.00**
.... Paperback 0-19-963095-X **$45.00**
67. Gel Electrophoresis of Proteins (2/e) Hames, B.D. & Rickwood, D. (Eds)
.... Spiralbound hardback 0-19-963074-7 **$75.00**
.... Paperback 0-19-963075-5 **$50.00**
66. Clinical Immunology Gooi, H.C. & Chapel, H. (Eds)
.... Spiralbound hardback 0-19-963086-0 **$69.95**
.... Paperback 0-19-963087-9 **$50.00**
65. Receptor Biochemistry Hulme, E.C. (Ed)
.... Spiralbound hardback 0-19-963092-5 **$70.00**
.... Paperback 0-19-963093-3 **$50.00**
64. Gel Electrophoresis of Nucleic Acids (2/e) Rickwood, D. & Hames, B.D. (Eds)
.... Spiralbound hardback 0-19-963082-8 **$75.00**
.... Paperback 0-19-963083-6 **$50.00**
63. Animal Virus Pathogenesis Oldstone, M.B.A. (Ed)
.... Spiralbound hardback 0-19-963100-X **$68.00**
.... Paperback 0-19-963101-8 **$40.00**
62. Flow Cytometry Ormerod, M.G. (Ed)
.... Paperback 0-19-963053-4 **$50.00**
61. Radioisotopes in Biology Slater, R.J. (Ed)
.... Spiralbound hardback 0-19-963080-1 **$75.00**
.... Paperback 0-19-963081-X **$45.00**
60. Biosensors Cass, A.E.G. (Ed)
.... Spiralbound hardback 0-19-963046-1 **$65.00**
.... Paperback 0-19-963047-X **$43.00**
59. Ribosomes and Protein Synthesis Spedding, G. (Ed)
.... Spiralbound hardback 0-19-963104-2 **$75.00**
.... Paperback 0-19-963105-0 **$45.00**
58. Liposomes New, R.R.C. (Ed)
.... Spiralbound hardback 0-19-963076-3 **$70.00**
.... Paperback 0-19-963077-1 **$45.00**
57. Fermentation McNeil, B. & Harvey, L.M. (Eds)
.... Spiralbound hardback 0-19-963044-5 **$65.00**
.... Paperback 0-19-963045-3 **$39.00**
56. Protein Purification Applications Harris, E.L.V. & Angal, S. (Eds)
.... Spiralbound hardback 0-19-963022-4 **$54.00**
.... Paperback 0-19-963023-2 **$36.00**
55. Nucleic Acids Sequencing Howe, C.J. & Ward, E.S. (Eds)
.... Spiralbound hardback 0-19-963056-9 **$59.00**
.... Paperback 0-19-963057-7 **$38.00**
54. Protein Purification Methods Harris, E.L.V. & Angal, S. (Eds)
.... Spiralbound hardback 0-19-963002-X **$60.00**
.... Paperback 0-19-963003-8 **$40.00**
53. Solid Phase Peptide Synthesis Atherton, E. & Sheppard, R.C.
.... Spiralbound hardback 0-19-963066-6 **$58.00**
.... Paperback 0-19-963067-4 **$39.95**
52. Medical Bacteriology Hawkey, P.M. & Lewis, D.A. (Eds)
.... Spiralbound hardback 0-19-963008-9 **$69.95**
.... Paperback 0-19-963009-7 **$50.00**
51. Proteolytic Enzymes Beynon, R.J. & Bond, J.S. (Eds)
.... Spiralbound hardback 0-19-963058-5 **$60.00**
.... Paperback 0-19-963059-3 **$39.00**
50. Medical Mycology Evans, E.G.V. & Richardson, M.D. (Eds)
.... Spiralbound hardback 0-19-963010-0 **$69.95**
.... Paperback 0-19-963011-9 **$50.00**
49. Computers in Microbiology Bryant, T.N. & Wimpenny, J.W.T. (Eds)
.... Paperback 0-19-963015-1 **$40.00**
48. Protein Sequencing Findlay, J.B.C. & Geisow, M.J. (Eds)
.... Spiralbound hardback 0-19-963012-7 **$56.00**
.... Paperback 0-19-963013-5 **$38.00**
47. Cell Growth and Division Baserga, R. (Ed)
.... Spiralbound hardback 0-19-963026-7 **$62.00**
.... Paperback 0-19-963027-5 **$38.00**
46. Protein Function Creighton, T.E. (Ed)
.... Spiralbound hardback 0-19-963006-2 **$65.00**
.... Paperback 0-19-963007-0 **$45.00**
45. Protein Structure Creighton, T.E. (Ed)
.... Spiralbound hardback 0-19-963000-3 **$65.00**
.... Paperback 0-19-963001-1 **$45.00**
44. Antibodies: Volume II Catty, D. (Ed)
.... Spiralbound hardback 0-19-963018-6 **$58.00**
.... Paperback 0-19-963019-4 **$39.00**
43. HPLC of Macromolecules Oliver, R.W.A. (Ed)
.... Spiralbound hardback 0-19-963020-8 **$54.00**
.... Paperback 0-19-963021-6 **$45.00**
42. Light Microscopy in Biology Lacey, A.J. (Ed)
.... Spiralbound hardback 0-19-963036-4 **$62.00**
.... Paperback 0-19-963037-2 **$38.00**
41. Plant Molecular Biology Shaw, C.H. (Ed)
.... Paperback 1-85221-056-7 **$38.00**
40. Microcomputers in Physiology Fraser, P.J. (Ed)
.... Spiralbound hardback 1-85221-129-6 **$54.00**
.... Paperback 1-85221-130-X **$36.00**
39. Genome Analysis Davies, K.E. (Ed)
.... Spiralbound hardback 1-85221-109-1 **$54.00**
.... Paperback 1-85221-110-5 **$36.00**
38. Antibodies: Volume I Catty, D. (Ed)
.... Paperback 0-947946-85-3 **$38.00**
37. Yeast Campbell, I. & Duffus, J.H. (Eds)
.... Paperback 0-947946-79-9 **$36.00**
36. Mammalian Development Monk, M. (Ed)
.... Hardback 1-85221-030-3 **$60.00**
.... Paperback 1-85221-029-X **$45.00**
35. Lymphocytes Klaus, G.G.B. (Ed)
.... Hardback 1-85221-018-4 **$54.00**
34. Lymphokines and Interferons Clemens, M.J., Morris, A.G. & others (Eds)
.... Paperback 1-85221-035-4 **$44.00**
33. Mitochondria Darley-Usmar, V.M., Rickwood, D. & others (Eds)
.... Hardback 1-85221-034-6 **$65.00**
.... Paperback 1-85221-033-8 **$45.00**

32. **Prostaglandins and Related Substances** Benedetto, C., McDonald-Gibson, R.G. & others (Eds)
.... Hardback 1-85221-032-X **$58.00**
.... Paperback 1-85221-031-1 **$38.00**

31. **DNA Cloning: Volume III** Glover, D.M. (Ed)
.... Hardback 1-85221-049-4 **$56.00**
.... Paperback 1-85221-048-6 **$36.00**

30. **Steroid Hormones** Green, B. & Leake, R.E. (Eds)
.... Paperback 0-947946-53-5 **$40.00**

29. **Neurochemistry** Turner, A.J. & Bachelard, H.S. (Eds)
.... Hardback 1-85221-028-1 **$56.00**
.... Paperback 1-85221-027-3 **$36.00**

28. **Biological Membranes** Findlay, J.B.C. & Evans, W.H. (Eds)
.... Hardback 0-947946-84-5 **$54.00**
.... Paperback 0-947946-83-7 **$36.00**

27. **Nucleic Acid and Protein Sequence Analysis** Bishop, M.J. & Rawlings, C.J. (Eds)
.... Hardback 1-85221-007-9 **$66.00**
.... Paperback 1-85221-006-0 **$44.00**

26. **Electron Microscopy in Molecular Biology** Sommerville, J. & Scheer, U. (Eds)
.... Hardback 0-947946-64-0 **$54.00**
.... Paperback 0-947946-54-3 **$40.00**

25. **Teratocarcinomas and Embryonic Stem Cells** Robertson, E.J. (Ed)
.... Hardback 1-85221-005-2 **$62.00**
.... Paperback 1-85221-004-4 **$0.00**

24. **Spectrophotometry and Spectrofluorimetry** Harris, D.A. & Bashford, C.L. (Eds)
.... Hardback 0-947946-69-1 **$56.00**
.... Paperback 0-947946-46-2 **$39.95**

23. **Plasmids** Hardy, K.G. (Ed)
.... Paperback 0-947946-81-0 **$36.00**

22. **Biochemical Toxicology** Snell, K. & Mullock, B. (Eds)
.... Paperback 0-947946-52-7 **$40.00**

19. **Drosophila** Roberts, D.B. (Ed)
.... Hardback 0-947946-66-7 **$67.50**
.... Paperback 0-947946-45-4 **$46.00**

17. **Photosynthesis: Energy Transduction** Hipkins, M.F. & Baker, N.R. (Eds)
.... Hardback 0-947946-63-2 **$54.00**
.... Paperback 0-947946-51-9 **$36.00**

16. **Human Genetic Diseases** Davies, K.E. (Ed)
.... Hardback 0-947946-76-4 **$60.00**
.... Paperback 0-947946-75-6 **$34.00**

14. **Nucleic Acid Hybridisation** Hames, B.D. & Higgins, S.J. (Eds)
.... Hardback 0-947946-61-6 **$60.00**
.... Paperback 0-947946-23-3 **$36.00**

13. **Immobilised Cells and Enzymes** Woodward, J. (Ed)
.... Hardback 0-947946-60-8 **$0.00**

12. **Plant Cell Culture** Dixon, R.A. (Ed)
.... Paperback 0-947946-22-5 **$36.00**

11a. **DNA Cloning: Volume I** Glover, D.M. (Ed)
.... Paperback 0-947946-18-7 **$36.00**

11b. **DNA Cloning: Volume II** Glover, D.M. (Ed)
.... Paperback 0-947946-19-5 **$36.00**

10. **Virology** Mahy, B.W.J. (Ed)
.... Paperback 0-904147-78-9 **$40.00**

9. **Affinity Chromatography** Dean, P.D.G., Johnson, W.S. & others (Eds)
.... Paperback 0-904147-71-1 **$36.00**

7. **Microcomputers in Biology** Ireland, C.R. & Long, S.P. (Eds)
.... Paperback 0-904147-57-6 **$36.00**

6. **Oligonucleotide Synthesis** Gait, M.J. (Ed)
.... Paperback 0-904147-74-6 **$38.00**

5. **Transcription and Translation** Hames, B.D. & Higgins, S.J. (Eds)
.... Paperback 0-904147-52-5 **$38.00**

3. **Iodinated Density Gradient Media** Rickwood, D. (Ed)
.... Paperback 0-904147-51-7 **$36.00**

Sets

Essential Molecular Biology: Volumes I and II as a set Brown, T.A. (Ed)
.... Spiralbound hardback 0-19-963114-X **$118.00**
.... Paperback 0-19-963115-8 **$78.00**

Antibodies: Volumes I and II as a set Catty, D. (Ed)
.... Paperback 0-19-963063-1 **$70.00**

Cellular and Molecular Neurobiology Chad, J. & Wheal, H. (Eds)
.... Spiralbound hardback 0-19-963255-3 **$133.00**
.... Paperback 0-19-963254-5 **$79.00**

Protein Structure and Protein Function: Two-volume set Creighton, T.E. (Ed)
.... Spiralbound hardback 0-19-963064-X **$114.00**
.... Paperback 0-19-963065-8 **$80.00**

DNA Cloning: Volumes I, II, III as a set Glover, D.M. (Ed)
.... Paperback 1-85221-069-9 **$92.00**

Molecular Plant Pathology: Volumes I and II as a set Gurr, S.J., McPherson, M.J. & others (Eds)
.... Spiralbound hardback 0-19-963354-1 **$0.00**
.... Paperback 0-19-963353-3 **$0.00**

Protein Purification Methods, and Protein Purification Applications, two-volume set Harris, E.L.V. & Angal, S. (Eds)
.... Spiralbound hardback 0-19-963048-8 **$98.00**
.... Paperback 0-19-963049-6 **$68.00**

Diagnostic Molecular Pathology: Volumes I and II as a set Herrington, C.S. & McGee, J. O'D. (Eds)
.... Spiralbound hardback 0-19-963241-3 **$0.00**
.... Paperback 0-19-963240-5 **$0.00**

Receptor Biochemistry; Receptor-Effector Coupling; Receptor-Ligand Interactions Hulme, E.C. (Ed)
.... Spiralbound hardback 0-19-963096-8 **$193.00**
.... Paperback 0-19-963097-6 **$125.00**

Signal Transduction Milligan, G. (Ed)
.... Spiralbound hardback 0-19-963296-0 **$60.00**
.... Paperback 0-19-963295-2 **$38.00**

Human Cytogenetics: Volumes I and II as a set (2/e) Rooney, D.E. & Czepulkowski, B.H. (Eds)
.... Hardback 0-19-963314-2 **$130.00**
.... Paperback 0-19-963313-4 **$90.00**

Peptide Hormone Secretion/Peptide Hormone Action Siddle, K. & Hutton, J.C. (Eds)
.... Spiralbound hardback 0-19-963072-0 **$135.00**
.... Paperback 0-19-963073-9 **$90.00**

ORDER FORM for USA and Canada

Qty	ISBN	Author	Title	Amount
			S&H	
			CA and NC residents add appropriate sales tax	
			TOTAL	

Please add shipping and handling: $2.50 for first book, ($1.00 each book thereafter)

Name ....

Address ....

....

.... Zip ....

[ ] Please charge $ .... to my credit card
Mastercard/VISA/American Express (circle appropriate card)

Acct. .... Expiry date ....

Signature ....

Credit card account address if different from above:

....

.... Zip ....

[ ] I enclose a cheque for $...........

Mail orders to: Order Dept. Oxford University Press, 2001 Evans Road, Cary, NC 27513